# A BREVIARY OF SEISMIC TOMOGRAPHY

## Imaging the Interior of the Earth and Sun

This is the first textbook to cover all the major aspects of seismic tomography at a level accessible to students. While focusing on applications in solid earth geophysics, the book also includes numerous excursions into helioseismology in order to demonstrate the strong affinity between the two fields.

The book presents a comprehensive introduction to seismic tomography including the basic theory of wave propagation, the ray and Born approximations required for interpretation of amplitudes, travel times and phases, eigenvibrations and surface waves, observational methods, model parametrization, finite-frequency methods, inversion, error and resolution analysis, and seismic anisotropy. It presents in-depth consideration of observational aspects of the subject, as well as practical recommendations for implementing numerical models using publicly available software.

Written by one of the leaders in the field, and containing numerous student exercises, this textbook is appropriate for advanced undergraduate and graduate courses. It is also an invaluable guide for seismology research practitioners in geophysics and astronomy. Solutions to the exercises, and a link to the author's tomographic software and user manual are available online from www.cambridge.org/9780521882446.

GUUST NOLET is the George J. Magee Professor of Geophysics Emeritus at Princeton University and currently teaches at the University of Nice/Sophia Antipolis in France. His research interests include seismic body wave tomography and investigations of the structure of the mantle and its role in shaping the Earth's surface. Professor Nolet is the winner of the 2006 Beno Gutenberg Medal of the European Geophysical Union, the 1983 Vening Meinesz prize, and the 1980 Prix Lagrange. He is a Fellow of the American Geophysical Union and is a member of the Academia Europea, the Royal Netherlands Academy of Sciences, and the American Academy of Arts and Sciences.

# A BREVIARY OF SEISMIC TOMOGRAPHY

## Imaging the Interior of the Earth and Sun

Guust Nolet
*University of Nice/Sophia Antipolis, France*

CAMBRIDGE
UNIVERSITY PRESS

CAMBRIDGE UNIVERSITY PRESS
Cambridge, New York, Melbourne, Madrid, Cape Town, Singapore, São Paulo, Delhi

Cambridge University Press
The Edinburgh Building, Cambridge CB2 8RU, UK

Published in the United States of America by Cambridge University Press, New York

www.cambridge.org
Information on this title: www.cambridge.org/9780521882446



First published 2008

Printed in the United Kingdom at the University Press, Cambridge

*A catalogue record for this publication is available from the British Library*

*Library of Congress Cataloguing in Publication data*
Nolet, Guust, 1945–
A breviary of seismic tomography : imaging the interior of the earth and sun / Guust Nolet.
p. cm.
Includes bibliographical references and index.
ISBN 978-0-521-88244-6 (hardback)
1. Seismic tomography. 2. Geodynamics. I. Title.
QE538.5.N65 2008
551.1′10284 – dc22 2008020508

ISBN 978-0-521-88244-6 hardback

---



---

To Tony Dahlen (1942–2007)

Bre"vi*a*ry, n. [F. bréviarie, L. breviarium summary, abridgment, neut. noun fr. breviarius abridged, fr. brevis short.]

1. An abridgment; a compend; an epitome; a brief account or summary. A book entitled the abridgment or breviary of those roots that are to be cut up or gathered.

2. A book containing the daily public or canonical prayers of the Roman Catholic or of the Greek Church for the seven canonical hours, namely, matins and lauds, the first, third, sixth, and ninth hours, vespers, and compline; – distinguished from the missal.

*(Webster Dictionary, 1913, Page: 180)*

# Contents

# Preface

Avec tout ce que je sais, on pourrait faire un livre. Il est vrai qu'avec tout ce que je ne sais pas, on pourrait faire une bibliothèque.

*Sacha Guitry*

After working on research topics related to seismic tomography for a quarter of a century, I decided it was time to write down all I know about the topic – but not in a grand unifying tome that covers everything from first principles to numerical applications. First of all, I have little patience for mathematical niceties; second, and more importantly, I wrote this book for the *practitioners* of the craft of seismic tomography. Those who go out into the field to collect data usually have no time for proofs of convergence or existence. The intended reader of this book is therefore an observational seismologist or helioseismologist who is not interested in lengthy derivations nor in the subtleties that fascinate the theoreticians, but who wants to understand the assumptions behind algorithms, even if these are mathematically intricate, and develop an understanding of the conditions for their validity, which forms the basis of that priceless commodity: scientific intuition. The level is such that it could be used for a one-semester course at upper undergraduate or beginning graduate level, perhaps following up on an introductory course based on Shearer [307] or Stein and Wysession [343]. Despite covering a wide range of topics, I have tried to keep it short (hence the title), while not economizing on references that may provide more detail if needed. As for references, choice is inevitable, and I have generally given preference to easily accessible papers in the English language. I realize that this gives short shrift to articles from Eastern Europe, Japan and China. This is unfortunate and I offer my apologies to colleagues in these countries.

The book is roughly divided into three parts. The basic theory of wave propagation and scattering needed to compute Fréchet (sensitivity) kernels for seismic tomography is expanded in the first half of the book. Although I made an attempt to write this as a self-contained part, even here I often steer away from lengthy derivations that the interested reader can find in the more general seismological literature, and strive instead to make results at least intuitively acceptable. I then discuss observations, paying attention to both the ambient noise and the capabilities of modern, digital, broadband instrumentation. The last part of the book is devoted to the tomographic inversion and imaging itself. I restrict the material to *transmission* tomography. The nonlinearities associated with reflection seismology form a topic apart, worthy of a monograph of equal or greater length than this book.

As a geophysicist, I have written this book from a 'terrestrial' viewpoint. Where the links with helioseismology are obvious, I have ventured onto the playing field of solar astronomy as well, mainly to demonstrate the large affinity between the two research fields and to help astronomers to recognize parallel developments more easily. However, geophysicists interested in the fascinating topics of helio- and astroseismology do well to consult other sources for an expert introduction into these fields. The lecture notes by Christensen-Dalsgaard [58] provide a general and very readable introduction to the theory of stellar oscillations. The 'living review' by Gizon and Birch [118] gives an up-to-date account of methods and results in local helioseismology. A special issue of Astronomische Nachrichten edited by Thierry Corbard, Laurent Gizon and Markus Roth [65] is an excellent source of information on current techniques and future plans and provides a wealth of further references.

If you are a student of the field, and undertake the journey of reading this book from first to last chapter, I hope that at the end you will not feel as though you have done the proverbial grand tour of six European cities in seven days. But I cannot deny that I try to cover a very large range of topics, each with its own jargon and notation. I have tried to stay close to notations one commonly finds in the literature. This implies that the same symbol is sometimes used with different meaning in different chapters. Usually, that meaning is clear from the context, but occasionally I have felt the need to explicitly comment on peculiarities in notation.

A special case is the notation of vectors, tensors and matrices. Throughout the book we deal with physical vectors such as the force $\boldsymbol{f} = (f_x, f_y, f_z)$, that have tensor-properties and are conceptually different from $N$-tuples such as the data 'vector' $\boldsymbol{d} = (d_1, d_2, ..., d_N)$ that we encounter in the last few chapters. I use the same bold font for both, and avoid the transpose notation (e.g. $\boldsymbol{d}^T$) that toggles between row and column vectors. A dot product $\boldsymbol{a} \cdot \boldsymbol{b}$ denotes $\sum_i a_i b_i$, without the complex conjugation (if it is needed, as in the definition of an inner product for normal modes in Chapter 9, it is explicitly used in the notation, e.g. $\boldsymbol{a} \cdot \boldsymbol{b}^*$). A matrix vector product such as $\sum_j A_{ij}b_j$ is written as $\boldsymbol{Ab}$, and quadratic products are therefore written as $\boldsymbol{a} \cdot \boldsymbol{Ab}$. The transpose of $\boldsymbol{A}$ is written as $\boldsymbol{A}^T$.

During the writing of this book, I have become painfully aware that the half life of a typo or sign convention or even error in math must be measured in weeks or months, not days. In view of the many that have been found before submitting the manuscript, some will undoubtedly remain. I shall be very grateful for readers to contact me about these.

Software for the computation of finite-frequency kernels is publicly available at the software website of CIG (Computational Infrastructure for Geodynamics).† In the text I refer to this as the *Software repository*.

Much of the research described in this book could never have been accomplished without the steady support of science foundations in a number of countries; in my own case the Dutch science foundation NWO and, over the last 15 years, the National Science Foundation in the US. Program directors like Robin Reichlin at NSF, who remain largely anonymous and rarely share in the glory, play a crucial role in allowing science to advance in the best way possible and making sure taxpayers' money is well spent. The ESA/NASA Solar and Heliospheric Observatory was instrumental in the acquisition of very high quality data for solar seismology as witnessed by some of the illustrations in this book.

Both as a researcher and as a teacher of the topic, I always felt a strong need for one text that covers all important aspects of the multidisciplinary science of seismic tomography. I started to write this book during a sabattical in 2005 with the intent to defy Richard's law (that one should multiply the expected time until submission – two years in my case – by $\pi$). That I succeeded is largely due to the help I received from many people. I wish to thank my colleagues at Geoazur of the Université de Nice/Sophia Antipolis, the Laboratoire de Geophysique Interne et Tectonophysique of the Université Joseph Fourier in Grenoble and the Institut de Physique du Globe in Paris, who all provided hospitable hiding space during the various stages of writing this book. A number of geoscientists and astronomers provided figures, valuable information, or commented on parts or all of earlier drafts of this book: Sebastien Chevrot, Jon Claerbout, Huub Douma, Adam Dziewonski, Bob Engdahl, Jim Fowler, Laurent Gizon, Brad Hindman, Shu-Huei Hung, Eystein Husebye, Alexander Kosovichev, Gabi Laske, Suzan van der Lee, Will Levandowski, Tolya Levshin, Guy Masters, Jean-Paul Montagner, Tarje Nissen-Meyer, Mark Panning, Jeroen Ritsema, Barbara Romanowicz, Génevieve Roult, Frederik Simons, Karin Sigloch, Roel Snieder, Toshiro Tanimoto, Albert Tarantola, Yue Tian, Jean Virieux, Cecily Wolfe, and Ying Zhou. I am very grateful to them and wish to make clear that the responsibility for any errors that survive is mine and mine alone. My beloved

† `http://geodynamics.org/cig/software/packages/seismo/`

Julia Frey corrected more than a few prepositions and other peculiarities in my use of the English language. But mostly I am indebted to my close friend Tony Dahlen with whom I collaborated intensively at Princeton and who died before he could see the final version of this book. Without his sharp theoretical insight and intellectual driving force the field of seismic tomography would never have evolved as rapidly and actively as it has.

# 1
# Introduction

Early in the 1970s, seismologists realized that a full three-dimensional (3D) interpretation was needed to satisfy variations in the observed seismic travel times. The starting point of modern seismic tomography (from $\sigma\epsilon\iota\sigma\mu\acute{o}\varsigma$ = quake and $\tau\acute{o}\mu o\varsigma$ = slice) is probably the 1974 AGU presentation by MIT's Keiti Aki (Aki et al. [3]) in which arrival times of P-waves were for the first time formally interpreted in terms of an 'image' as opposed to a simple one-dimensional graph of seismic velocity versus depth. That Aki's co-authors came from NORSAR – the Norwegian array to monitor nuclear test ban treaties – was caused by a quirk of history: Aki had originally planned a sabbatical in Chile, but when a military coup d'état brought the Allende government down in 1973 he changed plans and accepted an invitation from his former MIT student Eystein Husebye for a short sabbatical at NORSAR, which was equipped with a state-of-the-art digital seismic network and computing facilities. Even so, in the twenty-first century it is easy to underestimate the difficulties faced by early tomographers, who had to invert matrices of size 256 × 256 using a CPU with 512 Kbyte of memory. The collaboration between Aki, Husebye and Christoffersson was continued in 1975 at Lincoln Labs in Massachussets (Aki et al. [2, 4]).

The name that was later given to the new imaging technique is more than an accidental reference to medical tomography, because the earliest radiologic tomograms also attempted to get a scan of the body that focuses on a plane of interest, albeit using X-rays rather than seismic waves. This was obtained by moving the X-ray source and the photographic plate in opposite directions, such that objects outside the target plane would be blurred, but those in the plane would always illuminate the photographic plate at the same spot, a technique known as 'backprojection'. This enabled radiologists to reconstruct the tissue or bone density on that plane from observed X-ray intensities. In the 1970s the photographic plate was replaced by sensors, which feed a computer that reconstructs the density from discrete observations (computerized tomography). In seismic travel time tomography, we attempt to reconstruct the local seismic velocity in the Earth from arrival times of body waves observed with seismic sensors. Though very similar to the case of computerized medical tomography, the challenges are far greater. Added complications are that seismic waves do not follow paths that are straight, and that we have little or no control over the experiment – we have to use the earthquakes that are dealt to us, and seismic networks are virtually confined to the continents and a few islands in the oceans.

## 1.1 Early efforts at seismic tomography

The field of seismic tomography blossomed quickly, being an outgrowth of several developments that matured in the 1970s: Backus and Gilbert [14, 12, 13], Jackson [144] and Wiggins [395] developed geophysical inversion theory and provided the necessary tools to deal with the inevitable underdetermined nature of geophysical inverse problems. Dense digital seismic networks, the largest of which were the LASA and NORSAR arrays in Montana and Norway, had been installed to monitor the testing of nuclear weapons, and supplemented the analogue World Wide Standardized Seismograph Network (WWSSN) which served the same purpose. The monitoring arrays also provided densely spaced data to global seismologists. Sengupta and Toksöz [305] read arrival times from WWSSN seismograms and adapted Aki's method in a first attempt at global tomography. But the International Seismological Centre (ISC, then in Edinburgh, UK) had been assembling many more arrival time data from thousands of station operators around the world since 1964, and, even though some of the early ISC data were archived on 7-track tapes that had become difficult to read, the painstakingly careful efforts by Dziewonski [93] to recover them led to his first try at global tomography and would soon give Harvard a dominant position among the institutions doing global imaging studies.

In the meantime, the MIT efforts resulted in a number of local tomography experiments using Aki's technique: Ellsworth and Koyagani [97] imaged the structure beneath the Kilauea volcano on Hawaii in 1977, and in the same year Mitchell et al. [208] published a tomographic study of the New Madrid seismic zone. This was soon followed by larger experiments on a regional scale: Menke [206], Romanowicz [286, 287] and Taylor and Toksöz [354] pioneered continental tomography with early studies of the upper mantle under the Himalayas, and under the North American and European continents. With time, such local and regional studies became more precise: in 1984, Thurber [360] imaged velocities at Kilauea that were low enough to be interpreted as the underlying magma complex.

## 1.2 Ocean acoustic tomography

In 1981 a first ocean acoustic tomography experiment was performed near Bermuda, and tomography is now an accepted tool to study sound speed (a proxy for temperature) and flow in the upper layers of the oceans. Because ocean acoustic tomography has already been extensively documented in the monograph by Munk, Worcester and Wunsch [223], this book will not often venture into the domain of oceanography. Though the experimental methods used by oceanographers differ greatly from those used by seismologists or astronomers, the theory behind tomographic interpretation is very much the same. Though high frequencies used in ocean acoustics allow for a broader validity of ray theory, the Fresnel zones in ocean acoustic tomography are often large, and a finite-frequency approach is called for (Skarsoulis and Cornuelle [321]). This has the added effect of reducing the effects of wave chaos in range-dependent ocean models.

It is noteworthy that Simons et al. [317] succeeded in recording a teleseismic P-wave using a hydrophone mounted on a freely floating diver submerged at 700 m depth – a development that may eventually bring ocean acoustics together with global seismic tomography and open up the Southern oceans for dense seismometry and acoustic monitoring (see Chapter 17).

## 1.3 Global tomography

In solid Earth sciences, seismic tomography soon extended its reach to the inversion of the Earth's eigenfrequencies and dispersion properties of long-period surface waves, using perturbation theory on normal modes (see Chapter 9). In 1982, Masters et al. [199] at UCSD discovered the strong degree-2 pattern in the Earth's heterogeneity. This anomaly was later correlated by Cazenave et al. [43] to the surface topography of the Earth, topography being the dynamic response of the surface to mantle convection processes, confirming an earlier prediction by Hager et al. [128]. It became evident that the 3D seismic structure of the Earth was strongly linked to the deep dynamics of the planet, and that it was important to resolve that structure.

By 1984, Dziewonski's work on ISC delay times resulted in the first reliable global modal of long-wavelength P-wave velocity variations [90], and Woodhouse and Dziewonski [400] began to use long-period data to image the upper mantle using the lowest order spherical harmonics. This led to an immensely fruitful era in which the large scale heterogeneous structure of the Earth was mapped in increasingly finer detail.

In early global tomography it was customary to parametrize the velocity model in terms of a small number of spherical harmonics. This was not just a preference of the early pioneers; a parametrization in terms of a small number of coefficients was also mandated by the limited computer capacity available at the time, and spherical harmonics provide an excellent means of doing so while forcing velocity anomalies to vary smoothly. By parametrizing the Earth in terms of just a few hundred coefficients the least squares matrix – which is $N \times N$ in size for $N$ model parameters – could be diagonalized and inverted using singular value decomposition or other regularization techniques. However, the spherical harmonic expansion made little sense for the investigation of more localized structures, and for this purpose existing iterative matrix solvers were soon introduced into seismic tomography by Clayton and Comer [61], Nolet [233, 234], and Neumann-Denzau and Behrens [229]. These iterative solvers are very flexible in the type of parametrization and impose virtually no limit to the size of the set of model coefficients or data. With iterative solvers, the lack of a formal inverse hampers formal estimation of error or resolution, but sensitivity tests try to make up for that shortcoming (Spakman and Nolet [338]). In a separate development, Jackson [145], Tarantola and Valette [353], and Tarantola and Nercessian [352] circumvented the problem of model parametrization by imposing a-priori correlation properties on the model and introduced the tools of Bayesian inference – which allows for the assignment of *probabilities* to scientific truth – into the tomographic inversion.

Waveform inversions remained restricted to modelling the low frequency wavefield, though, independent from global tomography, 'waveform tomography' was formulated for exploration purposes by Tarantola [350], who used first order scattering theory, finite difference modelling, and an adjoint approach to attack the strongly nonlinear inverse problem for high frequency waveforms. This approach – which is related to seismic migration as used in the oil industry – stands at the basis of some of the more recent developments in seismic tomography. For example, in regional waveform tomography, Bostock and coworkers [31] succeeded in imaging the subducting lithosphere under the Pacific Northwest using converted wave energy to map discontinuities.

## 1.4 Some major discoveries

I refer the reader to the excellent reviews by Romanowicz [288, 291] for the later history of global tomography but note several of the major discoveries: In 1988, Spakman et al. [340] showed that the Hellenic subduction zone extends far deeper than the cut-off depth for seismicity, thereby opening up a vista that up to then seemed a mere hypothesis: could slabs sink deeper than the depth of the deepest located earthquakes near 700 km? Such penetration of slabs into the lower mantle had earlier been inferred by Creager and Jordan [69] from the statistics of travel time anomalies without the benefit of tomography. I still remember the day that

Suzan van der Lee, then an undergraduate at the University of Utrecht working on her senior thesis with Wim Spakman, produced a stunning image of the Aegean slab subducting into the lower mantle. This was in 1989 and her thesis was met with some scepticism – including from me – but by the time it found its way into the refereed literature [339] there were other, very strong indications from regions with deep seismicity for lower mantle subduction. A first image in which both the Farallon plate and the Tonga subduction zone enter the lower mantle came from global tomography by Inoue et al. [141]. In this early study, Inoue also imaged the major low velocity regions near the core–mantle boundary that we now call 'superplumes'. But it was van der Hilst et al. [372, 373], who used P and surface-reflected P-waves to construct the first tomographic images of the subducting slabs under the Northwest Pacific that contained sufficient detail to convince the geophysical community that several slabs indeed penetrate into the lower mantle. When Grand's S-wave models [123, 125] showed very similar cross-sections of slabs sinking into the lower mantle, any remaining doubts about the reliability of seismic tomography quickly melted away and the hypothesis of two-layered convection in the Earth was dealt a serious blow.

The first global model of topography on the 660 km discontinuity was constructed by Shearer and Masters [310] in 1992. The first global attenuation model was derived by Romanowicz [289], but to date little agreement exists among different attenuation models for the Earth, and this remains one of the major challenges of tomographic research. In 1997, Su and Dziewonski [345] discovered a negative correlation between anomalies in bulk and shear modulus in the lowermost mantle where the Pacific and African superplumes are located. Ishii and Tromp [142, 143] determined that the Pacific superplume has a higher density than the surrounding mantle, indicating that these are chemically distinct entities, though this finding is somewhat controversial.

The hypothesized – much thinner – thermal mantle plumes remained elusive in tomographic images, apart from a few contested interpretations. In this context, the debate over the origin of the Iceland plume is very illustrative. Following two field campaigns with portable seismometers in Iceland, seismic tomography by Wolfe et al. [396] and Allen et al. [7] left little doubt that Iceland caps a strong upper mantle plume. At the same time, Bijwaard and Spakman [22] presented tomographic evidence for a lower mantle extension of this plume and argued that the plume originates from a broad upwelling at the base of the mantle. This interpretation was strongly contested by Foulger et al. [104]. In 2004 Montelli et al. [215] used the enhanced resolving power of finite-frequency tomography to image more than a dozen plume-like anomalies in the lower mantle in detail, but argued that the Iceland plume disappears at mid-mantle depth, an interpretation that was supported by ray-theoretical tomographic imaging by Zhao [414]. More

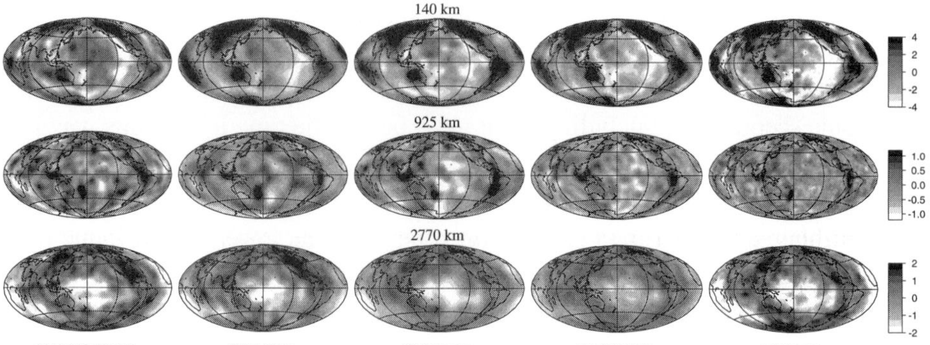


Fig. 1.1. A comparison of perturbations in S-wave velocity in five recent global tomographic models permits judgement of the degree of agreement at three depth levels (top to bottom): upper, mid- and lower mantle. The greyscale reflects perturbations in per cent. Figure courtesy Barbara Romanowicz, reprinted, with permission, from [291] © 2003 by Annual Reviews www.annualreviews.org.

recent finite-frequency tomography using S-waves by Montelli et al. [214] seems to re-open the question of the depth of the Iceland plume since the S-wave images suggest a weak connection between low velocity anomalies in the upper- and lowermost mantle.

The long-wavelength structure of the Earth is very similar across recent tomographic models, as can be judged from Figure 1.1, even though there is disagreement on the amplitudes of even the largest structures. But for smaller length scales the correlation between the models becomes less. Disagreements such as for Iceland abound for many weakly resolved features. Yet, barring undiscovered errors in the analyses, we must assume that all tomographic models have an acceptable fit to the data. The lack of agreement comes only partly from the fact that different investigators use different data sets, because in fact all global data sets rely heavily on the network of broadband, digital seismometers that has been growing over the past 20 years and there is a significant amount of overlap between the information used to construct each one of these models. A major factor that creates differences between tomographic solutions is that the inverse problem is underdetermined and requires regularization, which leaves the choice of 'optimum' model among all models that satisfy the data to the investigator. Apart from that (or perhaps because this aspect of arbitrariness has not forced us to face the issue) there is at present not much of a tradition among tomographers to use the same statistical criteria for goodness of fit. We console our scientific conscience with the argument that 'resolution tests' will convey how much wriggle room there is for other solutions. But such resolution tests are almost always limited in scope, depend on the misfit criterion used, are subject to the same theoretical approximations used in the inversion and usually give no quantitative information on the covariance of the solution.

## 1.5 Helioseismology

Helioseismology began in 1961, when Leighton et al. [178] reported the discovery of 'a striking repetitive correlation with a period $T = 296 \pm 3$ sec' in Doppler shifts of solar spectral lines. An explanation of such 'five minute oscillations' as standing waves trapped in the solar interior was provided in 1970 by Ulrich [370], who also predicted that discrete spectral lines in a frequency–wavenumber diagram would be observed if the spectral resolution could be improved. Deubner [85] was the first to observe such dispersion lines, using observations lasting several hours from the Anacapri observatory on Capri, Italy.

The results from helioseismology have been spectacular. Since 1995, the Global Oscillations Network Group (GONG) makes it possible to obtain 24-hour coverage of the solar observations through a network of six telescopes around the globe, leading to an important increase in the precision of the Doppler measurements. Even better, after December 1995 the Michelson Doppler Imager (MDI) began sending back data from the Solar and Heliospheric Observatory (SOHO), circling the Sun in a nearly stationary position with respect to Earth, around the $L_1$ Lagrangian point. These observations are crucial in the testing of models of energy generation in the Sun. Early determinations of the sound speed as a function of depth in the Sun seemed to indicate broad agreement with the standard model (Bahcall et al. [16]) – and thus provided no relief for the discrepancy in neutrino flux exhibited at the time by that model: the observed flux was about a third of that predicted (the discrepancy was later resolved when particle physicists realized that some of the neutrinos change their nature as they flow out of the Sun and were unobservable with the available detectors).

Recent reviews on helioseismology are given by Christensen-Dalsgaard [58] and Gizon and Birch [118]. The first 'sunquake' – oscillations in response to a flare of gas above the solar surface – was observed on July 9, 1996 by Kosovichev and Zharkova [167], but normally the Sun's interior is imaged using random waves. Helioseismic observations have also led to the discovery of 'weather patterns' in the upper layers of the convective zone, which are due to convective motions and flows associated with concentrations of magnetic activity (e.g. Kosovichev et al. [166]).

Originally, helioseismology relied on the inversion of dispersion relationships of surface waves. But in 1993 time–distance helioseismology was developed by Duvall et al. [89] using cross-correlation techniques to retrieve the Green's function for a fixed distance on the solar surface (Claerbout's conjecture [278]). This helped greatly the development of true 3D imaging of the Sun's convective zone. The helioseismic inversion methods are inspired by the work in terrestrial seismic tomography and there are many similarities.

## 1.6 Finite-frequency tomography

This book is for a large part motivated by the most recent development in the theory of body wave and surface wave tomography: the recognition that progress in sharpening the images requires that we step away from the shortcomings of ray theory. Ray theory is an infinite-frequency (or zero wavelength) approximation, but in reality seismic waves have wavelengths that range from 10 to 1000 km or even more, and wavelengths and Fresnel zones are often much larger than heterogeneities of considerable interest. As the models became more and more detailed, scepticism about the use of ray theory grew. To be valid, ray theory requires that wavelengths are short and Fresnel zones narrow. We now have simple but powerful diagnostics – Equation (7.2) for example – for the correctness of ray theory. Though much of the tomographic work so far made acceptable use of the ray-theoretical approximation, such tests also tell us that more sophisticated analytical techniques are called for if we want to move beyond the current limits on resolution.

For surface waves, the complexity of the finite-frequency sensitivity of waveforms to the heterogeneity in the Earth became apparent in an early study by Woodhouse and Girnius [401] who used a first order perturbation theory. Snieder [325, 327, 334] used first order perturbation theory for high frequency surface waves and was the first to attempt to use these in regional tomography [329]. Because first order perturbation methods were originally developed by the German – later naturalized British – physicist Max Born (1882–1970) within the context of quantum mechanics, first order perturbation theory is often referred to as 'Born theory'. For waves, perturbations lead to scattered waves and yet another term enters the jargon: first order or 'single scattering'. In this book I shall freely mix such terminology.

For body waves, the fact that the sensitivity of travel times and amplitudes extends over the full width of the Fresnel zone and that this extended sensitivity influences seismic tomography was pointed out 20 years ago by Wielandt [394] and Nolet [235, 237]. However, they did not recognize the power of first order perturbation theory to provide a quantitative analysis. Earlier, Hudson [137] had modelled the coda of P-waves using Born scattering theory. Coates and Chapman [63] showed that Born theory adequately models the changes in arrival time and amplitude of body waves using ray theory, opening up the possibility that one could improve on ray theory using ray theory! A formalism based on a first-order perturbation treatment of the wave equation was provided by Woodward [403] and Luo and Schuster [190], who proposed a waveform inversion method and compared this to travel time inversion for acoustic waves in a 2D (cross-well) tomographic experiment. Yomogida [409] was the first to recognize the power of paraxial ray theory in combination with first order perturbation theory. At the same time, Jin et al. [149] developed ray–Born inversion methods that were extended to attenuating media by Ribodetti et al. [276] and applied to invert the results of laboratory experiments. By the mid 1990s, the time seemed also ripe for a change in 3D global tomography, which was still relying heavily on pure ray theory for the inversion of body waves – the type of waves that, at least in theory, provide the best resolution.

Not surprisingly, the first attempts to provide a finite-frequency treatment of teleseismic body waves followed up on the earlier Born theory for normal modes and surface waves. Li and Tanimoto [186] studied the effect of coupling of normal modes on the waveforms of body waves, and Li and Romanowicz [185] used it to develop a tomographic algorithm, that Zhao and Jordan [417] extended to anisotropy. Tanimoto [346] used the slope of the waveform to translate the waveform perturbation directly into a travel time shift. These studies, however, were still restricted to inversions in a two-dimensional setting in which the Earth's properties are assumed independent of the third, horizontal, coordinate. However, Zhao and Dahlen [416] derived Fréchet kernels for body–waveform perturbations in a 3D Earth using asymptotic normal mode theory. At the same time, Snieder and Lomax [333] tried to incorporate finite-frequency effects directly to travel times by applying a smoothing function to the velocity medium.

A full 3D treatment of finite-frequency effects on travel times of elastic waves in the global Earth was given by Marquering et al. [196, 195], who coupled higher modes of surface waves for the case of three-dimensional perturbations and investigated the effect on the cross-correlograms of body waves, summing a large number of surface wave modes to synthesize the S-wave. Zhao et al. [418] subsequently developed three-dimensional kernels using a discrete normal mode formalism.

Because of the surprising result that the travel time of a body wave is insensitive to a perturbation located exactly on the ray itself, Marquering labelled the sensitivity kernels 'banana-doughnut' kernels, in reference to their banana-like shape with a sensitivity 'hole' in the centre. For the computation of three-dimensional kernels, Marquering's surface wave mode summation is considerably more efficient than the summation of discrete normal modes used by Li Zhao et al. [418] in the terrestrial case, and by Birch and Kosovichev [24] in solar tomography, but still not fast enough for routine application in seismic tomography with large data sets. The breakthrough that made finite-frequency tomography a practical possibility was provided by Dahlen et al. [76], who used dynamic ray tracing to make the computation of the sensitivity (or Fréchet) kernels efficient enough for application in large, global tomography inversions, and who provided a comprehensive theory for all types of teleseismic body waves, including the effects of caustics.

# 2

# Ray theory for seismic waves

In this chapter we introduce some basic concepts of the mathematics of wave propagation for acoustic and elastic waves. We shall often use heuristic or intuitive arguments rather than formal proofs, since our primary aim is to provide the necessary minimum background to readers not familiar with the fundamentals of continuum mechanics. Readers eager to educate themselves more extensively on the topics that we touch upon only briefly should consult the advanced seismology textbooks by Aki and Richards [5], Kennett [159] or Dahlen and Tromp [78]. Červený [45] and Chapman [50] have written more specialized books on ray theory and its extensions.

We limit ourselves to wave propagation in isotropic media. This means that the elastic properties of the medium do not depend on its orientation in space: to shear a cube in the $x$-direction requires the same force as in the $y$- or $z$-directions. Even if the real Earth is locally anisotropic in regions of fine layering or of crystal alignment because of solid state flow, the *background* model – the model with respect to which heterogeneity is defined – is usually defined to be isotropic (and spherically symmetric). As we shall see in Chapter 16, anomalies with respect to the background model can be anisotropic. In the Sun, the magnetic field introduces anisotropy, but here too the background model is assumed to be isotropic (a gas).

## 2.1 The stress tensor

The force acting on the side of a volume element is called a 'traction'. Since the force scales with the size of the surface, tractions are always specified per unit area. Figure 2.1 shows the tractions operating on the surfaces of a small volume in a continuous medium. Only the $x$- and $y$-directions are shown, but the concepts easily generalize to the third dimension $z$. The force on any particular surface element in the Earth is a linear function of the orientation of that surface. This dependence is defined by the *stress tensor*. Readers unfamiliar with tensors may

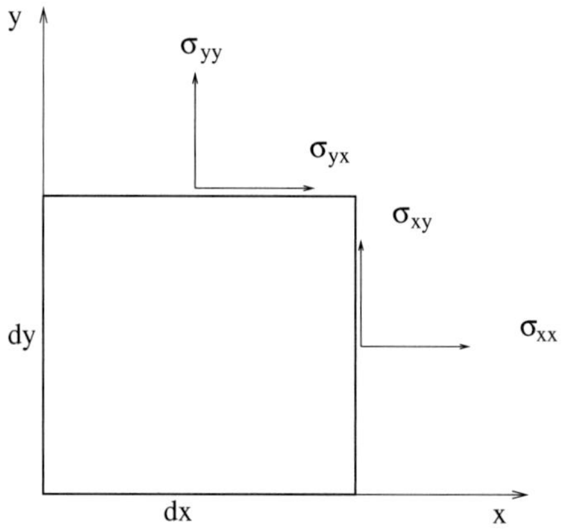


Fig. 2.1. Definition of stress tensor elements as tractions on the sides of a cube.

view the stress tensor as simply a $3 \times 3$ matrix of coefficients, such that the force in the $i$-direction on a unit surface area whose direction is given by a normal unit vector $\hat{\boldsymbol{n}}$ is given by:

$$F_i = \sum_{j=1}^{3} \sigma_{ji} \hat{n}_j \, .$$

When considering the tractions on a patch of the surface of a volume element or layer – as we shall almost always do – we choose the direction of $\hat{\boldsymbol{n}}$ by convention such that it points outward. The first index of the stress tensor gives the direction of the normal to a surface, whereas the second one identifies the force direction. Note that we number $i$ and $j = 1, 2, 3$ for the three Cartesian coordinates $(x, y, z)$. We shall use both $(x, y, z)$ and when convenient the alternative notation, $(x_1, x_2, x_3)$.

Once the stress tensor is known for a point in space we can compute the traction on an arbitrary surface through that point if we know its orientation $\hat{\boldsymbol{n}}$. Since the physical unit for stress is force *per unit area* or N/m$^2$, we obtain a force by multiplying with the area over which the stress operates. One commonly uses Pascal (Pa) instead of N/m$^2$. In general, the stress tensor is time and space dependent: when a wave propagates through a medium the deformations of a three-dimensional medium give rise to local distortions that generate stress, just as they do in a one-dimensional string spring.

## 2.2 Forces in continuous media

Consider a volume element $\mathrm{d}x\,\mathrm{d}y\,\mathrm{d}z$ within a continuum of mass $\mathrm{d}m = \rho\,\mathrm{d}x\,\mathrm{d}y\,\mathrm{d}z$ if $\rho$ is its density. For a surface area $\mathrm{d}x\,\mathrm{d}y$ on such a volume element (with normal in the $z$-direction) the definition of the stress tensor implies the force $F_z$ in the $z$-direction is given by $\sigma_{zz}\,\mathrm{d}x\,\mathrm{d}y$, or in the $x$-direction by $F_x = \sigma_{zx}\,\mathrm{d}x\,\mathrm{d}y$. This is the key to finding the total force on a volume. For example, to find the force in the $x$-direction we sum all forces in that direction on the six sides:

$$\begin{aligned} F_x = & [\sigma_{xx}(x+\mathrm{d}x, y, z) - \sigma_{xx}(x, y, z)]\,\mathrm{d}y\,\mathrm{d}z \\ & + [\sigma_{yx}(x, y+\mathrm{d}y, z) - \sigma_{yx}(x, y, z)]\,\mathrm{d}x\,\mathrm{d}z \\ & + [\sigma_{zx}(x, y, z+\mathrm{d}z) - \sigma_{zx}(x, y, z)]\,\mathrm{d}x\,\mathrm{d}y\,. \end{aligned} \tag{2.1}$$

The volume $\mathrm{d}x\,\mathrm{d}y\,\mathrm{d}z$ or $\mathrm{d}x_1\,\mathrm{d}x_2\,\mathrm{d}x_3$ is denoted by $\mathrm{d}V$, and one easily sees that for example $\mathrm{d}y\mathrm{d}z = \mathrm{d}V/\mathrm{d}x$, so that the first term in $F_x$ can be written as $\mathrm{d}V\,\partial\sigma_{xx}/\partial x$, etc., to give (replacing $x$ by 1):

$$F_1 = \sum_{j=1}^{3} \frac{\partial \sigma_{j1}}{\partial x_j}\mathrm{d}V\,.$$

The forces in the $y$- or $z$-direction may receive the same treatment, and one obtains the general expression:

$$F_i = \sum_{j=1}^{3} \frac{\partial \sigma_{ji}}{\partial x_j}\mathrm{d}V\,. \tag{2.2}$$

Note that forces on opposite sides of the volume have opposite direction. This is easy to understand when one imagines a cube clamped between two hands, one hand pushes left, the other right. There is an intuitive argument to show that the traction on one side of the cube must be the negative of that on the other side: try to shear a cube, with a stress $\sigma_{xy}$. Then imagine the cube to shrink to a very flat box as $\mathrm{d}y \to 0$. The surface area $\mathrm{d}x\mathrm{d}z$ does not go to zero, so that a finite net traction will not go to zero because of the shrinking $\mathrm{d}y$. But when $\mathrm{d}y$ goes to zero the cube has zero mass and we would end up with an infinite acceleration. Thus, no net traction should operate, i.e. the two tractions on either side should cancel exactly and because the normal vectors $\hat{\boldsymbol{n}}$ are of opposite sign we conclude that $\lim_{\mathrm{d}y\to 0} \sigma_{yx}(x, y+\mathrm{d}y, z) = \sigma_{yx}(x, y, z)$. A similar argument can be used to show that the stress tensor must be symmetric to avoid an infinite angular momentum for rotations of the volume, thus

$$\sigma_{ij}(x, y, z) = \sigma_{ji}(x, y, z)\,.$$

## 2.3 Newton's law and the elastodynamic equations

Now that we have the expression for the force (2.2) it is easy to formulate the *elastodynamic equations*, the balance of forces that is simply a restatement of Newton's second law: force = mass × acceleration, formulated for a continuum. When a wave passes through the continuum, the volume elements will be slightly displaced from their equilibrium position. Let $\boldsymbol{u}$ be the very small displacement of the volume element $\mathrm{d}V$ with mass $\mathrm{d}m = \rho \mathrm{d}V$ with respect to the static situation. Its acceleration is then $\partial^2 \boldsymbol{u}/\partial t^2$. Equation (2.2) reflects the forces in the absence of any sources. A source is usually localized in a small region and zero elsewhere. Adding a volume force $\boldsymbol{f}\mathrm{d}V$ as the source term, inserting (2.2) in Newton's law and dividing by $\mathrm{d}V$ yields the elastodynamic equations:†

$$\rho \frac{\partial^2 u_i}{\partial t^2} = \sum_j \frac{\partial \sigma_{ji}}{\partial x_j} + f_i \,. \tag{2.3}$$

Note that the physical unit of $f_i$ is $\mathrm{N/m^3}$, a consequence of the fact that we chose the force term to scale with the volume $\mathrm{d}V$. One well known volume force is gravity. However, we referred the displacement with respect to the static situation, in which the gravitational force is in equilibrium with the compression of the Earth under its own gravitational force. Because of this, the acceleration of gravity does not enter into (2.3). Only in case the displacements are of such long wavelength that the gravitational field itself is noticeably changed, such as for the Earth's lowest eigenvibrations or for g-modes in stars and the Sun, do we need to extend (2.3) with a *perturbation* in the gravity force. This we shall ignore here.

With (2.3) we have three equations, one for each direction, but nine unknowns (three displacements $u_i$ and the six independent elements of the symmetric tensor $\sigma_{ji}$). Empirically, we know that a solid deforms in response to stress in a unique way, so we shall wish to remedy the underdetermined nature of (2.3) by introducing empirical laws that relate stress to displacement. Such laws can be very complicated, but are greatly simplified when we ignore the hysteresis caused by anelastic effects and when we confine ourselves to very small displacements. In that case the medium deforms approximately linearly with the applied stress. We know from elementary physics that the extension of a spring is linearly proportional to the force applied to the spring, a relationship known as Hooke's law. In a similar way, we expect the deformation of a solid to be proportional to the stresses applied. A constant $\boldsymbol{u}$ would represent a simple translation of a solid, and not be accompanied by any stress – so what really interests us is the *deformation* of the medium. The deformation is given by the spatial derivatives of $\boldsymbol{u}$, which can be written as $\partial u_i/\partial x_j$, or in shorthand notation: $\partial_j u_i$. However, even if $\partial_j u_i \neq 0$ we cannot conclude that the volume is deforming, since pure rotations of the volume also cause nonzero $\partial_j u_i$. Such displacement fields are easily recognizable, though. Figure 2.2 shows that for a pure rotation, $\partial_x u_y = -\partial_y u_x$. This holds for all directions,

† Throughout this book the notation is sometimes simplified by leaving out the limits of summation signs, either where these are obvious – as it is here – or sometimes when a subjective choice is involved: e.g. a sum that theoretically runs to infinity as in $\sum_{k=1}^{\infty}$ will have to be cut-off in practical implementations, and the shorter notation $\sum_k$ reflects this ambiguity.

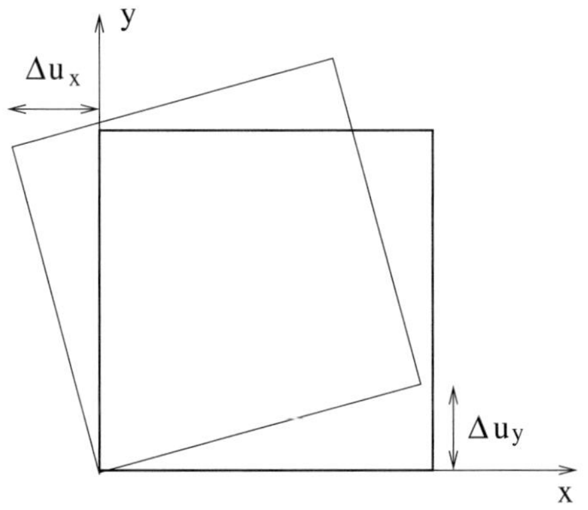


Fig. 2.2. A rotation leads to $\Delta u_y = u_y(x + \mathrm{d}x, y) - u_y(x, y) = -\Delta u_x = -[u_x(x, y + \mathrm{d}y) - u_x(x, y)]$.

so in general: $\partial_i u_j = -\partial_j u_i$. This we can use to remove any rotational component from the expression of the strain tensor. To this end we define a rotation tensor:

$$\Omega_{ij} = \frac{1}{2}\left(\frac{\partial u_j}{\partial x_i} - \frac{\partial u_i}{\partial x_j}\right),$$

as well as a (symmetric) strain tensor as:

$$\epsilon_{ij} = \frac{1}{2}\left(\frac{\partial u_j}{\partial x_i} + \frac{\partial u_i}{\partial x_j}\right), \tag{2.4}$$

such that

$$\frac{\partial u_j}{\partial x_i} = \Omega_{ij} + \epsilon_{ij} .$$

Hooke's law for a spring in the $x$-direction with spring constant $k$ states that $\epsilon_{xx} = k\sigma_{xx}$. There is no other choice for a linear relationship because in a one-dimensional situation $\epsilon_{xx}$ and $\sigma_{xx}$ are the only existing tensor elements (in fact, in 1D it is a little silly to use a tensor formalism). A simple experiment with a sponge makes clear that the simplest linear relationship between stress and strain in 3D, of the form $\sigma_{ij} = c\epsilon_{ij}$ does not hold: if we apply a vertical stress $\sigma_{zz}$ the sponge will shrink in the $z$-direction, but expand in $x$- and $y$-directions even though $\sigma_{xx}$ and $\sigma_{yy}$ are 0. For the same reason, simply varying the constant $c$ with direction ($\epsilon_{ij} = c_{ij}\sigma_{ij}$) does not work. Instead, we must use the most general expression of linearity between stress and strain that allows for a stress field in one direction to influence the displacement in all other directions:

$$\sigma_{ij} = \sum_{kl} c_{ijkl}\epsilon_{kl} . \tag{2.5}$$

The material constants $c_{ijkl}$ form the elasticity tensor. Since $\epsilon_{kl} = \epsilon_{lk}$, the elasticity tensor has the simple symmetry $c_{ijkl} = c_{ijlk}$, so that:

$$\sigma_{ij} = \sum_{kl} c_{ijkl} \frac{\partial u_l}{\partial x_k} , \tag{2.6}$$

where we use the definition of the strain tensor (2.4). The elasticity tensor exhibits more symmetries:

$$c_{ijkl} = c_{jikl} = c_{ijlk} = c_{lkij} . \tag{2.7}$$

The first two symmetries arise from the symmetries of the stress and strain tensor; for a proof of the last symmetry see Aki and Richards [5] or Dahlen and Tromp [78].

If we insert (2.6) into (2.3) we obtain the elastodynamic equations:

$$\rho \frac{\partial^2 u_i}{\partial t^2} = \sum_{jkl} \frac{\partial}{\partial x_j} \left( c_{ijkl} \frac{\partial u_l}{\partial x_k} \right) + f_i . \tag{2.8}$$

### *Exercise*

**Exercise 2.1** Show that a pure rotation gives rise to $\epsilon_{ij} \equiv 0$, i.e. no deformation.

## 2.4 The acoustic wave equation

The oceans and the outer core represent fluid areas in the Earth, where the simplest type of seismic wave, an acoustic wave, propagates. The gaseous layers of the Sun also transmit acoustic waves. By definition of a fluid (of zero viscosity), shear stresses such as $\sigma_{xy}$ are zero, and the remaining stress is isotropic, i.e. equal in all directions: $\sigma_{xx} = \sigma_{yy} = \sigma_{zz}$. In a gas or fluid we prefer to use the scalar pressure $P$ rather than the stress field tensor $\sigma$. We define the pressure as the *negative* value of these three stresses:

$$\sigma_{ij} = -P\delta_{ij} , \tag{2.9}$$

where $\delta_{ij}$, Kronecker's delta, equals 1 if $i = j$ and 0 otherwise.

Hydrophones are often used in the oceans to record pressure $P$ directly. Since pressure is a stress, it is again measured in Pascal (Pa). We show that the simplification of the stress tensor leads to one simple scalar equation rather than three equations involving the acceleration in each direction. For small (acoustic) motions in a fluid or gas Hooke's law becomes:

$$-P = \kappa \sum_j \frac{\partial u_j}{\partial x_j}, \tag{2.10}$$

where $\kappa$ is the incompressibility or bulk modulus, measured in N/m$^2$ or Pa. Note that $\sum_j \partial_j u_j$ is the divergence $\nabla \cdot \boldsymbol{u}$, i.e. the relative volume change of $dV$: pressure is linearly proportional to the change in volume of a fluid element. From here, the derivation of a wave equation is rather straightforward. Insert (2.9) into (2.3) :

$$\rho \frac{\partial^2 u_i}{\partial t^2} = -\frac{\partial P}{\partial x_i} + f_i\,,$$

divide by $\rho$, take the divergence and use (2.10) to eliminate $u_i$:

$$\frac{1}{\kappa}\frac{\partial^2 P}{\partial t^2} = \sum_i \frac{\partial}{\partial x_i}\left(\frac{1}{\rho}\frac{\partial P}{\partial x_i}\right) - \sum_i \frac{\partial}{\partial x_i}\left(\frac{f_i}{\rho}\right). \tag{2.11}$$

In a medium of constant density, (2.11) reduces to a second order partial differential equation of the hyperbolic type known as the wave equation:

$$\frac{\partial^2 P}{\partial t^2} = c^2 \sum_i \frac{\partial^2 P}{\partial x_i^2} - c^2 \frac{\partial f_i}{\partial x_i} = c^2 \nabla^2 P - c^2 \nabla \cdot \boldsymbol{f}\,, \tag{2.12}$$

where the acoustic wave velocity $c$ is defined by

$$c = \sqrt{\kappa/\rho}\,. \tag{2.13}$$

Though $c$ is a property of the wave, we shall often treat the velocity as a property of the medium itself by assigning the local velocity of a wave (e.g. as measured locally in a laboratory test). Where confusion between the two definitions could arise, we denote the local velocity of the medium as the 'intrinsic' velocity.

We modelled the source as a force distribution $\boldsymbol{f}$. For example, for an underwater explosion, $\boldsymbol{f}$ would represent the force of the air pushing against the wall of the bubble. This is usually called an 'equivalent force'. Alternatively, we can model the source as the divergence of a stress tensor, the 'stress glut' $\Gamma$: $f_i = \sum_j \partial_j \Gamma_{ij}$. Outside of the source region we may set $f_i = 0$ (we shall do this often without explicitly specifying this condition). The frequency domain solution for homogeneous media is obtained by assuming a harmonic time dependence of the form $P(\boldsymbol{r}, t) = P(\boldsymbol{r}, \omega)e^{i\omega t}$:†

$$c^2\nabla^2 P + \omega^2 P = 0\,. \tag{2.14}$$

In a homogeneous medium, it is easy to show (Exercise 2.3) that the solution to the wave equation is given by any wave of the form $g(x - ct)$. A special wave is the *harmonic* wave where $g$ is a sine, cosine, or most generally, a complex function e.g.:

$$P(x, t) = P(k, \omega)e^{ik(x-ct)} = P(k, \omega)e^{i(kx-\omega t)}\,, \tag{2.15}$$

where $P(k, \omega)$ is the amplitude of the harmonic wave and where the angular frequency $\omega$ and the wavenumber $k$ are related by what is often called a dispersion relationship:

$$k = \frac{\omega}{c}\,.$$

Individual harmonic waves are important because we can sum them in a Fourier series to form waves of more general waveshape. The fact that we use complex functions is for mathematical convenience only. Functions in nature are always real, and in fact the imaginary part of the term $e^{i(kx-\omega t)}$ cancels if we sum over both positive and negative components of $k$. For $P(x, t)$ to be real, we require

$$P(k, \omega) = P(k, -\omega)^* = P(-k, \omega)^*\,. \tag{2.16}$$

If we see $P(x, t)$ as a function of $x$ only (at fixed time, as in a snapshot) the distance between two peaks in the sine waves is called the wavelength $L$, and we see that:

$$\text{wavelength } L = \frac{2\pi}{k}\,,$$

where $k$ is known as the circular or *angular wavenumber*. On the other hand, if we watch how $P(x, t)$ changes as a function of time at one fixed $x$, the time between two maxima is the 'period' $T$ of the wave:

$$T = \frac{2\pi}{\omega}\,.$$

The direction of the $x$-axis is arbitrary as long as we have a homogeneous medium, and (2.15) can thus always be used in this form. For a system of homogeneous layers we can still use the homogeneous solution within each layer, but the wave will change direction as it passes from one layer to the next and it may be

† $P(\boldsymbol{r}, \omega)$ is just one component in a spectrum of frequencies. The sign in the exponent is defined by our time-to-frequency Fourier transform convention. Fourier transform conventions differ across different disciplines. In this book we define the Fourier transform $f(\omega)$ as: $f(\omega) = \int f(t)\exp(+i\omega t)dt$, which implies the inverse transform $f(t) = (2\pi)^{-1}\int f(\omega)\exp(-i\omega t)d\omega$. See Appendix A at the end of this chapter.

more convenient to introduce a wavenumber vector $\boldsymbol{k}$ in the direction of the wave propagation:

$$P(\boldsymbol{r}, t) = P(\boldsymbol{k}, \omega)\mathrm{e}^{\mathrm{i}(\boldsymbol{k}\cdot\boldsymbol{r}-\omega t)} . \tag{2.17}$$

A point source in a homogeneous medium will excite spherical rather than plane waves, and it is better to formulate the problem in spherical coordinates $(r, \theta, \phi)$. Since derivatives in the angular directions are zero for a spherical wave, the Laplacian $\nabla^2$ in spherical coordinates only involves differentiation with respect to the radius $r$ and the acoustic wave equation can be written:

$$\nabla^2 P = \frac{\partial^2 P}{\partial r^2} + \frac{2}{r}\frac{\partial P}{\partial r} = \frac{1}{c^2}\frac{\partial^2 P}{\partial t^2} . \tag{2.18}$$

## *Exercises*

**Exercise 2.2** Show that $p(t)$ is a real signal if its Fourier transform $P(\omega)$ satisfies $P(\omega) = P(-\omega)^*$

**Exercise 2.3** In one dimension, (2.12) reduces to the wave equation for a string:

$$c^2\frac{\partial^2 P}{\partial x^2} = \frac{\partial^2 P}{\partial t^2} . \tag{2.19}$$

Assume that the velocity $c$ is constant. Use the chain rule of differentiation to show that *any* differentiable function of the form

$$P(x, t) = g(x - ct) \tag{2.20}$$

satisfies the wave equation for the string. Explain why, therefore, this equation is named the ‘wave equation’.

**Exercise 2.4** In three dimensions, (2.12) is shorthand for

$$c^2\left(\frac{\partial^2 P}{\partial x^2} + \frac{\partial^2 P}{\partial y^2} + \frac{\partial^2 P}{\partial z^2}\right) \equiv c^2\nabla^2 P = \frac{\partial^2 P}{\partial t^2} .$$

Again use the chain rule of differentiation to show that *any* differentiable function of the form $P(x, y, z, t) = g(n_x x + n_y y + n_z z - ct)$ with $n_x^2 + n_y^2 + n_z^2 = 1$ satisfies the three-dimensional wave equation. Explain why we may interpret the solution as a plane wave in the direction given by the unit vector $\hat{\boldsymbol{n}}$.

**Exercise 2.5** Show that *any* differentiable function of the form

$$P(r, t) = \frac{1}{r}g(t - r/c) \tag{2.21}$$

satisfies the spherical wave equation (2.18). Explain why the amplitude has to diminish as $1/r$ to conserve energy (Hint: energy is proportional to the square of the pressure amplitude). For a wave in a two-dimensional membrane, how does the amplitude vary with $r$?

## 2.5 The ray approximation

What if the medium is not homogeneous? In that case we are forced to use approximations, short of using brute force and solving (2.11) numerically. The ray approximation, which allows us to predict wave propagation in smoothly varying media, is the most important tool we have to study seismic waves in the Earth, and much of this book is devoted to it (and to its shortcomings!). Again, it will be advantageous to use the frequency domain formulation of the wave equation rather than (2.12). This can be obtained by assuming a harmonic solution of the form $P(\boldsymbol{r}, t) = P(\boldsymbol{r}, \omega)\exp(-\mathrm{i}\omega t)$ and substituting into (2.11):

$$-\omega^2 P = \kappa\nabla \cdot \left(\frac{1}{\rho}\nabla P\right) - \kappa\nabla \cdot \left(\frac{1}{\rho} f\right), \tag{2.22}$$

where $\boldsymbol{f}(\omega)$ now denotes the spectrum of the source. If $P(\boldsymbol{r}, \omega)$ is a constant then the time signal has the shape of a delta function (see 2.62). Such a sharp pulse travels with a local velocity and will have a delay different from $r/c$ in (2.21). At location $\boldsymbol{r} = (x_1, x_2, x_3)$ let us denote the delay by $\tau(\boldsymbol{r})$:

$$P(\boldsymbol{r}, t) = A(\boldsymbol{r})\delta[t - \tau(\boldsymbol{r})] .$$

In an acoustic medium where the density $\rho$ varies with location, the simple wave equation (2.14) or (2.18) does not hold. However, if the change in $\rho$ is gradual, its derivatives will be small and we expect a solution that still behaves like a wave. Its velocity may be faster or slower depending on the location, but in the absence of sharp changes we expect an absence of reflections, such that we retain a wavefront, albeit not the spherical front given by $\tau = r/c$ that we encountered in the homogeneous case. According to (2.64), the delay in the time domain causes a phase shift in the frequency domain, This motivates us to try the following approximate solution for $P(\boldsymbol{r}, \omega)$ with a delay $\tau$ and amplitude $A$ that depend on the location $\boldsymbol{r}$:

$$P(\boldsymbol{r}, \omega) = A(\boldsymbol{r}) \exp[\mathrm{i}\omega\tau(\boldsymbol{r})] . \tag{2.23}$$

Inspection of (2.23) reveals some of the shortcomings of this approximation. The travel time $\tau(\boldsymbol{r})$ defines just one wavefront. Although one can handle a more complex wavefield by adding up rays for reflections and reverberations, such an approach quickly becomes computationally cumbersome. Also, $\tau$ does not depend on the frequency $\omega$, so that diffracted waves can not be handled adequately (diffracted waves are generally dispersive). But for clearly recognizable arrivals on

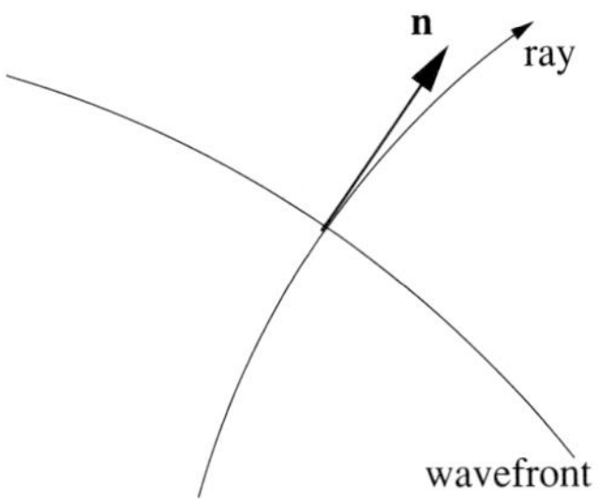


Fig. 2.3. Ray and wavefront geometry

a seismogram (2.23) generally does a good job of representing the major characteristics of the pulse. In Chapter 5 we shall see that the effects of attenuation – which are frequency dependent – can easily be incorporated into ray theory.

If we substitute (2.23) into (2.22), divide by $\omega^2$, set $f = 0$ outside the source region and retain only terms up to order $1/\omega$ we find:

$$-Ae^{i\omega\tau} = \sum_i \kappa \left[ \frac{\partial}{\partial x_i}\left(\frac{1}{\rho}\right) \frac{iA}{\omega} \frac{\partial\tau}{\partial x_i} + \frac{1}{\rho} \frac{iA}{\omega} \frac{\partial^2\tau}{\partial x_i^2} + \frac{2i}{\rho\omega} \frac{\partial A}{\partial x_i} \frac{\partial\tau}{\partial x_i} - \frac{1}{\rho}\left(\frac{\partial\tau}{\partial x_i}\right)^2 A \right] e^{i\omega\tau}.$$

After dividing out the harmonic term $e^{i\omega\tau}$ the two real terms of zero order yield:

$$\sum_i \left(\frac{\partial\tau}{\partial x_i}\right)^2 = |\nabla\tau|^2 = \frac{\rho}{\kappa} = \frac{1}{c^2}, \tag{2.24}$$

which is known as the *eikonal* equation for the wavefront $\tau(\boldsymbol{r})$. The eikonal equation implies that $c\nabla\tau$ is a unit vector (the vector $\hat{\boldsymbol{n}}$ in Figure 2.3), perpendicular to the surface where $\tau =$ constant (the wavefront) and therefore by definition parallel to the ray.

This allows us to derive an equation for the ray geometry itself. Let $d\boldsymbol{r}$ be a tangent along the ray with length $ds$. Thus $\hat{\boldsymbol{n}} = d\boldsymbol{r}/ds$, or, using $c\nabla\tau = \hat{\boldsymbol{n}}$:

$$\nabla\tau(\boldsymbol{r}) = \frac{1}{c}\frac{d\boldsymbol{r}}{ds}.$$

With $\boldsymbol{n} \cdot \nabla\tau = d\tau/ds = 1/c$ and $d(\nabla\tau)/ds = \nabla(d\tau/ds)$:

$$\frac{d}{ds}\left(\frac{1}{c}\frac{d\boldsymbol{r}}{ds}\right) = \nabla\left(\frac{1}{c}\right). \tag{2.25}$$

Computers are much better at solving first-order systems than second-order ones such as (2.25). It is not difficult to transform the system to a first-order system by defining the *slowness* vector $\boldsymbol{p}$:

$$\boldsymbol{p} = \frac{1}{c}\frac{\mathrm{d}\boldsymbol{r}}{\mathrm{d}s} . \tag{2.26}$$

Then

$$\frac{\mathrm{d}\boldsymbol{r}}{\mathrm{d}s} = c\boldsymbol{p}, \qquad \frac{\mathrm{d}\boldsymbol{p}}{\mathrm{d}s} = \nabla\left(\frac{1}{c}\right) . \tag{2.27}$$

Because of its simplicity, differential equations reduced to first order, as is (2.27), are often said to be in 'canonical form', after the Greek $\kappa\alpha\nu\omega\nu$ or 'rule'.

### *Exercise*

**Exercise 2.6** Show that the elements of the vector $\boldsymbol{p}$ are equal to the direction cosines $\gamma_i \equiv x_i/r$ of the ray path scaled by the velocity $c$.

## 2.6 Ray solutions in layered and spherical systems

In a medium with horizontal stratification, where $c(x, y, z) = c(z)$, we may without loss of generality choose our $x$-axis in the direction of propagation. This confines the ray to the $x - z$ plane (i.e. with $p_y = 0$). From (2.27) $\mathrm{d}p_x/\mathrm{d}s = 0$, hence:

$$p_x = \text{constant} = \frac{1}{c}\frac{\mathrm{d}x}{\mathrm{d}s} = \frac{\sin i}{c} , \tag{2.28}$$

which is known as Snel's law.† The angle $i$ is the angle the ray makes with the vertical (Figure 2.4). In the case of horizontal layering the subscript $x$ on $p_x$ is usually omitted. The horizontal slowness $p$ is measured in s/m or s/km. The constant $p$ is also known as the 'ray parameter'. At an interface with a jump in $c$, the slowness $p$ remains constant, which implies a jump in the incidence angle $i$ (refraction). From Figure 2.4b we see that $\mathrm{d}T = \mathrm{d}x\ \sin i/c$ or

$$p = \frac{\mathrm{d}T}{\mathrm{d}X} . \tag{2.29}$$

With the help of Figure 2.4a one finds that the horizontal distance travelled by a ray with slowness $p$ is:

$$X(p) = \int \mathrm{d}x = \int \tan i\ \mathrm{d}z = \int \frac{\sin i}{\sqrt{1 - \sin^2 i}} \mathrm{d}z = \int \frac{cp}{\sqrt{1 - c^2 p^2}} \mathrm{d}z , \tag{2.30}$$

† Though commonly spelled as Snell's law, the law was discovered by Willebrord Snel van Royen (1580–1626), a Dutch mathematician who latinized his name to Snellius. When using his original surname, the spelling with one l is more appropriate.

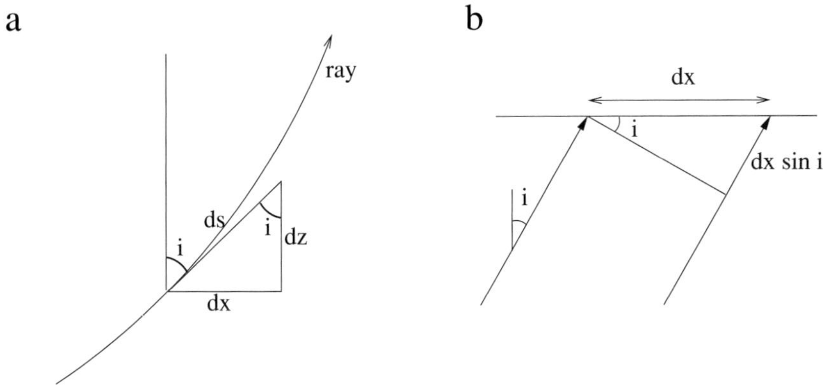


Fig. 2.4. Incidence angle $i$ and other quantities used in the derivation of $T(p)$ and $X(p)$.

and the time travelled is:‡

$$T(p) = \int \mathrm{d}t = \int \frac{\mathrm{d}s}{c} = \int \frac{\mathrm{d}z}{c\sqrt{1-c^2p^2}}, \tag{2.31}$$

where the integration limits are the start and end of the ray. If a diving ray reverses direction, the integrations of the $z$-variable require the integrals to be split up in a downgoing and upcoming part.

In a spherically symmetric body, we can assume without loss of generality that the ray is located in the equatorial plane ($\theta = \pi/2$). The ray equations can then be formulated in polar coordinates with radius $r$ and longitude (or epicentral distance) $\phi$. Let $i$ be the angle of incidence (the angle with the radial direction). From Figure 2.5 we see that:

$$\begin{aligned} \frac{\mathrm{d}r}{\mathrm{d}s} &= \cos i \\ \frac{\mathrm{d}\phi}{\mathrm{d}s} &= \frac{\sin i}{r}. \end{aligned} \tag{2.32}$$

As in the case of plane layers, we may define a constant 'slowness' or ray parameter in the spherically symmetric case in which $\partial c/\partial\theta = \partial c/\partial\phi = 0$. First, observe that for a unit vector $\hat{\boldsymbol{r}}$:

$$\frac{\mathrm{d}}{\mathrm{d}s}(\hat{\boldsymbol{r}} \times \boldsymbol{p}) = \hat{\boldsymbol{r}} \times \frac{\mathrm{d}\boldsymbol{p}}{\mathrm{d}s} + \frac{\mathrm{d}\hat{\boldsymbol{r}}}{\mathrm{d}s} \times \boldsymbol{p} = \hat{\boldsymbol{r}} \times \nabla(1/c) + c\boldsymbol{p} \times \boldsymbol{p} = \boldsymbol{0},$$

‡ Note on the different notations for the time variable: in general, we denote a travel time field with $\tau(\boldsymbol{r})$, but the travel time of a particular ray with $T$, whereas $t$ is reserved for time as a variable. The distinction between $\tau$ and $T$ is not always razor sharp, though.

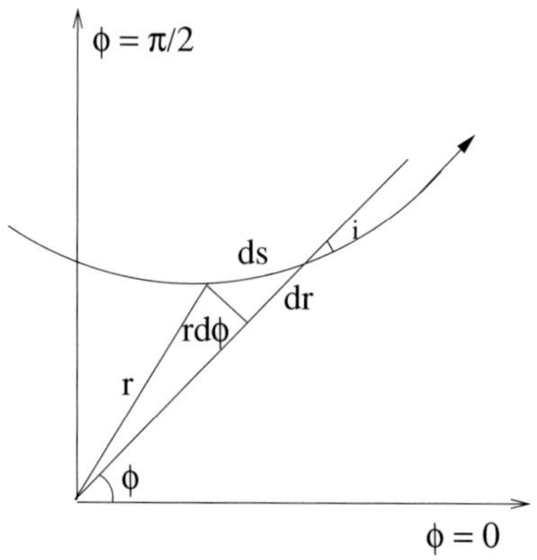


Fig. 2.5. The polar coordinate system for a ray in the equatorial plane.

since $\nabla(1/c)$ is oriented in the direction of $\hat{\boldsymbol{r}}$. Thus $\hat{\boldsymbol{r}} \times \boldsymbol{p}$ is a constant vector, the length of which we denote by $p$, as in the Cartesian case, but Snel's law now takes a different form:

$$p = \frac{r \sin i}{c} \quad \text{(spherical).} \tag{2.33}$$

Replacing d$x$ with $r$d$\Delta$ in Figure 2.4b we see that for the spherical slowness:

$$p = \frac{\mathrm{d}T}{\mathrm{d}\Delta}.$$

If we wish to integrate (2.32) with the pathlength $s$ as an independent variable, we still lack an equation for the angle of incidence $i$ as a function of $s$. This can be obtained by differentiating (2.33) with respect to $s$. Using the fact that $p$ is constant for a ray, thus independent of $s$, we obtain:

$$\frac{\mathrm{d}i}{\mathrm{d}s} = \frac{p}{r}\left(\frac{\mathrm{d}c}{\mathrm{d}r} - \frac{c}{r}\right). \tag{2.34}$$

Eliminating d$s$ from the three differential equations for $r$, $\phi$ and $i$ yields a more condensed system for $r$ and $i$ as a function of $\phi$:

$$\frac{\mathrm{d}r}{\mathrm{d}\phi} = r \cot i$$
$$\frac{\mathrm{d}i}{\mathrm{d}\phi} = \frac{\mathrm{d}c}{\mathrm{d}r}\frac{r}{c} - 1, \tag{2.35}$$

with continuity of $p$ at an interface with a jump in $c$. Both formulations, as well as one with $r$ as independent variable (Exercise 2.10) are used in ray tracing applications (Chapter 3).

For laterally heterogeneous Earth models, the equations (2.27) and (2.35) must be solved by numerical methods (e.g. Červený et al. [47] or Moser [221]) – to which we return in the next chapter – or by the perturbation methods developed by Snieder and Aldridge [331].

### *Exercises*

**Exercise 2.7** Verify that $|\hat{\boldsymbol{r}} \times \boldsymbol{p}| = r \sin i / c$.

**Exercise 2.8** Check that the correct units for the spherical slowness are seconds/radian (s/rad). For an Earth with surface radius 6371 km, how fast does a ray travel (in km/s) with $p = 637$ s/rad?

**Exercise 2.9** Where along the ray path is $p = r/c$? Give a geometrical interpretation by drawing a picture of the ray and its wavefront at the location where $p = r/c$.

**Exercise 2.10** Reformulate (2.35) in terms of a system that gives $\phi$ and $i$ as a function of $r$.

**Exercise 2.11** Show that the travel time $T$ and epicentral distance $\Delta$ for a ray in a spherically symmetric Earth are given by:

$$T(p) = \int \frac{\mathrm{d}r}{c\sqrt{1 - p^2c^2/r^2}} \tag{2.36}$$

$$\Delta(p) = \int \frac{\mathrm{d}r}{r\sqrt{r^2/p^2c^2 - 1}} \,. \tag{2.37}$$

**Exercise 2.12** For a ray that bottoms at $r = b$, the total travel time in (2.36) is given by a sum of two integrals, both starting at $b$. Give the complete expression (i.e. including integration bounds) for the travel time of a ray starting from an earthquake at depth $h$ and recorded at the Earth's surface where $r = a$.

## 2.7 Geometrical spreading

Consider the solid angle $\mathrm{d}\Omega$ which is spanned by seismic rays leaving the source within a small incidence angle $\mathrm{d}i_0$ (Figure 2.6). These rays span a band with circumference $2\pi\epsilon \sin i_0$ and width $\epsilon \mathrm{d}i_0$ on a small sphere with a total surface of $4\pi\epsilon^2$. The numerical value of the solid angle is by definition the surface area of the ray bundle crossing a unit sphere, so that:

$$\mathrm{d}\Omega = \frac{2\pi\epsilon \sin i_0 \epsilon \mathrm{d}i_0}{4\pi\epsilon^2} \times 4\pi = 2\pi \sin i_0 \mathrm{d}i_0 \,.$$

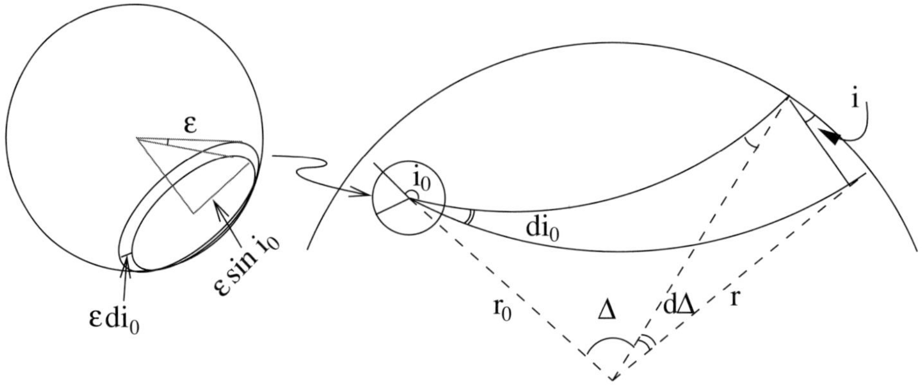


Fig. 2.6. The width of a ray bundle. The focal sphere is enlarged on the left, and shows the band of width $\epsilon \mathrm{d}i_0$ with a radius $\epsilon$ sin $i_0$.

When it arrives back at the Earth's surface the ray tube has a cross-section $S$ which is:

$$S = 2\pi r \sin\Delta \; r \, \mathrm{d}\Delta \; \cos i \equiv \mathcal{R}^2 \mathrm{d}\Omega \, ,$$

where the last term defines the geometrical spreading $\mathcal{R}$, so that:

$$\mathcal{R}^2 = \frac{r^2 \sin\Delta \cos i}{\sin i_0} \left( \frac{\mathrm{d}\Delta}{\mathrm{d}i_0} \right) .$$

Since $p = r_0 \sin i_0 / c_0$ we have

$$\frac{\mathrm{d}p}{\mathrm{d}\Delta} = \left( \frac{\mathrm{d}i_0}{\mathrm{d}\Delta} \right) \frac{r_0}{c_0} \cos i_0 \quad \rightarrow \quad \frac{\mathrm{d}i_0}{\mathrm{d}\Delta} = \frac{c_0}{r_0 \cos i_0} \left( \frac{\mathrm{d}^2 T}{\mathrm{d}\Delta^2} \right) ,$$

or

$$\mathcal{R}^2 = \frac{r^2 r_0 \cos i_0 \sin\Delta \cos i}{c_0 \sin i_0} \left( \frac{\mathrm{d}^2 T}{\mathrm{d}\Delta^2} \right)^{-1} . \tag{2.38}$$

This geometrical spreading factor $\mathcal{R}$ is equivalent to the distance $R$ from the source in the case of a homogeneous medium. Amplitude is proportional to $\mathcal{R}^{-1}$ in order to conserve energy, and this is seen to blow up at singular points where $\mathcal{R} = 0$ (caustics) in the travel time curve. This is a fundamental shortcoming of ray theory. Another important shortcoming is that effects of wave diffraction are not modelled by rays; we return to this in Chapter 7.

## 2.8 Rays in an isotropic, elastic Earth

The acoustic case was easy to handle because the equations only deal with a scalar quantity, the pressure $P$. The next level of complexity is encountered when we

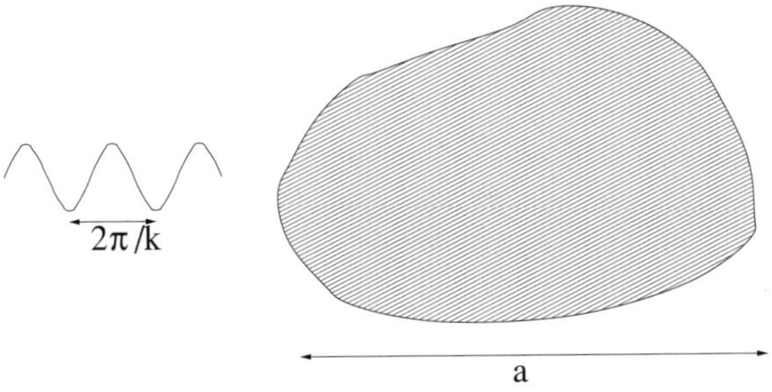


Fig. 2.7. A sketch of the assumption that $\mathcal{O}(\partial u_i/\partial x_j) \gg \mathcal{O}(\partial\mu/\partial x_j)$, or $ka \gg 1$

introduce an elastic Earth. To reduce the large number of elastic variables $c_{ijkl}$ we usually assume that the elastic properties of the Earth are not dependent on the orientation of the rock. In such an isotropic Earth the elasticity tensor involves only two constants, the bulk modulus $\kappa$ and the shear modulus $\mu$:

$$c_{ijkl} = \left(\kappa - \frac{2}{3}\mu\right)\delta_{ij}\delta_{kl} + \mu(\delta_{ik}\delta_{jl} + \delta_{il}\delta_{jk})\,. \tag{2.39}$$

A wave with a spatial dependence $u \sim \mathrm{e}^{ikx}$ will have derivatives $\partial u/\partial x \sim ku$. Similarly, variations in elastic parameters with a correlation distance† $a$ will give derivatives scaled by $a$, e.g. $\partial\mu/\partial x \sim \mu/a$. If $ka \gg 1$, we may neglect the spatial derivatives of the elastic parameters. This is equivalent to saying that the wavelength of the wave is much smaller than the scale length of the heterogeneities (Figure 2.7). Substituting (2.39) into (2.8), introducing the Lamé parameter, $\lambda \equiv \kappa - \frac{2}{3}\mu$, and going into the frequency domain by replacing $\partial^2 u_i/\partial t^2$ by $-\omega^2 u_i$:

$$-\rho\omega^2 u_i = \sum_{jkl} \frac{\partial}{\partial x_j}\left(c_{ijkl}\frac{\partial u_l}{\partial x_k}\right) \tag{2.40}$$

$$= \sum_{jkl} \frac{\partial}{\partial x_j}\left[\lambda\delta_{ij}\delta_{kl}\frac{\partial u_l}{\partial x_k}\right] + \frac{\partial}{\partial x_j}\left[\mu(\delta_{ik}\delta_{jl} + \delta_{il}\delta_{jk})\frac{\partial u_l}{\partial x_k}\right]$$

$$= \frac{\partial}{\partial x_i}\left[\lambda\sum_k \frac{\partial u_k}{\partial x_k}\right] + \frac{\partial}{\partial x_j}\left[\mu\left(\frac{\partial u_j}{\partial x_i} + \frac{\partial u_i}{\partial x_j}\right)\right]. \tag{2.41}$$

Or, in vector notation, writing $\nabla\boldsymbol{u}$ for $\partial u_i/\partial x_j$ and $\nabla\boldsymbol{u}^T$ for its transpose $\partial u_j/\partial x_i$:

$$-\rho\omega^2\boldsymbol{u} = \nabla(\lambda\nabla\cdot\boldsymbol{u}) + \nabla\cdot[\mu(\nabla\boldsymbol{u} + \nabla\boldsymbol{u}^T)]\,. \tag{2.42}$$

† For the purpose of this book it is sufficient to define the correlation distance as the distance over which a parameter still looks smooth.

If we can neglect the derivatives of $\lambda$ and $\mu$, i.e. if the variations in elastic properties are smooth:

$$-\rho\omega^2 u_i = \sum_j \lambda \frac{\partial^2 u_j}{\partial x_i \partial x_j} + \mu \frac{\partial}{\partial x_j}\left(\frac{\partial u_i}{\partial x_j} + \frac{\partial u_j}{\partial x_i}\right). \qquad (2.43)$$

The trial solution comparable to (2.23) is now:

$$\boldsymbol{u}(\boldsymbol{r}, \omega) = \boldsymbol{A}(\boldsymbol{r})\mathrm{e}^{\mathrm{i}\omega\tau(\boldsymbol{r})}. \qquad (2.44)$$

Substituting this into the elastodynamic equation, and collecting all terms with $\omega^2$:

$$-\rho A_i = \sum_j -(\lambda + \mu)\frac{\partial \tau}{\partial x_i}\frac{\partial \tau}{\partial x_j} A_j - \mu \left(\frac{\partial \tau}{\partial x_j}\right)^2 A_i ,$$

which is more easily interpretable in vector notation:

$$-\rho \boldsymbol{A} + (\lambda + \mu)\nabla\tau(\nabla\tau \cdot \boldsymbol{A}) + \mu|\nabla\tau|^2 \boldsymbol{A} = \boldsymbol{0}.$$

This equation contains three terms, two of which are parallel to $\boldsymbol{A}$, the other one is parallel to $\nabla\tau$. Obviously, the equation can only be satisfied if all nonzero terms are parallel, which implies either

$$\boldsymbol{A} = \text{constant } \nabla\tau \rightarrow |\nabla\tau|^2 = \frac{\rho}{\lambda + 2\mu} = \frac{\rho}{\kappa + \frac{4}{3}\mu} = \frac{1}{V_\mathrm{P}^2}, \qquad (2.45)$$

or

$$\boldsymbol{A} \cdot \nabla\tau = 0 \rightarrow |\nabla\tau|^2 = \frac{\rho}{\mu} = \frac{1}{V_\mathrm{S}^2}. \qquad (2.46)$$

(2.45) and (2.46) define the velocities of compressional or P-waves and shear or S-waves in the Earth. The formal equality of these equations with (2.24) implies that the ray tracing equations we derived for an acoustic medium are equally valid for elastic P- and S-waves, even though the polarizations of P- and S-waves differ fundamentally (Exercises 2.13 and 2.14).

## *Exercises*

**Exercise 2.13** How can you show that the direction of displacement of S-waves is perpendicular to the ray? The polarization of an S-wave may be in the (local) plane of propagation or perpendicular to it. Is the S-velocity dependent on its polarization?

**Exercise 2.14** Why can you state that the direction of displacement of a P-wave is parallel to the ray direction? Are there different polarizations of the P-wave?

**Exercise 2.15** In a fluid $\mu = 0$. Reduce the expressions for $V_P$ and $V_S$ for this case and compare to the acoustic velocity $c$.

**Exercise 2.16** The displacement $\boldsymbol{u}$ has a pressure $P$ associated with it given by (2.10). Does this pressure satisfy the wave equation in a solid? If so, how fast is it propagating?

**Exercise 2.17** Physically, $\kappa$ measures the resistance to volume change, while leaving the shape unchanged, whereas $\mu$ measures the resistance to change in shape with constant volume. Explain why the P-velocity is not independent of $\mu$.

## 2.9 Fermat's Principle

Fermat's Principle, originally formulated in the first century by Hero of Alexandria for mirror reflections, but named after the French lawyer/mathematician Pierre de Fermat (1601–1665) who generalized it for arbitrary light rays, states that the travel time of a ray between two given points in space must be stationary for small perturbations in the path followed by the ray. We shall prove the principle in this section for seismic rays. The travel time $\mathrm{d}t$ along a ray segment $\mathrm{d}\boldsymbol{r}$ is given by:

$$\mathrm{d}t = \frac{|\mathrm{d}\boldsymbol{r}|}{c} = \frac{\mathrm{d}s}{c},$$

and the total travel time by:

$$T = \int \mathrm{d}t = \int \frac{\mathrm{d}s}{c}, \tag{2.47}$$

where $\mathrm{d}s = |\mathrm{d}\boldsymbol{r}|$ is small enough so that we make only a small error by replacing the velocity $c$ along the segment by its average over $\mathrm{d}\boldsymbol{r}$. If we perturb the ray location such that $\mathrm{d}\boldsymbol{r} \to d(\boldsymbol{r} + \delta\boldsymbol{r})$, the ray encounters a different velocity $c + \delta c$ and:

$$\mathrm{d}t + \delta\mathrm{d}t = \frac{|\mathrm{d}\boldsymbol{r} + \mathrm{d}\delta\boldsymbol{r}|}{c + \delta c} \approx \frac{|\mathrm{d}\boldsymbol{r} + \mathrm{d}\delta\boldsymbol{r}|}{c}\left(1 - \frac{\delta c}{c}\right) \approx |\mathrm{d}\boldsymbol{r} + \mathrm{d}\delta\boldsymbol{r}|\left(\frac{1}{c} + \delta\left(\frac{1}{c}\right)\right).$$

By writing out, one verifies that:

$$|\mathrm{d}\boldsymbol{r} + \mathrm{d}\delta\boldsymbol{r}| = |\mathrm{d}\boldsymbol{r}| + \frac{\mathrm{d}\boldsymbol{r} \cdot \mathrm{d}\delta\boldsymbol{r}}{|\mathrm{d}\boldsymbol{r}|} = \hat{\boldsymbol{n}} \cdot (\mathrm{d}\boldsymbol{r} + \mathrm{d}\delta\boldsymbol{r}),$$

where $\hat{\boldsymbol{n}} = \mathrm{d}\boldsymbol{r}/|\mathrm{d}\boldsymbol{r}| = \mathrm{d}\boldsymbol{r}/\mathrm{d}s$, a unit vector along the ray, so that (to first order):

$$\delta\mathrm{d}t = \left(\frac{1}{c}\right)\hat{\boldsymbol{n}} \cdot \mathrm{d}\delta\boldsymbol{r} + \delta\left(\frac{1}{c}\right)\hat{\boldsymbol{n}} \cdot \mathrm{d}\boldsymbol{r}.$$

This is only the change in the travel time $\mathrm{d}t$ along the small segment. The change in total travel time $T$ between two points A and B is obtained by integrating $\mathrm{d}t$

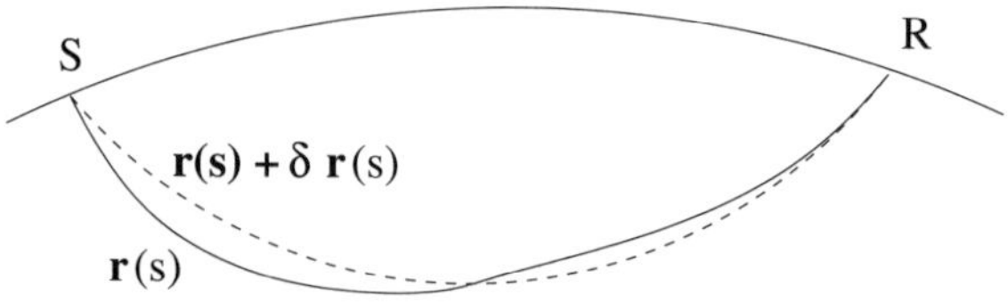


Fig. 2.8. Fermat's principle: perturbing the true minimum time ray from the actual trajectory $\boldsymbol{r}(s)$ (solid line) to a simplified path $\boldsymbol{r}(s) + \delta\boldsymbol{r}(s)$ (broken line) influences the time integral only to second order in $\delta\boldsymbol{r}$. If the perturbed ray is computed for a spherically symmetric Earth model, this greatly improves the efficiency of the calculations.

from A to B, so the *perturbation* in $T$ is:

$$\begin{aligned}\delta T &= \int_A^B \delta\left(\frac{1}{c}\right)\hat{\boldsymbol{n}}\cdot \mathrm{d}\boldsymbol{r} + \int_A^B \left(\frac{1}{c}\right)\hat{\boldsymbol{n}}\cdot \mathrm{d}\delta\boldsymbol{r} \\ &= \int_A^B \delta\left(\frac{1}{c}\right)\hat{\boldsymbol{n}}\cdot \frac{\mathrm{d}\boldsymbol{r}(s)}{\mathrm{d}s}\mathrm{d}s + \int_A^B \left(\frac{1}{c}\right)\hat{\boldsymbol{n}}\cdot \frac{\mathrm{d}\delta\boldsymbol{r}(s)}{\mathrm{d}s}\mathrm{d}s\,.\end{aligned}$$

We insert $\delta(1/c) = \delta\boldsymbol{r}\cdot\nabla(1/c)$ in the first integral, integrate the second one by parts and use $\hat{\boldsymbol{n}} = \mathrm{d}\boldsymbol{r}/\mathrm{d}s$:

$$\delta T = \int_A^B \delta\boldsymbol{r}\cdot\left(\nabla\left(\frac{1}{c}\right) - \frac{\mathrm{d}}{\mathrm{d}s}\left(\frac{1}{c}\frac{\mathrm{d}\boldsymbol{r}}{\mathrm{d}s}\right)\right)\mathrm{d}s\,,$$

where the end contributions disappear because we impose that the ray starts at A and ends at B so that $\delta\boldsymbol{r}_{\mathrm{A}} = \delta\boldsymbol{r}_{\mathrm{B}} = 0$. The term within the brackets is zero because of (2.25), so that $\delta T = 0$, to first order in the ray perturbation. This is very important when we formulate a perturbation theory for travel time tomography: when we perturb the Earth we adopt the ray path calculated for some standard, usually spherically symmetric Earth model, even though the outcome of the interpretation will be an Earth model full of three-dimensional heterogeneities. Fermat's Principle assures us that the resulting error is of second order.

## 2.10 Huygens, Fresnel and Green

We have seen that ray theory is really an infinite-frequency approximation, since we neglect terms of the order $\omega^{-1}$ and higher. In reality, seismic waves have a finite frequency. Seismic sources that are strong enough to emit waves that can be observed worldwide are generally of magnitude 5 or higher and have rupture surfaces of the order of a kilometre or more. That conforms to the wavelength of a P-wave of 5–10 Hz. Higher frequencies are not efficiently emitted because waves emanating from different parts of the rupture surface do not interfere constructively.

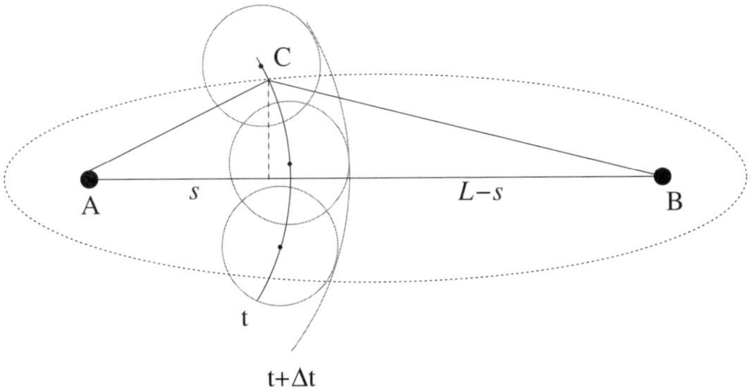


Fig. 2.9. Huygens' Principle: the wavefront at time $t$ hosts secondary sources (small grey circles), the envelope of which after $\Delta t$ seconds defines the wavefront at time $t + \Delta t$. Point C on the wavefront emits a secondary wave towards B with a certain delay. All secondary sources with the same delay are located on an ellipse through C (dotted line).

But even frequencies higher than 1 Hz are rarely observed at large distances because of anelastic attenuation.

What is the effect of finite-frequency on the seismic wave? A somewhat simplistic answer can be found by application of an important principle discovered by the Dutch scientist and inventor Christiaan Huygens (1629–1695). Huygens' Principle states that *every point on the wavefront acts as a secondary source emitting a new wavefront*. Figure 2.9 illustrates this principle. The secondary sources on the wavefront at time $t$ send out waves that, at time $t + \Delta t$, have arrived at the location given by the small circles. The envelope of these small wavefronts constitutes the wavefront at time $t + \Delta t$. In three dimensions, one can show that the backpropagating waves from the secondary sources cancel perfectly in a homogeneous medium – such that for example a $\delta$-like pulse keeps it shape as the wave propagates. This is not true in two dimensions, as one can easily observe by throwing a stone in a very shallow pond and observing a series of concentric circles spreading out. Note that this means that we can only talk to each other in 3D; in 2D any sound message would be garbled.

The French physicist Augustin Fresnel (1788–1827) showed how the secondary sources can give us an idea of the 'width' of a ray at finite-frequency. Consider the wave in Figure 2.9 travelling from A to B, and secondary source at C on the wavefront, at a distance $h$ from the direct path AB. Energy from C will arrive at B with a delay. If AB has length $L$, and C, when projected on AB is at a distance $s$ from A, then the difference in length is:

$$ACB - AB = \sqrt{s^2 + h^2} + \sqrt{(L-s)^2 + h^2} - L \approx \frac{h^2}{2}\left(\frac{1}{s} + \frac{1}{L-s}\right),$$

where we use a first-order Taylor expansion for small $h/s$ and $h/(L-s)$. The secondary wave will still interfere constructively with the direct arrival if the difference is less than half a wavelength $\lambda$ (some prefer a stricter criterion of $\lambda/4$). This region is called the 'Fresnel zone'. This criterion defines an ellipse with a maximum $h$ at the centre where $s = L/2$ and where $2h^2/L \leq \lambda/2$ or

$$h = \frac{1}{2}\sqrt{\lambda L}\,. \tag{2.48}$$

For the diameter we thus obtain the simple expression $\sqrt{\lambda L}$. For non-homogeneous media we may still use this to get a back-of-the envelope estimate of the Fresnel zone: a teleseismic P-wave with a typical wavelength of 40 km and a length of 8000 km thus has a Fresnel zone with a maximum diameter of 566 km. This is at the limit of the resolving power of current tomographic studies, and the motivation behind the development of *finite-frequency* tomography. For long-period P-waves the Fresnel zone easily exceeds a diameter of 1000 km.

The modern version of Huygens' Principle is the representation theorem for the wavefield $P(\boldsymbol{r}, \omega)$, which is a form of Green's theorem. We shall derive the principle for acoustic waves. We start from (2.22) which we reformulate for a field $P_1$ excited by a force $\boldsymbol{f}_1$:

$$\frac{\omega^2}{\kappa}P_1 + \nabla\cdot\left(\frac{1}{\rho}\nabla P_1\right) = S_1\,,$$

where $S_1 \equiv \nabla\cdot \boldsymbol{f}_1/\rho$. We multiply this equation by a field $P_2$ from a different source $S_1 = \nabla\cdot \boldsymbol{f}_2/\rho$:

$$\frac{\omega^2}{\kappa}P_2P_1 + P_2\nabla\cdot\left(\frac{1}{\rho}\nabla P_1\right) = P_2S_1\,.$$

We obtain a similar equation with indices 1 and 2 reversed if we start with $P_2$. Doing so, and subtracting:

$$\nabla\cdot\left[\frac{1}{\rho}(P_2\nabla P_1 - P_1\nabla P_2)\right] = P_2S_1 - P_1S_2\,.$$

Integrating over volume $V$ with surface $S$ and applying Gauss' theorem:

$$\int_V (P_2S_1 - P_1S_2)\mathrm{d}^3\boldsymbol{r} = \int_S \frac{1}{\rho}\hat{\boldsymbol{n}}\cdot(P_2\nabla P_1 - P_1\nabla P_2)\,\mathrm{d}^2\boldsymbol{r}\,. \tag{2.49}$$

The right-hand side disappears at the surface of the Sun or ocean, where the pressure is zero. In general, for homogeneous boundary conditions on $S$, the volume integral equals zero.

So far, we have simply manipulated equations for arbitrary wavefields and sources. The power of (2.49) only becomes evident when one introduces *point* sources. This was recognized by George Green (1793–1841), a British amateur mathematician, who published his essay privately in 1828 because he 'felt that it was presumptuous to submit his paper to a journal' and who was not fully appreciated until his work was rediscovered by Lord Kelvin [48]. The solution to a point force is now commonly referred to as a Green's function† and denoted as $G(\boldsymbol{r}, \omega; \boldsymbol{r}_s)$, if the source is located in $\boldsymbol{r}_s$.

Following this idea, we assume that the boundary conditions are such that the surface integral disappears and that $S_1$ is a point source at location $\boldsymbol{r}_1$ at time $t_1$:

$$S_1(\boldsymbol{r}, \omega) = \delta(\boldsymbol{r} - \boldsymbol{r}_1)\mathrm{e}^{\mathrm{i}\omega t_1} . \tag{2.50}$$

Similarly, for $S_2$ we assume a point source at location $\boldsymbol{r}_2$ at time $t_2$:

$$S_2(\boldsymbol{r}, \omega) = \delta(\boldsymbol{r} - \boldsymbol{r}_2)\mathrm{e}^{\mathrm{i}\omega t_2} .$$

The wavefields resulting from these point sources are by definition Green's functions: $P_1 \equiv G(\boldsymbol{r}, \omega; \boldsymbol{r}_1)$ and $P_2 \equiv G(\boldsymbol{r}, \omega; \boldsymbol{r}_2)$. Inserting this into (2.49) immediately gives an important result, the *reciprocity* relation:

$$G(\boldsymbol{r}_2, \omega; \boldsymbol{r}_1)\mathrm{e}^{\mathrm{i}\omega t_2} = G(\boldsymbol{r}_1, \omega; \boldsymbol{r}_2)\mathrm{e}^{\mathrm{i}\omega t_1} . \tag{2.51}$$

An even simpler symmetry becomes evident if we choose $t_1 = t_2$. The time delays $t_1$ and $t_2$ are added as phases in the spectra of $G$ through the exponentials, and not directly evident in the notation we adopt for $G$ in the spectral domain. But in the time domain we write the time dependence explicitly (using 2.65):

$$G(\boldsymbol{r}_2, t - t_2; \boldsymbol{r}_1, t_1) = G(\boldsymbol{r}_1, t - t_1; \boldsymbol{r}_2, t_2). \tag{2.52}$$

Equation (2.52) tells us that we can switch source and receiver and record the same signal, with the same delay and amplitude, no matter how heterogeneous the transmitting medium is.

The connection between Green's functions and Huygens' Principle becomes clear if we set $S_2 = 0$ but consider an arbitrary surface $S$ such that the surface integral in (2.49) does not disappear. Note that the wave equation has solutions even if the source term is zero, so that $P_2$ is not necessarily zero. We again assume a point source (2.50) for the first wavefield, and without loss of generality we choose the time origin at the timing of the source so that $t_1 = 0$. This allows us to write a wavefield $P_2$ in terms of its value at that surface – away from any of its

† A Green's function at location $\boldsymbol{r}$ and time $t$ from a source at $\boldsymbol{r}'$ at time $\tau$ is written as $G(\boldsymbol{r}, t; \boldsymbol{r}', \tau)$. This notation is somewhat cumbersome and not always needed. If the source is at the origin at time 0 , we shall often simplify notation and write $G(\boldsymbol{r}, t)$ or $G(\boldsymbol{r}, \omega)$. For arbitrary source location $\boldsymbol{r}'$ we write $G(\boldsymbol{r}, t; \boldsymbol{r}')$, or $G(\boldsymbol{r}, \omega; \boldsymbol{r}')$. If delayed by $\tau$ seconds, the Green's function becomes $G(\boldsymbol{r}, t; \tau) = G(\boldsymbol{r}, t - \tau)$.

sources – and the Green's function for sources on that surface:

$$\int_V (P_2(\boldsymbol{r},\omega)\delta(\boldsymbol{r}-\boldsymbol{r}_1))\mathrm{d}^3\boldsymbol{r} = P_2(\boldsymbol{r}_1,\omega)$$
$$= \int_S \frac{1}{\rho}\hat{\boldsymbol{n}}\cdot[P_2(\boldsymbol{r},\omega)\nabla G(\boldsymbol{r},\omega;\boldsymbol{r}_1) - G(\boldsymbol{r},\omega;\boldsymbol{r}_1)\nabla P_2(\boldsymbol{r},\omega)]\,\mathrm{d}^2\boldsymbol{r}\,. \qquad (2.53)$$

From a practical point of view, (2.53) is not very efficient since it would force us to compute the Green's function for many different source locations $\boldsymbol{r}_1$ if we wish to know $P_2$ at those locations. A look at the reciprocity (2.51) shows, however, that we only need source at the surface $S$ if we exchange $\boldsymbol{r}$ and $\boldsymbol{r}_1$ in the expressions for $G$. Also, the notation of (2.53) is unnecessarily cumbersome. We can do away with the subscript of $P_2$. This gives:

$$P(\boldsymbol{r}_1,\omega) = \int_S \frac{1}{\rho}\hat{\boldsymbol{n}}\cdot[P(\boldsymbol{r},\omega)\nabla G(\boldsymbol{r}_1,\omega;\boldsymbol{r}) - G(\boldsymbol{r}_1,\omega;\boldsymbol{r})\nabla P(\boldsymbol{r},\omega)]\,\mathrm{d}^2\boldsymbol{r}\,,$$

which one usually finds in the literature with a change in notation ($\boldsymbol{r}\to\boldsymbol{r}'$, $\boldsymbol{r}_1\to\boldsymbol{r}$):

$$P(\boldsymbol{r},\omega) = \int_S \frac{1}{\rho}\hat{\boldsymbol{n}}\cdot[P(\boldsymbol{r}',\omega)\nabla' G(\boldsymbol{r},\omega;\boldsymbol{r}') - G(\boldsymbol{r},\omega;\boldsymbol{r}')\nabla' P(\boldsymbol{r}',\omega)]\,\mathrm{d}^2\boldsymbol{r}'. \qquad (2.54)$$

Since multiplication in the frequency domain is equivalent to convolution in the time domain (see Appendix A), the time-domain version of (2.54) involves an integration over time delay $\tau$ as well:

$$P(\boldsymbol{r},t) = \int_S \frac{1}{\rho}\hat{\boldsymbol{n}}\cdot[P(\boldsymbol{r}',t) * \nabla' G(\boldsymbol{r},t;\boldsymbol{r}',0) - G(\boldsymbol{r},t;\boldsymbol{r}',0) * \nabla' P(\boldsymbol{r}',t)]\,\mathrm{d}^2\boldsymbol{r}'\,.$$

Huygens' Principle is obtained for the special case that $S$ is the wavefront of $P$. We see, however, that one important difference with Huygens' original concept arises: the source is composed of both a pressure source ($P$) and a dipole source ($\nabla P$).

Similar representation theorems can be established for the vector wavefield in the elastic case. In Chapter 4 we encounter ray-theoretical expressions for the Green's function.

## 2.11 Flow: solar p-waves or ocean acoustic waves

In the atmosphere and oceans, as well as in the Sun, the medium of interest is a gas or fluid *in flow*. In fact, in these cases the direction and magnitude of flow is of direct scientific interest. Figure 2.10 sketches the situation. The flow velocity is small with respect to the acoustic velocity $c$ – of the order of 0.1 m/s (versus 1500 m/s for the sound velocity) in the case of ocean flow (Munk et al. [223]). Typical wind velocities of 10 m/s are also well below the velocity of sound in the atmosphere (Virieux et al., [386]). The motions in the Sun's convective region are very complex. Gizon and Birch [118] summarize some of the observations.

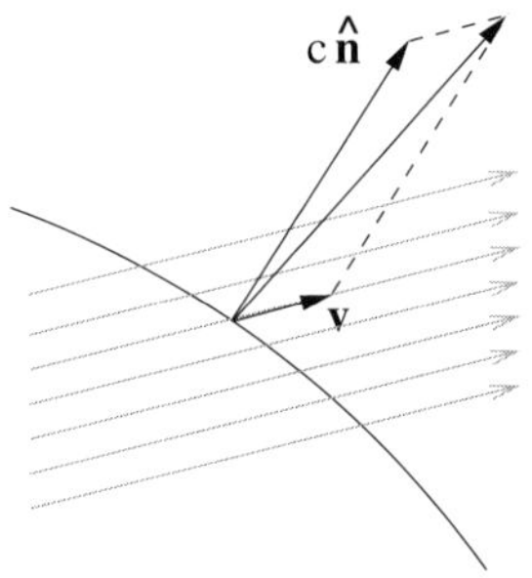


Fig. 2.10. A wavefront (solid line) in a flow field (grey vectors). A point on the wavefront travels in the direction given by the sum of the wavefront normal velocity $c\hat{\boldsymbol{n}}$ and the flow vector $\boldsymbol{v}$.

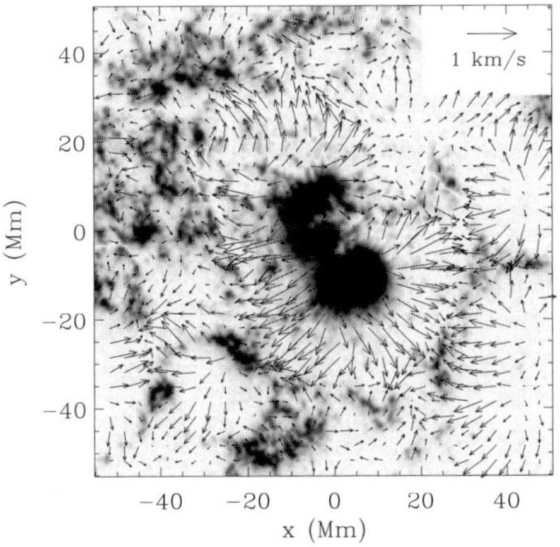


Fig. 2.11. Example of a flow map at 2 Mm (2000 km) beneath the solar surface. The magnitude of the magnetic field is depicted by the greyscale. The map origin is at a sunspot. Inversion of travel time differences shows strong outflows from the sunspot as well as from 'supergranules', short-lived convection patterns of unknown origin. Figure courtesy Laurent Gizon.

Flow speeds as large as a km/s are observed near the surface in active regions (Figure 2.11). Axisymmetric meridional flows in the direction of the poles, and flows at depth are much slower; all flow velocities are again well below the typical sound velocity of 10 km/s near the surface of the Sun, and can be handled as a perturbation to the sound velocity of the medium at rest.

The starting point is the realization that if we consider a coordinate system moving with the flow, nothing would change. Denote the material flow by a vector $\boldsymbol{v}$. For an observer in a fixed coordinate system, a point on the wavefront will move with a velocity $c\hat{\boldsymbol{n}} + \boldsymbol{v}$, where $\hat{\boldsymbol{n}}$ is the unit vector normal to the wavefront. Note that this is the 'ray' velocity, but in this case the ray is not perpendicular to the wavefront. But the velocity that concerns us is not the velocity normal to the wavefront (we don't really care which point from the wavefront arrives in our sensor!), but the velocity defined by the arrival time of the wavefront itself, which we find by dotting with $\hat{\boldsymbol{n}}$:

$$c_{\text{flow}} = c + \boldsymbol{v} \cdot \hat{\boldsymbol{n}} \,.$$

This is the velocity we need to use in the eikonal equation, and this logically leads to a travel time equal to:

$$T = \int_P \frac{\mathrm{d}s}{c + \boldsymbol{v} \cdot \hat{\boldsymbol{n}}} \,.$$

Fermat's Principle allows us to use the wavepath $P_0$ for a stationary medium if we assume that the perturbation imposed by $\boldsymbol{v}$ is small (i.e. $|\boldsymbol{v} \cdot \hat{\boldsymbol{n}}| \ll c$). It looks as though we end up with a complicated, anisotropic tomography problem because the velocity depends on the direction $\hat{\boldsymbol{n}}$. But a simple trick allows us to separate the effects of flow from the intrinsic acoustic velocity. To this end, helioseismologists as well as ocean acoustic tomographers work with the average of the travel time in two directions as well as with the difference (Munk et al. [223]); define

$$T^+ = \int_{P_0} \frac{\mathrm{d}s}{c + \boldsymbol{v} \cdot \hat{\boldsymbol{n}}} \approx \int_{P_0} \frac{\mathrm{d}s}{c} \left(1 - \frac{\boldsymbol{v} \cdot \hat{\boldsymbol{n}}}{c}\right) = T_0 - \int_{P_0} \frac{\boldsymbol{v} \cdot \hat{\boldsymbol{n}}}{c^2} \mathrm{d}s \,.$$

For the ray in the reversed direction we may use the same equation (using Fermat to neglect the small change in ray location) but reverse the sign of $\hat{\boldsymbol{n}}$:

$$T^- = T_0 + \int_{P_0} \frac{\boldsymbol{v} \cdot \hat{\boldsymbol{n}}}{c^2} \mathrm{d}s \,,$$

where $T_0$ is the travel time in the non-moving medium. We then retrieve both $T_0$ and a constraint on $\boldsymbol{v}$:

$$T_0 = \frac{1}{2}(T^+ + T^-) \tag{2.55}$$

$$T_{\text{dif}} = \frac{1}{2}(T^+ - T^-) = -\int_P \frac{\boldsymbol{v} \cdot \hat{\boldsymbol{n}}}{c^2} \mathrm{d}s \,. \tag{2.56}$$

(2.57)

## 2.12 Appendix A: Some elements of Fourier analysis

In this book we use the following convention for the time–frequency Fourier transform:

$$u(\omega) = \int_{-\infty}^{\infty} u(t) \mathrm{e}^{\mathrm{i}\omega t} \mathrm{d}t \,, \tag{2.58}$$

where the angular frequency $\omega = 2\pi f$ if $f$ is the frequency in Hz. The inverse transform is then:

$$u(t) = \frac{1}{2\pi} \int_{-\infty}^{\infty} u(\omega) \mathrm{e}^{-\mathrm{i}\omega t} \mathrm{d}\omega \,. \tag{2.59}$$

The power spectrum is defined as $|u(\omega)|^2 = u(\omega)^* u(\omega)$.

In seismology, pulse-like functions are very important. If the earthquake originates on a small rupture surface, the duration of the rupture can be very short, the resulting seismic wave has the shape of a brief pulse. We define Dirac's *delta-function* $\delta(t)$ as the mathematical abstraction of a pulse with zero time duration and infinite amplitude, such that:

$$\int_{-\infty}^{\infty} \delta(t)\mathrm{d}t = 1\,, \tag{2.60}$$

and, for an arbitrary time signal $f(t)$:

$$\int_{-\infty}^{\infty} \delta(t-\tau)f(t)\mathrm{d}t = f(\tau)\,. \tag{2.61}$$

The latter equation implies that the Fourier transform of $\delta(t)$ equals 1:

$$\delta(\omega) = \int_{-\infty}^{\infty} \delta(t)\mathrm{e}^{\mathrm{i}\omega t}\mathrm{d}t = \mathrm{e}^{\mathrm{i}\omega 0} = 1\,, \tag{2.62}$$

and the inverse transform leads to the important equality

$$\frac{1}{2\pi}\int_{-\infty}^{\infty} \mathrm{e}^{-\mathrm{i}\omega t}\mathrm{d}\omega = \frac{1}{2\pi}\int_{-\infty}^{\infty} \mathrm{e}^{\mathrm{i}\omega t}\mathrm{d}\omega = \delta(t)\,. \tag{2.63}$$

Similarly, (2.61) implies that a delta function delayed by $\tau$ seconds acquires a positive phase $\omega\tau$ in its spectrum:

$$\int_{-\infty}^{\infty} \delta(t-\tau)\mathrm{e}^{\mathrm{i}\omega t}\mathrm{d}t = \mathrm{e}^{\mathrm{i}\omega\tau}\,. \tag{2.64}$$

As a consequence, delaying an arbitrary signal by $\tau$ seconds will increase the phase in its spectrum by $\omega\tau$:

$$\int_{-\infty}^{\infty} f(t-\tau)\mathrm{e}^{\mathrm{i}\omega t}\mathrm{d}t = F(\omega)\mathrm{e}^{\mathrm{i}\omega\tau}\,. \tag{2.65}$$

Heaviside's step function $h(t)$ is defined as a function that is 0 for $t < 0$ and 1 for $t > 0$ and has a Fourier transform

$$h(\omega) = \int_{0}^{\infty} \mathrm{e}^{\mathrm{i}\omega t}\mathrm{d}t = \left.\frac{\mathrm{e}^{\mathrm{i}\omega t}}{\mathrm{i}\omega}\right|_0^{\infty} = \frac{\mathrm{e}^{\mathrm{i}\omega\infty}-1}{\mathrm{i}\omega}\,.$$

With a little bit of damping, modelled by a small positive imaginary component of $\omega$, $\mathrm{e}^{\mathrm{i}\omega\infty} \to 0$ and

$$h(\omega) = -\frac{1}{\mathrm{i}\omega} = \frac{\mathrm{i}}{\omega}\,. \tag{2.66}$$

Finally, if $v(t) = \mathrm{d}u(t)/\mathrm{d}t$ or $\dot{u}(t)$ for short, then we can use integration by parts to find the spectrum of $v$ in terms of that of $u$, assuming $u(t) \to 0$ for $|t| \to \infty$:

$$v(\omega) = \int_{-\infty}^{\infty} \dot{u}(t)\mathrm{e}^{\mathrm{i}\omega t}\mathrm{d}t = 0 - \mathrm{i}\omega \int_{-\infty}^{\infty} u(t)\mathrm{e}^{\mathrm{i}\omega t}\mathrm{d}t = -\mathrm{i}\omega u(\omega)\,. \tag{2.67}$$

We shall use equations (2.61)–(2.67) repeatedly in this book.

An important equality is given by Parseval's theorem. Let $u(t)$ and $v(t)$ be two time series, which we here assume to be real. The product of the two is proportional to amplitude squared, or energy:

$$\begin{aligned} E &= \int_{-\infty}^{\infty} u(t)v(t)\mathrm{d}t \\ &= \int_{-\infty}^{\infty} u(t)\frac{1}{2\pi}\int_{-\infty}^{\infty} v(\omega)\mathrm{e}^{-\mathrm{i}\omega t}\mathrm{d}\omega\mathrm{d}t \\ &= \frac{1}{2\pi}\int_{-\infty}^{\infty}\left[\int_{-\infty}^{\infty} u(t)\mathrm{e}^{-\mathrm{i}\omega t}\mathrm{d}t\right] v(\omega)\mathrm{d}\omega \\ &= \frac{1}{2\pi}\int_{-\infty}^{\infty} u^*(\omega)v(\omega)\mathrm{d}\omega\,. \end{aligned} \tag{2.68}$$

A special case is whcn $u(t) = v(t)$:

$$\int_{-\infty}^{\infty} u(t)^2\mathrm{d}t = \frac{1}{2\pi}\int_{-\infty}^{\infty} u(\omega)u^*(\omega)\mathrm{d}\omega\,. \tag{2.69}$$

For real signals $u(t)$ the spectrum satisfied $u(-\omega) = u^*(\omega)$ (see Exercise 2.2). With that one easily establishes a simplified version of Parseval for real signals $u(t)$ and $v(t)$, involving only the positive spectrum:

$$\int_{-\infty}^{\infty} u(t)v(t)\mathrm{d}t = \frac{1}{\pi}\mathrm{Re}\int_{0}^{\infty} u^*(\omega)v(\omega)\mathrm{d}\omega\,. \tag{2.70}$$

The energy spectral density of the signal $u(t)$ is defined as $(2\pi)^{-1}u(\omega)u^*(\omega)$. For a stationary time series (such as microseismic noise), the total energy would be infinite, so this definition does not work. If $n(t)$ is a noise signal of infinite duration we can define the average power:

$$P = \lim_{T\to\infty}\frac{1}{2T}\int_{-T}^{T} n(t)^2\mathrm{d}t\,.$$

Since $n(t)$ has infinite total energy, its Fourier transform does not exist. But if we analyse a finite time window, we can still determine the relative contribution of different frequencies within this window. Let $n_T(\omega)$ be the spectrum of $n(t)$

truncated to the window $-T \leq t \leq T$. Using Parseval's theorem (2.69):

$$P = \lim_{T\to\infty} \frac{1}{2T} \int_{-T}^{T} n(t)^2 \mathrm{d}t = \lim_{T\to\infty} \frac{1}{2\pi} \int_{-\infty}^{\infty} \frac{1}{2T} |n_T(\omega)|^2 \mathrm{d}\omega \,.$$

The function

$$P(\omega) = \lim_{T\to\infty} \frac{1}{2T} |n_T(\omega)|^2 \,, \tag{2.71}$$

defines the power density spectrum. Another way to define this is by using the autocorrelation of the noise:

$$C_n(t) = \lim_{T\to\infty} \frac{1}{2T} \int_{-T}^{T} n(\tau)n(\tau - t)\mathrm{d}\tau \,. \tag{2.72}$$

With a little algebra one establishes that $P(\omega)$ is the Fourier transform of $C_n(t)$:

$$C_n(t) = \frac{1}{2\pi} \int_{-\infty}^{\infty} P(\omega)\mathrm{e}^{-\mathrm{i}\omega t} \mathrm{d}\omega. \tag{2.73}$$

This result is known as the Wiener–Khintchine theorem.

Another important equality between the two domains concerns convolution. The convolution $c(t)$ between two signals $u(t)$ and $v(t)$ arises in filter theory and in physical systems with a memory, where the past value of the signal (or function) $v(t)$ is multiplied with the current value of the signal $u(t)$. Convolution is often denoted by an asterisk ($*$) and is defined as:

$$c(t) = u(t) * v(t) \equiv \int_{-\infty}^{\infty} u(\tau)v(t - \tau)\mathrm{d}\tau \,. \tag{2.74}$$

The Fourier transform of $c(t)$ is:

$$\begin{aligned} c(\omega) &= \int_{-\infty}^{\infty} c(t)\mathrm{e}^{\mathrm{i}\omega t} \mathrm{d}t \\ &= \int_{-\infty}^{\infty} \int_{-\infty}^{\infty} u(\tau)v(t - \tau)\mathrm{d}\tau \mathrm{e}^{\mathrm{i}\omega t} \mathrm{d}t \\ &= \int_{-\infty}^{\infty} u(\tau)\mathrm{e}^{\mathrm{i}\omega\tau} \mathrm{d}\tau \int_{-\infty}^{\infty} v(t - \tau)\mathrm{e}^{\mathrm{i}\omega(t-\tau)} \mathrm{d}t \\ &= u(\omega)v(\omega), \end{aligned} \tag{2.75}$$

showing that convolution in the time domain is equivalent to multiplication in the frequency domain. Because of the symmetry of the Fourier transform, the reverse is also true, i.e. multiplication in the time domain transforms to convolution in the frequency domain:

$$\int u(t)v(t)\mathrm{e}^{\mathrm{i}\omega t} \mathrm{d}t = u(\omega) * v(\omega) \,. \tag{2.76}$$

# 3
# Ray tracing

To find the correct geometry of a ray in realistic models of the Earth or Sun, we need to solve (2.27) numerically. This is comparatively easy in the case of layered or spherically symmetric media. On the other hand, if the seismic velocity is also a function of one or two horizontal coordinates, it may be very difficult. Fortunately, Fermat's Principle often allows us to use background models with lateral homogeneity, as I discussed in Section 2.9. In extreme cases, however, the seismic velocities may change sufficiently fast that the ray computed for a layered Earth is too far away from the ray in the true, heterogeneous Earth. In that case we must use full 3D ray tracing. In this chapter we take a look at the most promising algorithms available for both cases but warn the reader that accurate ray tracing in 3D is still an active area of research that has not yet converged to one 'ideal' method. In fact all methods still have shortcomings.

## 3.1 The shooting method

To find the correct ray geometry between a given source and receiver location we not only need an accurate solver for the differential equations such as (2.34) and (2.32), but also a way to determine which initial condition (ray orientation at the source) satisfies the end condition (ray arriving in the receiver). The term 'shooting' refers to the latter: we aim, compute and aim again until we 'hit' the receiver.

We recast the equations (2.32)–(2.34) for a ray in the equatorial plane $\theta = \pi/2$ in the form:

$$\frac{\mathrm{d}i}{\mathrm{d}s} = \frac{\sin i}{c}\left(\frac{\mathrm{d}c}{\mathrm{d}r} - \frac{c}{r}\right) \qquad \text{(2.34 again)}$$

$$\frac{\mathrm{d}r}{\mathrm{d}s} = \cos i$$

$$\frac{\mathrm{d}\phi}{\mathrm{d}s} = \frac{\sin i}{r} \qquad \text{(2.32 again)}$$

$$\frac{\mathrm{d}T}{\mathrm{d}s} = c^{-1},$$

to which we added a simple equation for the travel time $T$. In the case of elastic waves, the velocity $c$ may denote either $V_P$ or $V_S$.

This system is of the form $\mathrm{d}\boldsymbol{f}/\mathrm{d}s = \boldsymbol{F}[\boldsymbol{f}]$ and can be solved using well established numerical methods for the solution of ordinary differential equations, such as the second-order Runge-Kutta method (Press et al. [269]):

$$\boldsymbol{f}(s+\Delta s) = \boldsymbol{f}(s) + \Delta s \boldsymbol{F}\left[\boldsymbol{f}(s) + \frac{1}{2}\Delta s \boldsymbol{F}(s)\right].$$

Often, a fourth-order Runge-Kutta method is used. However, when the velocity model contains local sharp increases, a second-order Runge-Kutta with a smaller step size in $\mathrm{d}s$ may be preferable. A convenient diagnostic for the precision of a solution is offered by Snel's law (2.33): at every point of the ray $r \sin i/c$ needs to equal the constant ray parameter $p$. At the surface, and wherever the model contains a discontinuity at a fixed radius $r_d$, one should switch to differentiation with respect to $r$ rather than pathlength $s$ (see Exercise 2.10). This way one can ensure that the end of the step coincides exactly with the discontinuity, before the discontinuity is crossed (or reflected, depending on which ray is chosen) in the next step.

The shooting method consists of trying out starting values for the angle $i_0$ at the source until the ray crosses the receiver location at $r = a$, $\phi = \phi_r$ within a specified precision. Subroutines solving the differential equations often assume the source to be located at $\phi = 0$ in the equatorial plane, in which case the longitudinal coordinate $\phi$ is equivalent to the epicentral distance $\Delta$. Because of the spherical symmetry, we can simply rotate the ray to fit a specific source–receiver pair.

The epicentral distance $\Delta$ between a receiver at longitude $\varphi_r$ and co-latitude† $\vartheta_r$ and a source at $(\varphi_s, \vartheta_s)$ is found from the dot product between the (unit) vectors $\hat{\boldsymbol{r}}_r$ and $\hat{\boldsymbol{r}}_s$ that point from the Earth's centre to the receiver and source location, respectively:

$$\hat{\boldsymbol{r}}_s \cdot \hat{\boldsymbol{r}}_r = \cos\Delta \,.$$

With the expression for spherical coordinates

$$\begin{aligned} x &= r \sin\varphi \sin\vartheta \\ y &= r \cos\varphi \sin\vartheta \\ z &= r \cos\vartheta \,, \end{aligned}$$

† The co-latitude is the spherical coordinate, i.e. measured from the North Pole ($\vartheta = 0$) on a spherical Earth. Thus colatitude = $\pi/2$ − latitude. We denote geographical longitude and colatitude by $(\varphi, \vartheta)$ but use $(\phi, \theta)$ for arbitrary spherical coordinate systems. The symmetry of the Earth provides extra flexibility. For body waves the coordinate system is usually rotated such that the ray is in the equatorial plane and starts in ($\phi = 0, \theta = \pi/2$), so that $\phi$ equals epicentral distance. On the other hand, when we deal with normal modes we often define $\theta = 0$ as the source location, so that $\theta$ is the epicentral distance.

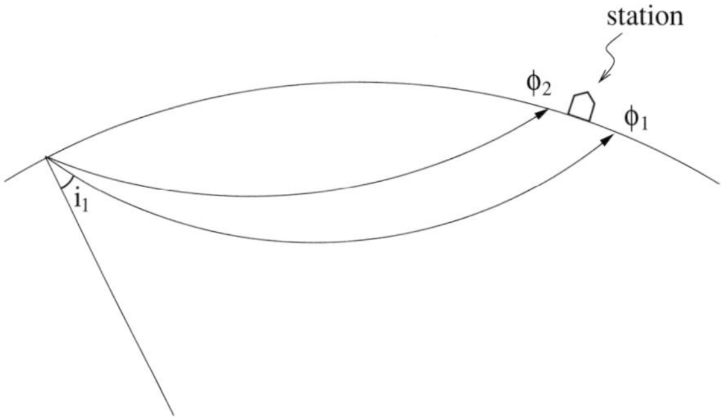


Fig. 3.1. The shooting method: the first try uses angle $i_1$ but overshoots the seismograph in the station, whereas $i_2$ gives a ray that arrives at a distance that is too short compared to the epicentral distance $\phi$ of the station. The next try would be given by (3.2).

we can work out the dot product $\hat{\boldsymbol{r}}_s \cdot \hat{\boldsymbol{r}}_r$ with $r = 1$ and obtain:

$$\cos\Delta = \cos\vartheta_r \cos\vartheta_s + \sin\vartheta_r \sin\vartheta_s \cos(\varphi_r - \varphi_s)\,. \tag{3.1}$$

More than one ray may arrive in a receiver. A suitable strategy is to compute a table with epicentral distances for a specific source depth, with dynamic increments in incidence angle $\Delta i$ such that $\Delta\phi < \epsilon$, where $\epsilon$ is a specified precision, e.g. 0.3°. A Newton interpolation scheme is usually effective in quickly locating the ray arriving at $\phi_1 < \phi < \phi_2$ by iterating:

$$i_{\text{next}} = i_1 + \frac{\phi - \phi_1}{\phi_2 - \phi_1}(i_2 - i_1), \tag{3.2}$$

and narrowing down the interval $(i_1, i_2)$ such that the interval $(\phi_1, \phi_2)$ always contains $\phi$ (see Figure 3.1). Note that $\phi$ is not necessarily a monotonously decreasing function of $i$, and that the iteration may fail if the interval contains a point where $d\phi/di = 0$. In tomographic inversions with $10^6$ or more data it is almost inevitable that some rays fail to converge. Rejecting them usually has no significant effect on the resulting image, while trying to salvage them may consume an inordinate amount of time.

## 3.2 Ray bending

Tracing rays between fixed sources and receivers in three-dimensional media is harrowingly difficult. If the velocity $c$ depends on the horizontal coordinates as well as on $z$ or $r$, the shooting method may so often fail to converge that it becomes useless. For this reason we prefer to use a ray bending method. As the name indicates, we set up a set of equations to 'bend' the ray until its travel time is stationary as prescribed by Fermat's principle.

The only method that is guaranteed to yield the fastest raypath in every circumstance is Dijkstra's method, first used in seismic tomography by Nakanishi and Yamaguchi [224] but better known in the efficient variant developed by Moser [221] as the 'shortest path method'. Unfortunately, the stability of the shortest path method is offset by two important disadvantages: its travel time is only approximate and in case more than one ray arrives in the receiver, only the fastest ray can be found. Whereas the first disadvantage can be mitigated using the bending methods that we shall describe, the second can only be circumvented if the later arrivals have clearly different properties from the first arrival, e.g. when they reflect from some surface and their path can be broken down into different components. Yet the shortest path method has its use in seismic tomography, because we are often only interested in the first arriving wave. For that reason we shall give a brief introduction to the basic principles of the method.

The shortest path method makes use of a weighted graph. Imagine that we place markers inside the Earth, much like road signs in a town, and we impose the restriction that a seismic ray can only travel by going from marker to marker. There are roads between the marker and several of its neighbouring markers, and the positive travel time between a marker and these neighbours is specified. In graph theory, the markers are called 'nodes' or 'vertices', the roads between a pair of markers are 'edges' or 'arcs'. The collection of all connected neighbours of a marker is denoted as its 'forward star'. The collection of all nodes constitutes a 'graph'. Suppose we wish to find the shortest travel time to go from a source node A to a receiver node B. We could try to find all permutations, compute the time needed, and select the shortest one (note that it makes no sense to visit one node twice, since eliminating the loop in between two visits will always result in a shorter time). This, however, would take too much time. A faster method was discovered by the Dutch mathematician Edsger Dijkstra (1930–2002).

Dijkstra's algorithm uses the fact that once the shortest path from the source node to a particular node is known, we can use this information when searching for the shortest path to other nodes. Let $\mathcal{Q}$ be the set of nodes for which we have not yet determined the shortest path, though some preliminary travel times may have been calculated already. We define the set $\mathcal{P}$ as the complement of $\mathcal{Q}$, i.e. it contains all nodes with known shortest paths to the source. Initially, all nodes are in $\mathcal{Q}$ with infinite travel time. The algorithm is then described by the following steps:

(i) Set the travel time to the source node to 0
(ii) If $\mathcal{Q}$ is empty, stop
(iii) Choose the node $s$ with the smallest travel time in $\mathcal{Q}$
(iv) For each node $i$ in the forward star of $s$, update the travel time such that $T_i \leftarrow \min\{T_i, T_s + \Delta T_{si}\}$
(v) Move node $s$ from set $\mathcal{Q}$ to set $\mathcal{P}$
(vi) Goto (ii)

A simple Fortran implementation of the algorithm is available in the software repository.† Note that the 'path' itself is fully prescribed by assigning to each node the preceding node along its shortest path to the source. For a proof that the algorithm works, see a textbook in discrete mathematics, e.g. Johnsonbaugh [150]. The algorithm can be speeded up considerably by sorting the nodes in $\mathcal{Q}$ into a 'heap', which is defined as a set of $N$ travel times $T_j$ such that:

$$T_j \geq T_{j/2} \text{ for } 1 \leq j/2 < j \leq N\,.$$

A heap is a comparatively fast way to find the fastest path among a set of paths without having to sort the whole set. For details, including how to set up the heap in the first place, see Press et al. [269] or subroutine `cheap.f` in the software repository.

The precision of the shortest path algorithm depends on the complexity of the forward star, but is never very great. Because the ray is represented by straight line segments between nodes, its travel time is not very accurate and, because of Fermat's Principle, always an overestimate of the shortest time that is possible for an arbitrary ray. To obtain a more precise estimate of the travel time, we 'bend' the resulting ray to get rid of the angle discretization imposed by the structure of the forward star. Moser et al. [222] present a method to bend this ray, using the graph nodes as beta spline supports that can be moved around. An alternative (and sometimes more stable) bending can be effected by discarding the spline interpolation and using linear connections instead. The node locations $x_i$, $y_i$ and $z_i$ satisfy a set of linear equations if they are to minimize the total travel time, given by

$$T = \sum_{i=2}^{N} \frac{L_i}{\bar{c}_i}\,,$$

for a ray consisting of $N$ segments of length $L_i$:

$$L_i = \sqrt{(x_i - x_{i-1})^2 + (y_i - y_{i-1})^2 + (z_i - z_{i-1})^2}\,,$$

† `http://geodynamics.org/cig/software/packages/seismo/`.

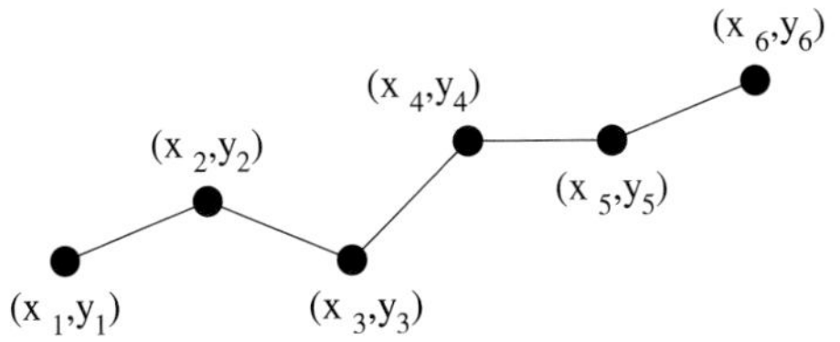


Fig. 3.2. The shortest path algorithm gives rays in which individual segments are oriented in a finite set of discrete angles. Bending involves changing the segment coordinates $(x_i, y_i)$ to minimize the travel time $T$. The coordinates of the endpoints are held fixed.

where we defined the average velocity $\bar{c}_i = \frac{1}{2}(c_i + c_{i-1})$ between two nodes (see Figure 3.2). According to Fermat's Principle $\partial T/\partial x_k = 0$ for $k = 2, \ldots, N-1$ (the end nodes are fixed), and similarly for $y_k$ and $z_k$:

$$\frac{\partial T}{\partial x_k} = \frac{\partial L_k}{\bar{c}_k \partial x_k} + \frac{\partial L_{k+1}}{\bar{c}_{k+1} \partial x_k} - \frac{L_k}{\bar{c}_k^2} \frac{\partial \bar{c}_k}{\partial x_k} - \frac{L_{k+1}}{\bar{c}_{k+1}^2} \frac{\partial \bar{c}_{k+1}}{\partial x_k}, \tag{3.3}$$

where

$$\frac{\partial L_k}{\partial x_k} = \frac{x_k - x_{k-1}}{L_k},$$

$$\frac{\partial L_{k+1}}{\partial x_k} = \frac{x_k - x_{k+1}}{L_{k+1}},$$

and

$$\frac{\partial \bar{c}_k}{\partial x_k} = \frac{\partial \bar{c}_{k+1}}{\partial x_k} = \frac{1}{2} \frac{\partial c_k}{\partial x_k}.$$

This gives a system of equations of the form:

$$\alpha_k x_{k-1} + \beta_k x_k + \gamma_k x_{k+1} = r_k,$$

with similar equations for $y_k$ and $z_k$. Since it is tri-diagonal it is efficient to solve. Equation (3.3) is not exactly linear, since a relocation of the nodes may change the average velocity $\bar{c}_k$, but it generally iterates quickly to a minimum. The coordinate changes must be orthogonalized to the local ray direction to avoid the problem of the travel time being optimized by collocating many nodes in regions of high velocity (making lower velocities invisible to the interpolation).

The precision of travel times and amplitudes can be tested by applying the principle of reciprocity (2.52), which implies that the travel time from $A \to B$ should be the same as the time from $B \to A$. The reciprocity test is not conclusive – for example in the case of the shortest path algorithm the travel time reciprocity is satisfied automatically. One should therefore always test a ray tracing algorithm against simple models for which analytical solutions are available, generally layered models. If reciprocity is not satisfied, something is clearly wrong. For bent ray travel times or for amplitudes in 3D media reciprocity is usually a very good diagnostic. In my own experience, relative precision in travel times is usually better than $10^{-4}$, i.e. an error of 0.1 s on a typical teleseismic travel time of 1000 s, or an error of one ms in a crustal reflection that arrives after 10 s. This is sufficient precision in view of the measurement accuracy.

### *Exercises*

**Exercise 3.1** The following is a very simple graph of only four nodes with specified travel times for five connections:

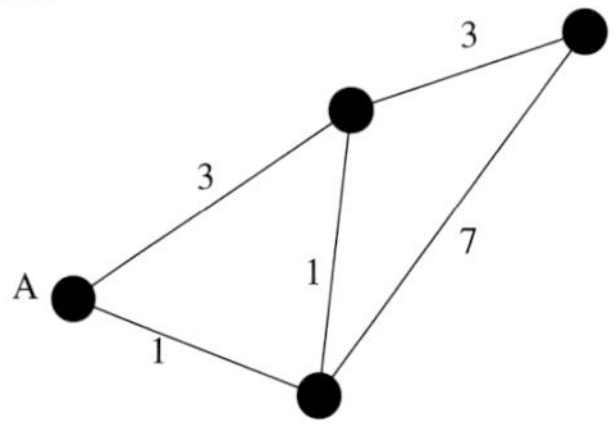


Use the shortest path algorithm to find the travel times and ray paths from node A to the three other nodes.

The following problems require Fortran routines from the software repository.

**Exercise 3.2** In subroutine `shpath` identify each of the steps in the algorithm by the line number(s) in the subroutine.

**Exercise 3.3** Write a small program that calls subroutine `shpath` for a $10 \times 10$ two-dimensional homogeneous model. Use the given example `forstar` subroutine. Plot the paths from a node near the centre to all other nodes in the model. Compare the computed travel times with the theoretically correct ones.

**Exercise 3.4** Extend the forward star routine to include more nodes, further away from the source, and repeat the calculations of the previous problem.

**Exercise 3.5** Adapt subroutine `forstar`, such that the velocity increases linearly with one of the coordinate axes. Repeat the calculations from the previous problem for the new model and plot.

## 3.3 Other raytracing algorithms for 3D media

A number of other algorithms exist for heterogeneous, 3D media, in which the eikonal equation (2.24) is solved by some form of finite differencing. These are

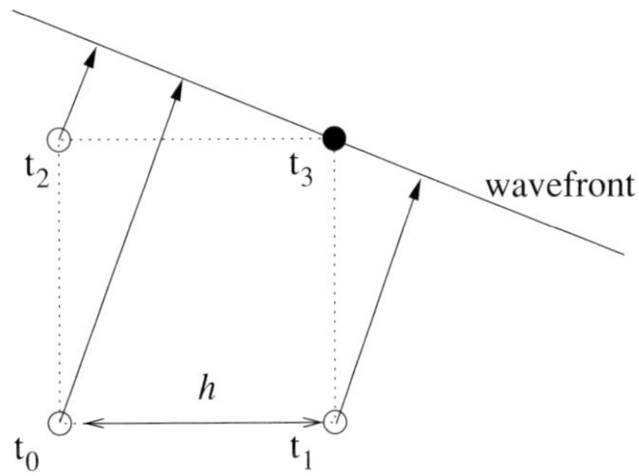


Fig. 3.3. A simple finite difference grid for the eikonal equation. The arrival times $t_0$, $t_1$ and $t_2$ are known and define a flat (or spherical) wavefront that arrives at $t_3$ in the fourth node.

not strictly 'ray tracing' methods, since they compute the travel time field $\tau(\boldsymbol{r})$ without calculating the ray trajectory. Rays need to be reconstructed following the gradient of $\tau$. For example, Vidale [383] solves the eikonal equation (in 2D but the extension to 3D is trivial):

$$\left(\frac{\partial \tau}{\partial x}\right)^2 + \left(\frac{\partial \tau}{\partial y}\right)^2 = \frac{1}{c^2},$$

on a square grid (Figure 3.3) with grid spacing $h$ we can approximate the derivatives of $\tau$ at the centre of each element with finite differences:

$$\frac{\partial \tau}{\partial x} = \frac{1}{2h}(t_1 + t_3 - t_0 - t_2)$$
$$\frac{\partial \tau}{\partial y} = \frac{1}{2h}(t_2 + t_3 - t_0 - t_1).$$

Assuming a wavefront comes in from the lower left corner, $t_0$, $t_1$ and $t_2$ are known and $t_3$ can be determined by solving the eikonal equation with the finite difference approximations:

$$t_3 = t_0 + \sqrt{2(h/c)^2 - (t_2 - t_1)^2},$$

the method is very inaccurate but can be improved by going to higher order approximations as shown by Podvin and Lecomte [263], Qin et al. [270] and Van Trier and Symes [379]. Rawlinson and Sambridge [271] show how to combine a regular grid intersected by interfaces defined by more arbitrary node locations, and use a related, but even less accurate, finite difference solver known as the Fast Marching Method.

More recently, attention has been directed toward 'essentially non-oscillatory' (ENO) algorithms to solve the 'Hamilton–Jacobi' form of the eikonal equation, in

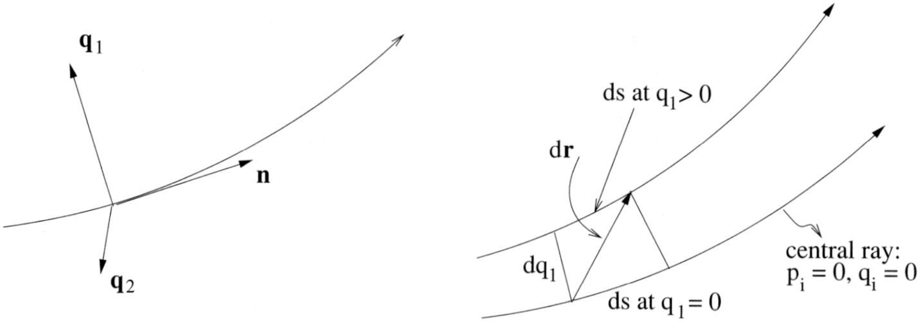


Fig. 3.4. Ray-centred coordinates. Left: unit vectors $\hat{\boldsymbol{n}}$, $\hat{\boldsymbol{q}}_1$ and $\hat{\boldsymbol{q}}_2$; right: a volume element spanned by a perturbation d$\boldsymbol{r}$.

which one coordinate is isolated, e.g. (in Cartesian coordinates):

$$\frac{\partial \tau}{\partial z} = \sqrt{\frac{1}{c^2} - \left(\frac{\partial \tau}{\partial x}\right)^2 - \left(\frac{\partial \tau}{\partial y}\right)^2}.$$

Errors in finite-differences occur when neighbouring gridpoints are separated by a cusp in the wavefront. Osher and Sethian [248] use the locally smoothest finite-difference approximation to avoid this and improve stability. Kim and Cook [164] use a post-processing scheme ('Post Sweeping') in which the travel time solution is checked for accuracy and iteratively adapted to increase accuracy.

Finite difference solvers to the wave equation in 3D media have the advantage that they do not require the shooting approach, which may lead to severe convergence difficulties, especially if the ray position varies rapidly with changing initial angles. They are therefore the algorithm of choice for small, strongly heterogeneous models. However, all of the difference methods have difficulties reaching the required precision of the order of $10^{-4}$ for long raypaths, and appear to be rather useless in global tomography without an additional step in which one reconstructs raypaths from the wavefronts, and re-computes the travel time with (2.47).

A relatively untested method, proposed by Nichols [230] and Shin et al. [311], is to solve the wave equation numerically for a damped, pulse-like source and to decompose the resulting time signal into wavefront arrivals. This has the advantage of calculating both arrival times and amplitudes that incorporate many finite-frequency effects. It is likely, however, that precision problems at longer distances will also hamper applications in global tomography.

## 3.4 Ray-centred coordinates

In numerical applications, it is often useful to work with ray centred rather than spherical or Cartesian coordinates. Figure 3.4 shows how we define ray-centred coordinates in a spherically symmetric Earth or Sun. The first coordinate is the pathlength $s$ along the ray, the local direction indicated by the unit vector $\hat{\boldsymbol{n}}$, tangent to the ray. A second coordinate, $q_1$ in direction $\hat{\boldsymbol{q}}_1$ is in the plane of propagation of the ray and orthogonal to $\hat{\boldsymbol{n}}$, whereas the third coordinate $q_2$ in direction $\hat{\boldsymbol{q}}_2$ is perpendicular to the first two such that

$$\hat{\boldsymbol{q}}_2 = \hat{\boldsymbol{n}} \times \hat{\boldsymbol{q}}_1 \qquad (\text{or } \hat{\boldsymbol{q}}_1 \times \hat{\boldsymbol{q}}_2 = \hat{\boldsymbol{n}}) .$$

Both $\hat{\boldsymbol{n}}$ and $\hat{\boldsymbol{q}}_1$ change direction as a function of $s$. In a general heterogeneous medium without lateral heterogeneity, the ray moves out of the plane spanned by $\hat{\boldsymbol{n}}$ and $\hat{\boldsymbol{q}}_1$ and $\hat{\boldsymbol{q}}_2$ also changes direction (the ray experiences 'torsion'). To simplify the – already considerable – algebra, we only treat the case of layered models or spherical symmetry. However, the extension to full heterogeneity is straightforward and the important expressions we derive are valid in the more general case. In a medium with lateral homogeneity, the change in $\hat{\boldsymbol{q}}_1$ is perpendicular to $\hat{\boldsymbol{q}}_1$ and in the ray propagation plane, hence proportional to $\hat{\boldsymbol{n}}$. We define this proportionality as the ray curvature $K$:

$$\frac{\mathrm{d}\hat{\boldsymbol{q}}_1}{\mathrm{d}s} = -K\hat{\boldsymbol{n}} .$$

Ray coordinates allow one to define a field of rays. The central ray is defined by $q_1 = q_2 = 0$. By varying $q_1$ and/or $q_2$, we move on to a neighbouring ray. The pathlength coordinate $s$ defines where we are in the ray field. As usual with curvilinear coordinates, we must be careful with the length properties of coordinate increments. A well known example in spherical coordinates is that an element $\mathrm{d}\phi$ has length $r\mathrm{d}\phi$, not $\mathrm{d}\phi$. If the element moves closer to the origin, its length decreases despite the fact that we keep $\mathrm{d}\phi$ constant. In the case of ray-centred coordinates we can write the position of any point close to the ray as

$$\boldsymbol{r}(s, q_1, q_2) = \boldsymbol{r}(s, 0, 0) + q_1\hat{\boldsymbol{q}}_1 + q_2\hat{\boldsymbol{q}}_2 .$$

However, a length element $\mathrm{d}s$ on the central ray does not have the same length when projected on a neighbouring ray since the ray curves (see Figure 3.4). We can easily find the appropriate scaling factor by calculating the length $|\boldsymbol{r}|^2$ with a

Taylor expansion for $\boldsymbol{r}$:

$$d\boldsymbol{r} = \boldsymbol{r}(s + \mathrm{d}s, \mathrm{d}q_1, \mathrm{d}q_2) - \boldsymbol{r}(s, 0, 0)$$
$$= \left(\frac{\partial \boldsymbol{r}(s,0,0)}{\partial s} + q_1 \frac{\mathrm{d}\hat{\boldsymbol{q}}_1}{ds}\right) \mathrm{d}s + \mathrm{d}q_1 \hat{\boldsymbol{q}}_1 + \mathrm{d}q_2 \hat{\boldsymbol{q}}_2 \,,$$

where we used that $\partial \hat{\boldsymbol{q}}_2/\partial s = 0$ and $\partial \boldsymbol{r}/\partial q_1 \equiv \hat{\boldsymbol{q}}_1$. Recognizing that $\partial \boldsymbol{r}/\partial s = \hat{\boldsymbol{n}}$ and $\hat{\boldsymbol{n}} \cdot (\mathrm{d}\hat{\boldsymbol{q}}_1/\mathrm{d}s) = -K$:

$$|\mathrm{d}\boldsymbol{r}|^2 = \mathrm{d}\boldsymbol{r} \cdot \mathrm{d}\boldsymbol{r} = (1 - Kq_1)^2 \mathrm{d}s^2 + \mathrm{d}q_1^2 + \mathrm{d}q_2^2 \,,$$

showing a scaling factor $1 - Kq_1$, usually denoted by $h$: $|\mathrm{d}\boldsymbol{r}|^2 = h^2 \mathrm{d}s^2 + \mathrm{d}q_1^2 + \mathrm{d}q_2^2$.

## 3.5 Dynamic ray tracing

Ray-centred coordinates allow us to extend the ray tracing to include nearby ('paraxial') rays. This procedure is called dynamic ray tracing. The term 'dynamic' in this context has nothing to do with forces – it has apparently been introduced because the method yields amplitudes as well as travel times. We shall closely follow the key paper by Červený and Hron [46].

The expression for the gradient $\nabla\tau$ in ray-centred coordinates can be found from the defining expression $\mathrm{d}\tau = \nabla\tau \cdot \mathrm{d}\boldsymbol{r}$. Since the step $\mathrm{d}s$ for the wavefront on the central ray reduces by a factor $h = 1 - Kq_1$ in length on the neighbouring ray, the scale factor $h$ appears in front of the $\partial\tau/\partial s$ term in the gradient:

$$\mathrm{d}\tau = \frac{1}{h}\frac{\partial \tau}{\partial s}\mathrm{d}s + \frac{\partial \tau}{\partial q_1}\mathrm{d}q_1 + \frac{\partial \tau}{\partial q_2}\mathrm{d}q_2 \,,$$

from which

$$\nabla\tau = \left(\frac{1}{h}\frac{\partial \tau}{\partial s}, \frac{\partial \tau}{\partial q_1}, \frac{\partial \tau}{\partial q_2}\right) , \tag{3.4}$$

so that the eikonal equation (2.24) is:

$$|\nabla\tau|^2 = \frac{1}{h^2}\left(\frac{\partial \tau}{\partial s}\right)^2 + \left(\frac{\partial \tau}{\partial q_1}\right)^2 + \left(\frac{\partial \tau}{\partial q_2}\right)^2 = \frac{1}{c^2} \,,$$

or

$$\frac{\partial \tau}{\partial s} = \frac{h}{c}\sqrt{1 - c^2\left(p_1^2 + p_2^2\right)} \equiv -\mathcal{H}(q_i, p_i) \,, \tag{3.5}$$

where $p_i = \partial\tau/\partial q_i$ and $\mathcal{H}$ is very similar to the 'Hamiltonian' known from mechanics, where the $p_i$ take the role of generalized momenta. The value of $\mathcal{H}$ depends on the $q_i$ as well as the $p_i$ because velocity $c$ varies with $q_i$. Since the ray is parallel to $\nabla\tau$, the derivatives of $\tau$ with respect to $q_1$ and $q_2$ are zero on the ray itself, so that

$$q_1 = q_2 = p_1 = p_2 = 0 \quad \text{on the ray.}$$

The stationarity of $\tau$ on the ray also implies that the travel time field for paraxial rays can be expressed, to second order, as:

$$\tau(s, q_1, q_2) = \tau(s, 0, 0) + \frac{1}{2}\boldsymbol{q} \cdot \boldsymbol{H}(s)\boldsymbol{q}\,, \tag{3.6}$$

where $\boldsymbol{q} = (q_1, q_2)$ and $\boldsymbol{H}$ is a $2 \times 2$ matrix with $H_{ij} = \partial^2\tau/\partial q_i \partial q_j$, the 'Hessian' matrix. Differentiation with respect to $s$ gives:

$$\begin{aligned}
\frac{\partial\tau}{\partial s} &= \frac{1}{c} + \frac{1}{2}\boldsymbol{q} \cdot \boldsymbol{H}'\boldsymbol{q} \\
\frac{\partial\tau}{\partial q_1} &= H_{11}q_1 + H_{12}q_2 \\
\frac{\partial\tau}{\partial q_2} &= H_{12}q_1 + H_{22}q_2\,,
\end{aligned} \tag{3.7}$$

where the prime $'$ denotes differentiation with respect to $s$. Similarly, we can expand (3.5) in a Taylor series; to second order this gives (see Exercise 3.6):

$$\begin{aligned}
&\mathcal{H}(q_1, q_2, p_1, p_2) \\
&= \mathcal{H}(0,0,0,0) + \frac{1}{2}c(p_1^2 + p_2^2) + \frac{1}{2}\frac{\partial^2\mathcal{H}}{\partial q_1^2}q_1^2 + \frac{\partial^2\mathcal{H}}{\partial q_1 \partial q_2}q_1 q_2 + \frac{1}{2}\frac{\partial^2\mathcal{H}}{\partial q_2^2}q_2^2 \\
&= -\frac{1}{c} + \frac{1}{2}c(p_1^2 + p_2^2) + \frac{1}{2c^2}\frac{\partial^2 c}{\partial q_1^2}q_1^2\,.
\end{aligned} \tag{3.8}$$

Equations (3.5) and (3.8) now give on the paraxial ray at distance $\boldsymbol{q}$:

$$\frac{\partial\tau}{\partial s} + \frac{1}{2}c(p_1^2 + p_2^2) + \frac{1}{2c^2}\frac{\partial^2 c}{\partial q_1^2}q_1^2 = \frac{1}{c}\,,$$

and with (3.7):

$$\begin{aligned}
&\frac{1}{2}\left(H'_{11} + c(H_{11}^2 + H_{12}^2) + \frac{1}{c^2}\frac{\partial^2 c}{\partial q_1^2}\right)q_1^2 \\
&\quad + \frac{1}{2}\left(H'_{22} + c(H_{22}^2 + H_{12}^2) + \frac{1}{c^2}\frac{\partial^2 c}{\partial q_2^2}\right)q_2^2 \\
&\quad + \frac{1}{2}\left(H'_{12} + cH_{12}(H_{11} + H_{22}) + \frac{1}{c^2}\frac{\partial^2 c}{\partial q_1 \partial q_2}\right)q_1 q_2 = 0\,.
\end{aligned}$$

Since this has to be satisfied for each small $(q_1, q_2)$, it follows that each of the terms is zero, or:

$$\frac{d\boldsymbol{H}}{ds} + c\boldsymbol{H}^2 = -\frac{1}{c^2}\boldsymbol{V}\,, \tag{3.9}$$

where we use the symmetry of $\boldsymbol{H}$ and where $V_{ij} = \partial^2 c/\partial q_i \partial q_j$.

Equation (3.9) is a Riccati equation that can be solved numerically (see Section 3.7). Though we have derived it here for a spherically symmetric Earth in which only $V_{11} \neq 0$, its extension to more generally heterogeneous media is straightforward; Červený and Hron [46] show that it leads to the same Riccati equation, though the scaling factor $h$ now becomes dependent on the orientation angle $\theta$ of the ray plane, given by its torsion: $h = 1 - K(q_1 \cos\theta + q_2 \sin\theta)$. However, neither $h$ nor $\theta$ occur in (3.9) which helps in the efficient construction of the matrix equations for finite-frequency tomography (Chapter 7).

Once we have calculated $\boldsymbol{H}$ we can find the travel time in the neighbourhood of a ray by using the expansion (3.6). In Chapter 5 we shall also establish a relationship between $\boldsymbol{H}$ and the ray amplitude.

### *Exercise*

**Exercise 3.6** In this problem you shall develop the steps to derive (3.8).

a. Use the fact that $p_i = \partial\tau/\partial q_i = 0$ on the ray, to show that also $\partial\mathcal{H}/\partial q_i = 0$ (hint: use $\mathcal{H} = -\partial\tau/\partial s$ and switch the order of differentiation); substitute this into (3.5) to show that

$$\frac{\partial\mathcal{H}}{\partial q_1} = \frac{\partial}{\partial q_1}\left(-\frac{h}{c}\right) = \frac{h\partial_1 c - c\partial_1 h}{c^2} = 0 \quad \text{on the ray,}$$

where we use the shorthand notation $\partial_i$ for $\partial/\partial q_i$.

b. Show that $h = 1$ on the ray

c. Show that

$$\frac{\partial\mathcal{H}}{\partial q_i} = \frac{h\partial_i c - c\partial_i h[1 - c^2(p_1^2 + p_2^2)]}{c^2\sqrt{1 - c^2(p_1^2 + p_2^2)}}\,,$$

and check that this is zero on the ray.

d. Derive a similar expression for $\partial\mathcal{H}/\partial p_i$ and check that this is zero on the ray.

e. Prove that $\partial^2\mathcal{H}/\partial q_i \partial p_i = 0$ by writing $\mathcal{H}$ as a quadratic function in $p_i$ and $q_i$ and imposing $\partial\mathcal{H}/\partial q_i = \partial\mathcal{H}/\partial p_i = 0$ on the ray.

f. Differentiate once more to find $\partial_{11}\mathcal{H} = \partial_{11}c/c^2$. Hint: use $h = 1 - Kq_1$.

## 3.6 Ray tracing on the sphere

For seismic surface waves in a spherically symmetric Earth, the raypaths are simply segments of great circles, i.e. lines defined by the intersection between the surface of the sphere and a plane through the centre of the Earth. For a laterally heterogeneous Earth, the ray approximation for surface waves implies that we assume a local phase velocity $c(\vartheta, \varphi)$ that is a function of location in geographical coordinates: co-latitude $\vartheta$ and longitude $\varphi$. We shall denote by $\vartheta(\varphi)$ the location of the ray on the sphere at longitude $\varphi$. Dahlen and Tromp [78] give the following system of differential equations for $\vartheta$ and ray azimuth $\zeta$ (but here $\zeta$ is defined as measured North over East as is common in observational seismology):

$$\frac{\mathrm{d}\vartheta}{\mathrm{d}\varphi} = -\sin\vartheta\ \cot\zeta$$
$$\frac{\mathrm{d}\zeta}{\mathrm{d}\varphi} = \cos\vartheta - \sin\vartheta\frac{\partial \ln c}{\partial\vartheta} - \cot\zeta\frac{\partial \ln c}{\partial\varphi}.$$

The geometrical spreading is given by

$$\mathcal{R} = \sin\zeta\left|\frac{\partial\vartheta}{\partial\zeta_s}\right|,$$

where $\zeta_s$ is the azimuth[†] at the source. We find the source azimuth from a convenient spherical trigonometric identity:

$$\sin\zeta_s = \frac{|(\hat{\boldsymbol{s}}\times\hat{\boldsymbol{r}})\times(\hat{\boldsymbol{s}}\times\hat{\boldsymbol{z}})|}{|(\hat{\boldsymbol{s}}\times\hat{\boldsymbol{r}})||(\hat{\boldsymbol{s}}\times\hat{\boldsymbol{z}})|} \tag{3.10}$$

for unit vectors $\hat{\boldsymbol{s}}$, $\hat{\boldsymbol{r}}$ and $\hat{\boldsymbol{z}}$ to source, receiver and North Pole, respectively. Ferreira and Woodhouse [102] give very much reduced expressions for the geometrical spreading and the phase of a surface wave:

$$\mathcal{R} = -\frac{a\sin^2\vartheta}{\sin\vartheta_s}\frac{1+\nu_s^2\sin^2\vartheta_s}{\sqrt{1+\nu^2\sin^2\vartheta}}\frac{\partial\cot\vartheta}{\partial\nu_s},$$

where $a$ is the Earth's radius and $\nu = -\mathrm{d}\cot\vartheta/\mathrm{d}\varphi = -\cot\zeta/\sin\vartheta$. As we shall see in Chapter 10, the 'phase' $\psi$ of the surface wave takes on a similar role to that of the travel time for body waves. When advancing a distance d$s$ along the ray, the phase at frequency $\omega$ advances by $(\omega/c)\mathrm{d}s$. From this:

$$\psi(\omega,\varphi_x) = \omega a\int_{\varphi_s}^{\varphi_x}\frac{1}{c(\vartheta,\varphi)}\left[\frac{\nu(\varphi)^2}{[1+\cot^2\vartheta(\varphi)]^2}+\frac{1}{1+\cot^2\vartheta(\varphi)}\right]^{\frac{1}{2}}\mathrm{d}\varphi.$$

† Observational seismologists define azimuth clockwise from North. Even though we normally define angles counterclockwise, we prefer to make an exception for azimuth and adhere to the clockwise convention that prevails in practice.

### 3.7 Computational aspects

In this section we summarize some computational aspects that need to be addressed when solving the Riccati equation (3.9), following the reduction of Červený and Hron's results to a spherically symmetric medium by Dahlen et al. [76]).

To obtain initial conditions for $\boldsymbol{H}$ we assume that the medium is homogeneous in a small region around the source (which we can always do by defining the region small enough, unless the source is located on a discontinuity). In that region the wavefront is spherical so that

$$\boldsymbol{H} \rightarrow \frac{1}{cs}\boldsymbol{I} \tag{3.11}$$

(see Exercise 3.7). Because of the spherical symmetry, derivatives with respect to $q_2$ are zero, and $\boldsymbol{H}$ is diagonal.

The Riccati equation is nonlinear, and not solvable by any of the standard solution methods for ordinary differential equations. Usually, one solves the Riccati equation by substitution of solutions that have some known characteristics. In our case we shall wish to make sure the singularity (3.11) is part of the solution. To this end, we write $\boldsymbol{H}$ as the product of two diagonal matrices $\boldsymbol{H} = \boldsymbol{P}\boldsymbol{Q}^{-1}$ where we impose $\mathrm{d}\boldsymbol{Q}/\mathrm{d}s = c\boldsymbol{P}$. Substituting this in (3.9) we obtain the following system:

$$\begin{aligned} \frac{\mathrm{d}\boldsymbol{Q}}{\mathrm{d}s} &= c\boldsymbol{P} \\ \frac{\mathrm{d}\boldsymbol{P}}{\mathrm{d}s} &= -\frac{1}{c^2}\boldsymbol{V}\boldsymbol{Q}\,. \end{aligned} \tag{3.12}$$

Note that the first of these equations assures that $\boldsymbol{H}$ satisfies the initial condition (3.11) if we start out with $\boldsymbol{P} = \boldsymbol{I}$. We denote the diagonal elements of $\boldsymbol{P}$ and $\boldsymbol{Q}$ by $P_1$, $P_2$, and $Q_1$. $Q_2$, so that $H_{11} = P_1/Q_1$, $H_{22} = P_2/Q_2$. In polar coordinates (see Figure 3.5) we have $\mathrm{d}\phi/\mathrm{d}s = \sin i/r$ and $\mathrm{d}i/\mathrm{d}q_1 = (\partial i/\partial\phi)(\partial\phi/\partial q_1) \approx \partial\phi/\partial q_1 = \cos i/r$. With that:

$$\begin{aligned} \frac{\mathrm{d}Q_1}{\mathrm{d}\phi} &= p^{-1}r^2 P_1 \\ \frac{\mathrm{d}Q_2}{\mathrm{d}\phi} &= p^{-1}r^2 P_2 \\ \frac{\mathrm{d}P_1}{\mathrm{d}\phi} &= -pc^{-1}(\partial_{rr}c + r^{-1}\partial_r c \cot^2 i)Q_1 \\ \frac{\mathrm{d}P_2}{\mathrm{d}\phi} &= -p^{-1}rc^{-3}\partial_r c\, Q_2\,. \end{aligned} \tag{3.13}$$

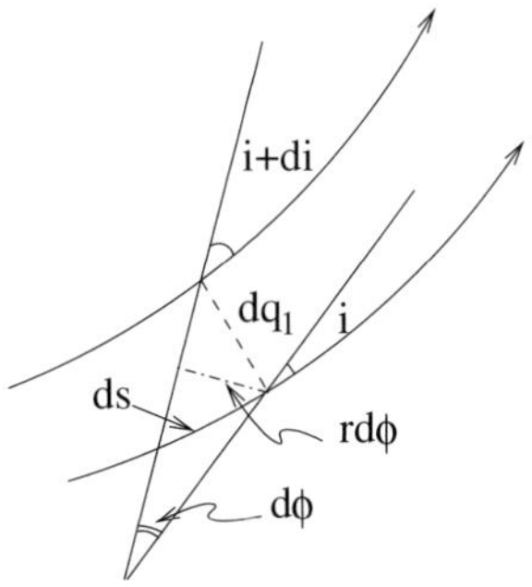


Fig. 3.5. Paraxial rays in a polar coordinate system.

The derivation of the continuity conditions at curved interfaces is quite lengthy and will not be repeated here. The continuity conditions are:

$$\begin{aligned}[\cos i P_1 + (p^2 r^{-2} \partial_r c - (rc)^{-1}) Q_1]_-^+ &= 0 \\ [P_2 - (rc)^{-1} \cos i\, Q_2]_-^+ &= 0 \\ [Q_1 / \cos i]_-^+ = [Q_2]_-^+ &= 0\,,\end{aligned} \tag{3.14}$$

and initial conditions follow from (3.11):

$$\begin{aligned} P_1(0) = P_2(0) &= 1 \\ Q_1(0) = Q_2(0) &= 0\,.\end{aligned} \tag{3.15}$$

The following analytical results can be used to check on the numerical accuracy of the integration:

$$P_2 = (pc)^{-1} r_s \sin(i + \phi) \tag{3.16}$$

$$Q_2 = p^{-1} r r_s \sin\phi \tag{3.17}$$

$$r_s \cos i_s = r \cos i\, P_1 + (p^2 r^{-1} \partial_r c - c^{-1}) Q_1\,, \tag{3.18}$$

where $r_s$ is the source location. Alternatively, one may forego the numerical integration of $P_2$ and $Q_2$ and use the analytical solutions.

The transformation from equatorial coordinates, where the ray starts in longitude $\phi = 0$ and $\theta = \pi/2 =$ constant, to a path between a source located in $(r_s, \vartheta_s, \varphi_s)$ and a receiver in $(r_r, \vartheta_r, \varphi_r)$ involves a pure rotation of the Cartesian coordinates. Let $\boldsymbol{T}$ be the matrix that transforms the Cartesian components of a vector $\boldsymbol{v}_e$ in the equatorial plane to the Cartesian components of a vector $\boldsymbol{v}_s = \boldsymbol{T}\boldsymbol{v}_e$ in the ray plane, the plane through the centre of the Earth or Sun spanned by the unit vectors $\hat{\boldsymbol{s}}$ and $\hat{\boldsymbol{r}}$ in source and receiver directions. Logical reasoning provides a shortcut to derive the rotation matrix $\boldsymbol{T}$ needed. In Cartesian coordinates, the equatorial source is located in (1,0,0) and moves to a location $\hat{\boldsymbol{r}}_s = \boldsymbol{T}(1, 0, 0)^T = (x_s, y_s, z_s)$ on the

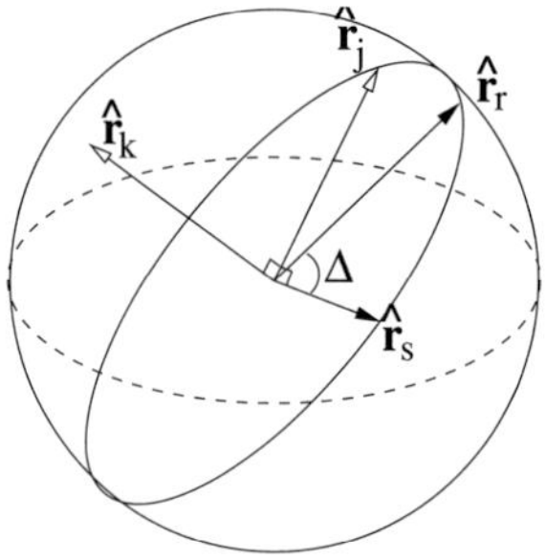


Fig. 3.6. The unit vectors that define the rotation matrix $\boldsymbol{T}$.

unit sphere. Thus the elements in the first column of $\boldsymbol{T}$ are:

$$x_s = \cos\varphi_s \sin\vartheta_s$$
$$y_s = \sin\varphi_s \sin\vartheta_s$$
$$z_s = \cos\vartheta_s\,.$$

The receiver is originally located in $(\cos\Delta, \sin\Delta, 0)$ and moves to $\hat{\boldsymbol{r}}_r = (x_r, y_r, z_r)$. The unit vectors $\hat{\boldsymbol{r}}_s$ and $\hat{\boldsymbol{r}}_r$ span the ray plane but are not orthogonal. We find an orthogonal unit vector by first defining a unit vector orthogonal to the ray plane:

$$\hat{\boldsymbol{r}}_k = \frac{\hat{\boldsymbol{r}}_s \times \hat{\boldsymbol{r}}_r}{\sin\Delta}\,.$$

and a third one orthogonal to both $\hat{\boldsymbol{r}}_k$ and $\hat{\boldsymbol{r}}_s$:

$$\hat{\boldsymbol{r}}_j = \hat{\boldsymbol{r}}_k \times \hat{\boldsymbol{r}}_s\,.$$

From Figure 3.6 we see that:

$$\hat{\boldsymbol{r}}_r = \hat{\boldsymbol{r}}_s \cos\Delta + \hat{\boldsymbol{r}}_j \sin\Delta\,, \tag{3.19}$$

which enables us to identify $\hat{\boldsymbol{r}}_j$ as the second column of $\boldsymbol{T}$. Since pure rotation matrices are orthonormal, the third column must be the unit vector orthogonal to both $\hat{\boldsymbol{r}}_s$ and $\hat{\boldsymbol{r}}_j$, i.e. $\hat{\boldsymbol{r}}_k$. Thus:

$$\boldsymbol{T} = \begin{pmatrix} x_s & x_j & x_k \\ y_s & y_j & y_k \\ z_s & z_j & z_k \end{pmatrix}. \tag{3.20}$$

The transpose of $\boldsymbol{T}$ is its inverse, which allows one to go back from geocentric to equatorial coordinates:

$$\boldsymbol{T}^{-1} = \begin{pmatrix} x_s & y_s & z_s \\ x_j & y_j & z_j \\ x_k & y_k & z_k \end{pmatrix}.$$

## *Exercises*

**Exercise 3.7** Consider a ray travelling along the z-axis in a homogeneous medium. Show that, for small deflections $x, y \ll z$ away from the $z$-axis, the travel time is:

$$\tau = \frac{z}{c} + \frac{1}{2c}\frac{x^2 + y^2}{z},$$

and derive the initial condition (3.11) for $\boldsymbol{H}$ by identifying $z$ with the ray coordinate $s$.

**Exercise 3.8** Diagonal matrices are commutative, i.e. $\boldsymbol{P}\boldsymbol{Q} = \boldsymbol{Q}\boldsymbol{P}$. However, (3.12) is also valid in the non-symmetric case when matrices do not commute. Derive the second equation in (3.12) by differentiating $\boldsymbol{P} = \boldsymbol{H}\boldsymbol{Q}$, and using both (3.9) and $\mathrm{d}\boldsymbol{Q}/\mathrm{d}s = c\boldsymbol{P}$.

# 4

# Wave scattering

The ray approximation that we used in the previous chapters is valid for high frequency waves travelling over rather short distances. Wu [405] distinguishes four wave propagation regimes, depending on the dimensionless variables $ka$, $L/a$ and the root-mean-square strength of the heterogeneities $v = \sqrt{\langle(\delta c/c)^2\rangle}$, where $k$ is the wavenumber, $a$ the scale length of heterogeneities, $L$ the propagation distance and $\langle.\rangle$ denotes the expected value (or average). At the microscopic level, the discriminating factor is $ka$, the ratio between scattering size and wavelength:

- $ka \lesssim 0.01$: the medium is quasi homogeneous, scattering is very weak,
- $0.01 \lesssim ka \lesssim 0.1$: Rayleigh scattering – small heterogeneities cause the wave to lose energy, with scattering in all directions (depending on wavetype) and scattered power proportional to $k^4$,
- $0.1 \lesssim ka \lesssim 10$: Mie scattering, with scattering in all directions but strongest in the forward direction and power only weakly dependent on wavelength,
- $ka \gtrsim 10$: Forward scattering, whereby most of the energy is in the forward direction, but causing focusing, diffraction and interference problems.

The difference between Rayleigh and Mie scattering is familiar to us from everyday life, looking at the sky. On a clear day with little humidity the sky above us is dark blue – this is because the air molecules are much smaller than the wavelength of visible light, so we are in the regime of Rayleigh scattering which is strongest for blue colours with short wavelength (it is the sensitivity of the retina in our eyes that poses a limit: if we were more sensitive to violet, the sky would look more purple). Small water droplets or dust particles are large enough to generate Mie scattering, which is why clouds are white and why the sky near the horizon becomes hazy – a result of the weak wavelength dependence of Mie scattering.

The subdivision in four scattering regimes is immediately applicable to seismic P- and S-waves in the Earth. The smallest $k = \omega/c$ is obtained for the lowest frequency, highest velocity waves, i.e. long period P-waves. A P-wave with a period

Table 4.1. *Maximum P ray length if $D \lesssim 1$*

| Period (s) | $a$ (km) | $L_{\max}$ (km) |
|---|---|---|
| 1 | 30 | 140 |
| 1 | 300 | 14,000 |
| 20 | 30 | 7 |
| 20 | 300 | 700 |

of 20 s and a typical propagation velocity of 10 km/s has $k \approx 0.03$ rad/km, so that forward scattering occurs for heterogeneities larger than about 300 km, whereas Mie scattering is optimal for $a = 30$ km. Although $V_S < V_P$ the largest wavenumbers belong also to P-waves because in general they reach higher frequencies than S-waves (see Chapter 5). A 1 Hz P-wave has $k \approx 0.6$ rad/km, making forward scattering dominant if $a > 16$ km.

However, whether the approximations of ray theory are good enough also depends on the macroscopic situation – the length of the raypath compared to the size $a$ of the heterogeneity. For the Earth, in which $v$ is small (a few per cent at most), $L/a$ is the defining quantity, or even better the ratio between $L/a$ and $ka$, known as the 'wave parameter' which can be directly related to the Fresnel zone (Exercise 4.1):

$$D = \frac{4L}{ka^2}.$$

The factor 4 has been chosen to make $D \approx 1$ a dividing threshold between regimes. If $D \lesssim 1$, the heterogeneities are larger than the Fresnel zone, diffraction around them is small and ray theory is applicable; but if $D \gtrsim 1$, diffracted waves cause amplitudes and travel times to be influenced by scattering, usually by reducing any travel time or amplitude anomaly, a phenomenon known as wavefront healing (see Chapter 6). Table 4.1 gives estimates for the maximum length of a P ray in a medium with a velocity of 10 km/s for ray theory to be valid. Thus, the 20 s P-wave and the 300 km heterogeneity fall in the strict ray-theoretical regime only as long as $L \lesssim \frac{1}{4}ka^2 = 700$ km. For short period P-waves the 30 km scatterers would only allow for $L \lesssim 140$ km; but a 300 km sized anomaly allows for a ray of 14000 km which includes most mantle waves of interest (see Figure 4.1).

Neither the strong diffraction regime nor the Mie scattering regime are at this moment easily treatable for the purposes of tomographic imaging. But for $ka \gtrsim 10$ we can analyse the diffraction effects using first-order (Born) scattering.

In the following sections we derive some fundamental results that are needed in the rest of this book. We limit the development to the key results needed to understand the role of scattering in finite-frequency tomography. For a more complete treatment of scattering see the book by Sato and Fehler [302]. This chapter relies

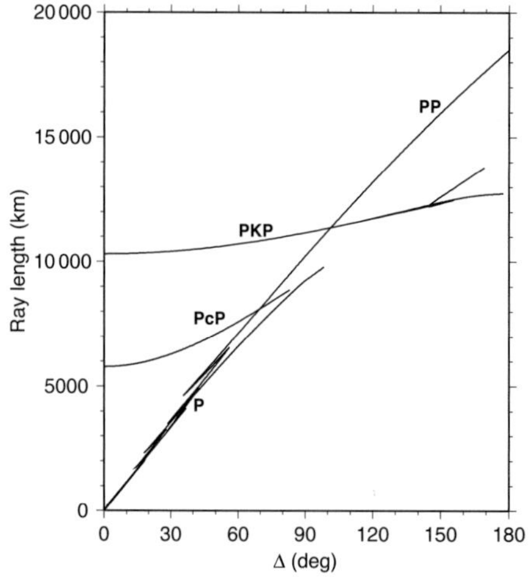


Fig. 4.1. Length of longitudinal (P) rays as a function of epicentral distance. The length of an S ray is comparable to that of a P ray at the same distance.

heavily on Wu and Aki's classical paper [406] and the scattering matrix formalism developed by Zhao and Dahlen [416].

### *Exercise*

**Exercise 4.1** If $d_{max}$ is the maximum diameter of the Fresnel zone in a homogeneous medium, show that

$$D = \frac{2}{\pi} \frac{d_{max}^2}{a^2} \,.$$

## 4.1 The acoustic Green's function

Central in any first-order perturbation theory is the concept of the response of the medium to a point force, the 'Green's function' $G(\boldsymbol{r}, t)$ of the system.† If we adopt a point source forcing of the form $c^2\nabla \cdot \boldsymbol{f} = \delta(\boldsymbol{r})$ to (2.22) we obtain the following equation for the Green's function in a homogeneous acoustic medium:

$$c^2\nabla^2 G + \omega^2 G = \delta(\boldsymbol{r}). \tag{4.1}$$

If we express $\nabla^2$ in spherical coordinates we need to retain derivatives with respect to $r$ only because the solution must have spherical symmetry:

$$c^2 \left( \frac{\partial^2 G}{\partial r^2} + \frac{2}{r}\frac{\partial G}{\partial r} \right) + \omega^2 G = \delta(\boldsymbol{r}) = 0 \quad \text{outside } r = 0\,. \tag{4.2}$$

† See footnote in Section 2.10 on the simplifications in notation. Note also that the $\delta$ function in (4.1) is not dimensionless.

A reasonable guess is that the amplitude will decrease as $1/r$ in order to conserve energy. This makes a substitution of the type $G(r) = \eta(r)/r$ logical. Doing so we find the following simple wave equation for $\eta$:

$$c^2 \frac{\partial^2 \eta}{\partial r^2} + \omega^2 \eta = 0 \quad (r \neq 0), \tag{4.3}$$

which is solved by any function of the form $\eta(\omega)\exp(\mathrm{i}\omega r/c)$, or, using (2.64): $\eta(t - r/c)$ in the time domain. We ignore the inward travelling wave of the form $\eta(t + r/c)$, which in all cases of interest is ruled out by the initial conditions. Hence the Green's function is of the form $\eta(t - r/c)/r$. We have derived this for $r > 0$; the solution is singular at $r = 0$. What is the shape of $f$? From physical experience (a gunshot is audible as a gunshot even at larger distance) we know that the shape of a pulse does not change as it propagates. Therefore a delta-function source must generate a wavefield (the 'impulse response') with $\delta$-function properties. It remains to scale the amplitude. The amplitude of the Green's function turns out to be $-1/4\pi rc^2$ (see Appendix), so that

$$G(\boldsymbol{r}, t) = -\frac{1}{4\pi rc^2}\delta(t - r/c), \tag{4.4}$$

or in the frequency domain:

$$G(\boldsymbol{r}, \omega) = -\frac{\mathrm{e}^{\mathrm{i}\omega r/c}}{4\pi rc^2}. \tag{4.5}$$

The use of Green's functions becomes apparent when we wish to solve the wave equation with more general force distributions:

$$c^2\nabla^2 P + \omega^2 P = F. \tag{4.6}$$

First, observe that (4.1) is invariant for translations of the coordinate system – it does not matter where we place the origin $\boldsymbol{r} = 0$, it gives the solution for a point source anywhere – all we need to do is replace $r$ with $|\boldsymbol{r} - \boldsymbol{r}'|$. We denote the Green's function at $\boldsymbol{r}$ for a point source at $\boldsymbol{r}'$ by $G(\boldsymbol{r}, \omega; \boldsymbol{r}')$. Because of the sampling property of the delta function (2.61), an arbitrary force distribution $F(\boldsymbol{r})$ can always be written as a sum of point sources: $F(\boldsymbol{r}) = \int F(\boldsymbol{r}')\delta(\boldsymbol{r} - \boldsymbol{r}')\mathrm{d}^3\boldsymbol{r}'$. Since the wave equation is linear, the solution for this force distribution can thus be obtained by summing (integrating) all the pulse responses:

$$P(\boldsymbol{r}, \omega) = \int G(\boldsymbol{r}, \omega; \boldsymbol{r}')F(\boldsymbol{r}')\mathrm{d}^3\boldsymbol{r}'. \tag{4.7}$$

## 4.2 An acoustic point scatterer

In this section we use the acoustic Green's function to derive an expression for the scattered wave from a point perturbation with velocity $c = c_0 + \delta c$ in an otherwise homogeneous acoustic medium with velocity $c_0$. We assume that the incoming wave field is a plane wave of the form:

$$P_0(\boldsymbol{r}, \omega) = e^{i\boldsymbol{k}_0(\omega)\cdot\boldsymbol{r}} ,$$

where $\boldsymbol{k}_0 = \omega\hat{\boldsymbol{n}}/c_0$ and $\hat{\boldsymbol{n}}$ a unit vector in the direction of wave propagation. Note that $\nabla^2 P_0 = -k_0^2 P_0 = -(\omega^2/c_0^2)P_0$. For the perturbed medium outside the source region we have:

$$(c_0 + \delta c)^2\nabla^2(P_0 + \delta P) + \omega^2(P_0 + \delta P) = 0 .$$

This equation is still exact. Its power lies in the fact that we clearly identify the wave perturbation. Assuming the scattered wave is small with respect to the unperturbed wave, we may neglect higher-order terms. Thus, equating first-order terms:

$$c_0^2\nabla^2\delta P + \omega^2\delta P = -2c_0\,\delta c\nabla^2 P_0 = 2\frac{\delta c}{c_0}\omega^2 P_0 .$$

This equation for $\delta P$ has the same form as the forced equation (4.6) for $P$, so the solution can be found using (4.5) and (4.7):

$$\delta P(\boldsymbol{r}, \omega) = -\int 2\frac{\delta c}{c_0}\omega^2 P_0(\boldsymbol{r}', \omega)\frac{e^{i\omega|\boldsymbol{r}-\boldsymbol{r}'|/c}}{4\pi c_0^2|\boldsymbol{r}-\boldsymbol{r}'|}d^3\boldsymbol{r}' .$$

We can now interpret this first-order scattering (or Born) approximation. The fact that $\delta P$ is ignored in the force term means that we neglect any contribution from an earlier scattered wave when this wave hits a scatterer: the incoming wavefield is assumed unchanged. The situation is sketched in Figure 4.2. For a point scatterer of strength $\Delta c$, i.e. $\delta c = \Delta c\delta(\boldsymbol{r}')$, at the origin where $P_0 = 1$ and $|\boldsymbol{r} - \boldsymbol{r}'| = r$ we find:

$$\delta P(\boldsymbol{r}, \omega) = -\omega^2\frac{\Delta c}{c_0^3}\frac{e^{i\omega r/c}}{2\pi r} . \tag{4.8}$$

Inspection of (4.8) shows two important characteristics of the acoustic scatterer. First, the amplitude of the scattered wave is proportional to $\omega^2$, so its energy depends on frequency as $\omega^4$, the characteristic frequency dependence of Rayleigh scattering. Second, the (velocity) point scatterer is isotropic – its energy radiates equally in all directions. In the next section we shall see that the first property

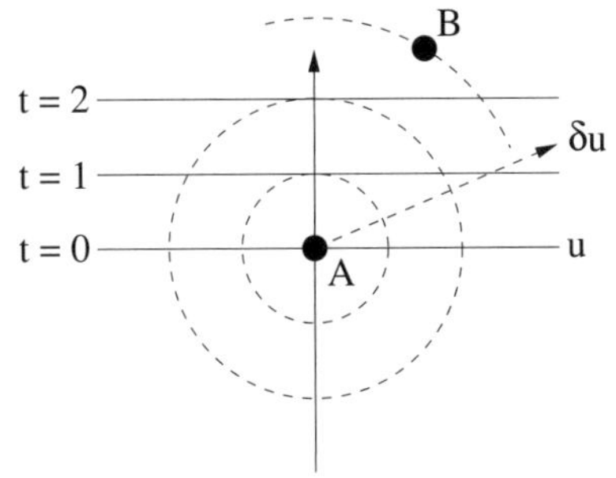


Fig. 4.2. A plane wave hitting scatterer A at time 0 (solid lines) generates a scattered wave (dashed lines). The Born approximation assumes that the plane wave itself is not noticeably perturbed by loss of energy to the scattered wave. It also ignores the interaction between the scattered wavefield and subsequent scatterers, such as here shown at B.

is shared with point scatterers in elastic media, but shear waves do not scatter isotropically.

### *Exercises*

**Exercise 4.2** Does the Born approximation give a wavefield in which energy is conserved?

**Exercise 4.3** If we have only one point scatterer in a homogeneous medium, multiple scattering is impossible. Does this mean that in this case the Born approximation yields the exact result?

## 4.3 Green's functions for elastic waves

The derivation of the Green's function for elastic waves in a homogeneous medium, though straightforward, is rather lengthy and will not be repeated here. But having derived the Green's function for the acoustic case, the result for a point force in an elastic medium can at least be made plausible and in this section we briefly summarize the main results. The essential difference with respect to the acoustic case is that the displacement $\boldsymbol{u}$ itself does not satisfy a simple wave equation like (4.6). This is logical, because we already know from Chapter 2 that the wavefield splits up into P- and S-waves that do not travel with the same velocity, and we have to deal with two elastic constants instead of one:

$$-\rho\omega^2\boldsymbol{u} = \nabla(\lambda\nabla\cdot\boldsymbol{u}) + \nabla\cdot[\mu(\nabla\boldsymbol{u}+\nabla\boldsymbol{u}^T)] \qquad \text{(2.42 again)}\,.$$

In a homogeneous medium we may write $\boldsymbol{u}$ as the sum of a rotation-free (i.e. P-wave) and divergence-free (S-wave) field by introducing Helmholtz potentials: $\boldsymbol{u} = \nabla\Phi + \nabla \times \boldsymbol{\Psi}$. It turns out that $\Phi$ and the components of the vector $\boldsymbol{\Psi}$ each satisfy a simple wave equation, with velocity $V_\mathrm{P}$ and $V_\mathrm{S}$, respectively. However, when we also develop the point force in a rotation-free and divergence-free field, we find that the potentials lose the point character of the force itself and spread out in space, giving rise to an extended wave that is especially strong in 'near field' at close distance to the source.

The wavefield emanating from a single force $f_j(t)$ in the (Cartesian) $x_j$-direction, situated at the origin $\boldsymbol{r} = 0$ has the following component in the $x_i$-direction (Aki and Richards [5]):

$$G_i^j(\boldsymbol{r}, t) = \frac{1}{4\pi\rho r^3}(3\gamma_i\gamma_j - \delta_{ij}) \int_{r/V_\mathrm{P}}^{r/V_\mathrm{S}} \tau f_j(t-\tau)\mathrm{d}\tau + \frac{1}{4\pi\rho V_\mathrm{P}^2 r}\gamma_i\gamma_j f_j(t - r/V_\mathrm{P})$$
$$- \frac{1}{4\pi\rho V_\mathrm{S}^2 r}(\gamma_i\gamma_j - \delta_{ij}) f_j(t - r/V_\mathrm{S}), \tag{4.9}$$

where $\gamma_i = x_i/r$, a direction cosine. The first term in this expression is the near-field term. The shape of $f_j(t)$ is not preserved in the near-field arrival, which changes with $r$ (we say this term is 'dispersive'), and its spreading in time reflects the spreading in space of the force potentials. Though it has a factor $r^{-3}$, its amplitude decreases more like $r^{-2}$ because the integral limits are proportional to $r$. The remaining ('far field') terms decrease like $r^{-1}$ and dominate for large $r$. For an arbitrary source location $\boldsymbol{r}'$ we denote the Green's function by $G_i^j(\boldsymbol{r}, t; \boldsymbol{r}')$; $\boldsymbol{G}$ is a tensor. One can also simply consider it to be a matrix of three columns, one for each of the wavefields excited by a force in $x$-, $y$- and $z$-directions, such that $\boldsymbol{G} \cdot \boldsymbol{f}$ gives the wavefield for an arbitrary force $\boldsymbol{f}$. We shall often use the Fourier transform of $G_i^j(\boldsymbol{r}, t; \boldsymbol{r}')$ in the far field, for a force $f_j(t)$ that is a delta function $\delta(t)$ in time. Using (2.65):

$$G_i^j(\boldsymbol{r}, \omega; \boldsymbol{r}') = \frac{\gamma_i\gamma_j}{4\pi\rho V_\mathrm{P}^2 r}\mathrm{e}^{\mathrm{i}k_\mathrm{P} r} - \frac{\gamma_i\gamma_j - \delta_{ij}}{4\pi\rho V_\mathrm{S}^2 r}\mathrm{e}^{\mathrm{i}k_\mathrm{S} r}, \tag{4.10}$$

where $r = |\boldsymbol{r} - \boldsymbol{r}'|$ and where $k_\mathrm{P} = \omega/V_\mathrm{P}$, $k_\mathrm{S} = \omega/V_\mathrm{S}$.

Equation (4.10) is for a homogeneous solid and needs to be understood in depth before we generalize it in the next section to smoothly varying media. The P-wave term is the easiest to grasp. We saw in Chapter 2 that the motion or polarity of the P-wave is in the direction of wave propagation, i.e. in the direction of $\boldsymbol{\gamma}$. This is indeed the case, because $G_i^j$ is proportional to $\gamma_i$. The factor $\gamma_j$ is multiplied by $f_j$, the component of the force. So far we have assumed the force is in either the $x$-,$y$- or $z$-direction, but we have already noted that we can sum three force components in these directions and obtain the response to an arbitrary force $\boldsymbol{f}$.

The dot product $\boldsymbol{\gamma} \cdot \boldsymbol{f}$ is known as the excitation factor for the wave. The product $\rho V_P$ is the impedance. The term impedance was originally used by Heaviside for the resistance of an electrical circuit to the input of an alternating current, but has been adopted in mechanics to describe the response of a solid to an applied force. A factor $V_P r$ captures the flux of energy with a speed $V_P$ across a surface of area proportional to $r^2$. Since the total flux needs to remain constant to conserve energy, this term describes the decrease of amplitude with distance $r$, and it is the same as the amplitude decrease for an acoustic wave which we derived in Section 4.1. Finally, the phase $k_P r$, or delay $(k_P r - \omega t)$ in the time domain, shows that this is a propagating wave.

The second expression, for the S-wave, has of course $V_S$ and $k_S$ instead of $V_P$ and $k_P$, but otherwise is the same except for the polarization factor, which is $\gamma_i \gamma_j - \delta_{ij}$. Since the motion of an S-wave is perpendicular to the ray direction $\boldsymbol{\gamma}$, the dot product of the motion with $\boldsymbol{\gamma}$ should be zero. To show that it is indeed so, first take the dot product of $G_i^j$ with an arbitrary force with components $f_j$: $\sum_j (\gamma_i \gamma_j - \delta_{ij}) f_j = \gamma_i (\boldsymbol{\gamma} \cdot \boldsymbol{f}) - f_i$. Dotting this with $\boldsymbol{\gamma}$ to find the motion in the direction of propagation gives $(\boldsymbol{\gamma} \cdot \boldsymbol{f}) - (\boldsymbol{\gamma} \cdot \boldsymbol{f}) = 0$.

The recognition that the numerators in (4.10) give the polarization of the wave allows us to write one simple dyadic expression for both P and S-waves:†

$$\boldsymbol{G}(\boldsymbol{r}, \omega; \boldsymbol{r}') = \frac{\boldsymbol{p}\boldsymbol{p}'}{4\pi \rho c^2 r} \mathrm{e}^{\mathrm{i}\omega r/c} . \tag{4.11}$$

For a P-wave in a homogeneous medium $\boldsymbol{p} = \boldsymbol{p}' = \boldsymbol{\gamma}$; but we shall soon generalize this to curved rays in smoothly varying media. In such media $\boldsymbol{p}'$ is the polarization at the source, and $\boldsymbol{p}$ at the point of observation, and we shall need to modify the expressions for amplitude and phase to reflect the heterogeneity.

Single forces are not common in the Earth, where all forces are created internally by stresses relating to slow convective motions. Such stresses build up slowly and are balanced by the elastic response of the rock, until the rock breaks. The forces on each side of a fractured surface are in opposite directions, forming a torque or moment. The forces are separated by a small distance across the fault, so that such a torque tries to rotate the fault. The unit of torque is force × distance, or Nm. Though the same SI unit is used for energy, the two are not the same: the torque needs to be multiplied by a (dimensionless) angle to produce the work done by a couple of forces. The torque with forces acting in direction $x_i$, offset in direction $x_j$, is denoted by $M_{ij}$. We usually rank the $M_{ij}$ in a $3 \times 3$ moment tensor. Its trace elements represent implosions or explosions, the off-diagonal elements refer to

† Notation: whereas $\boldsymbol{a} \cdot \boldsymbol{b}$ is a scalar, denoting the usual dot product $\sum_i a_i b_i$, the dyadic notation $\boldsymbol{ab}$ denotes a matrix with elements $a_i b_j$.

proper torques. The scalar moment is defined as:

$$M_0 = \sqrt{\frac{1}{2}} \left( \sum_{ij} M_{ij}^2 \right)^{\frac{1}{2}} \tag{4.12}$$

and serves as a measure of the strength of an earthquake. Near the locked ends of the fault, rock is strongly distorted and this creates a second torque that exactly balances the torque caused by forces in the fault direction: $M_{ij} = M_{ji}$ ('double couple'). Thanks to this symmetry of the moment tensor, the forces associated with earthquakes cannot speed up or slow down the rotation of the Earth, so that angular momentum is preserved, as expected for internal forces.

We thus need to generalize our Green's function expression to find the response to such torques. The response to a couple of forces of strength $f_j$ directed in the $\pm x_j$ directions but offset by $\Delta x'_k$ in the $x_k$ direction can be found by subtracting the two responses:

$$G_i^j(...; x'_k + \Delta x'_k, ...) f_j - G_i^j(...; x'_k, ...) f_j = \frac{\partial G_i^j}{\partial x'_k} \Delta x'_k f_j = \frac{\partial G_i^j}{\partial x'_k} M_{jk}\,, \tag{4.13}$$

which defines the moment tensor as $M_{jk} = \Delta x'_k f_j$.

Using $\partial r/\partial x'_k = -\partial r/\partial x_k = -x_k/r = -\gamma_k$ in (4.10), we find for the derivative of the Green's function in the far field, neglecting terms of $\mathcal{O}(r^{-2})$:

$$\begin{aligned} \partial G_i^j/\partial x'_k &= -\mathrm{i}k_\mathrm{P} \frac{\gamma_i \gamma_j \gamma_k}{4\pi\rho V_\mathrm{P}^2 r} \mathrm{e}^{\mathrm{i}k_\mathrm{P} r} + \mathrm{i}k_\mathrm{S} \frac{\gamma_i \gamma_j - \delta_{ij}}{4\pi\rho V_\mathrm{S}^2 r} \gamma_k \mathrm{e}^{\mathrm{i}k_\mathrm{S} r} \\ &= -\mathrm{i}\omega \frac{\gamma_i \gamma_j \gamma_k}{4\pi\rho V_\mathrm{P}^3 r} \mathrm{e}^{\mathrm{i}k_\mathrm{P} r} + \mathrm{i}\omega \frac{\gamma_i \gamma_j - \delta_{ij}}{4\pi\rho V_\mathrm{S}^3 r} \gamma_k \mathrm{e}^{\mathrm{i}k_\mathrm{S} r}\,. \end{aligned} \tag{4.14}$$

The multiplication with a factor $-\mathrm{i}\omega$ in (4.14) corresponds to time differentiation in the time domain, because the time dependence of this spectral component is given by $\mathrm{e}^{-\mathrm{i}\omega t}$. If we define the Green's function as the response to a step function, which is the time integral of the delta function, we can omit the factor $-\mathrm{i}\omega$ (see Exercise 4.4).

Earthquakes are caused by rupture, releasing stress. This is an irreversible process, and the movement at the source therefore looks very much like a step function. But the time derivative shows that we will observe this in the far field like the derivative of a step function, i.e. a delta function. Intuitively, this can be understood as the arrival of two waves with opposite motion, one from each side of the fault. The wave from the far side will arrive a little later than the near side, creating a time derivative in the seismogram. This is indeed what we observe in the real world.

The response to a single couple of forces at the origin, with moment $M_{jk}$, including the near-field terms, is given explicitly by:

$$
\begin{aligned}
u_i(\boldsymbol{r},t) = \sum_{jk} & \left(\frac{15\gamma_i\gamma_j\gamma_k - 3\gamma_i\delta_{jk} - 3\gamma_j\delta_{ik} - 3\gamma_k\delta_{ij}}{4\pi\rho r^4}\right)\int_{r/V_\mathrm{P}}^{r/V_\mathrm{S}} \tau M_{jk}(t-\tau)\mathrm{d}\tau \\
& + \left(\frac{6\gamma_i\gamma_j\gamma_k - \gamma_i\delta_{jk} - \gamma_j\delta_{ik} - \gamma_k\delta_{ij}}{4\pi\rho V_\mathrm{P}^2 r^2}\right) M_{jk}(t - r/V_\mathrm{P}) \\
& + \left(\frac{6\gamma_i\gamma_j\gamma_k - \gamma_i\delta_{jk} - \gamma_j\delta_{ik} - 2\gamma_k\delta_{ij}}{4\pi\rho V_\mathrm{S}^2 r^2}\right) M_{jk}(t - r/V_\mathrm{S}) \\
& + \frac{\gamma_i\gamma_j\gamma_k}{4\pi\rho V_\mathrm{P}^3 r}\dot{M}_{jk}(t - r/V_\mathrm{P}) - \frac{\gamma_i\gamma_j - \delta_{ij}}{4\pi\rho V_\mathrm{S}^3 r}\gamma_k\dot{M}_{jk}(t - r/V_\mathrm{S}),
\end{aligned}
\tag{4.15}
$$

where the dot denotes differentiation with respect to time.

## Exercises

**Exercise 4.4** Identify the far field response for a point moment tensor source in (4.15) and show that its spectrum can be written as

$$
u_i^{jk}(\boldsymbol{r},\omega) = \frac{\gamma_i\gamma_j\gamma_k}{4\pi\rho V_\mathrm{P}^3 r}\mathrm{e}^{\mathrm{i}k_\mathrm{P}r} - \frac{\gamma_i\gamma_j - \delta_{ij}}{4\pi\rho V_\mathrm{S}^3 r}\gamma_k \mathrm{e}^{\mathrm{i}k_\mathrm{S}r} \tag{4.16}
$$

if the source time function $\dot{M}(t)$ is a delta function. Explain the difference of a factor $-\mathrm{i}\omega$ with (4.14). Hint: see (2.66).

**Exercise 4.5** Verify that the far field response can be written as:

$$
\boldsymbol{u}(\boldsymbol{r},\omega) = \int \boldsymbol{G}(\boldsymbol{r},\omega;\boldsymbol{r}')\boldsymbol{f}(\boldsymbol{r}',\omega)\mathrm{d}^3\boldsymbol{r}' + \int \nabla'\boldsymbol{G}(\boldsymbol{r},\omega;\boldsymbol{r}') : \dot{\boldsymbol{M}}(\boldsymbol{r}',\omega)\mathrm{d}^3\boldsymbol{r}', \tag{4.17}
$$

where we write $\nabla'\boldsymbol{G}$ for $\partial'_k G_i^j$ and where ':' denotes contraction, summing over pairs $(j,k)$.

**Exercise 4.6** Discuss how the distinction between near and far field may depend on the time behaviour of $f_j(t)$. If $f_j(t)$ is the Heaviside function $H(t)$, which is 0 for $t < 0$ and 1 for $t > 0$, is then the 'near field' term still negligible in the 'far field'?

**Exercise 4.7** Show that for a unit force in the $z$-direction, the spherical components of the displacement are given by:

$$
\begin{aligned}
u_r &= \frac{\cos\theta}{4\pi\rho r}\left[\frac{2}{r^2}\int_{r/V_\mathrm{P}}^{r/V_\mathrm{S}} \tau f_z(t-\tau)\mathrm{d}\tau + \frac{1}{V_\mathrm{P}^2} f_z(t - r/V_\mathrm{P})\right], \\
u_\theta &= \frac{\sin\theta}{4\pi\rho r}\left[\frac{1}{r^2}\int_{r/V_\mathrm{P}}^{r/V_\mathrm{S}} \tau f_z(t-\tau)\mathrm{d}\tau - \frac{1}{V_\mathrm{S}^2} f_z(t - r/V_\mathrm{S})\right].
\end{aligned}
$$

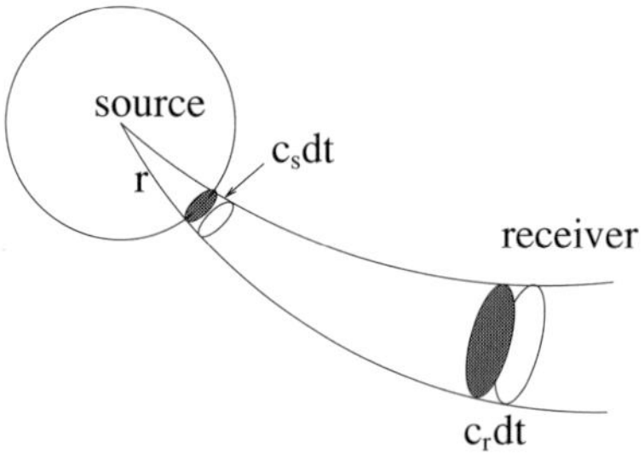


Fig. 4.3. Energy flux in a tube of rays.

Specify and sketch the behaviour of these expressions for $f_z(t) = \delta(t)$ and $f_z(t) = H(t)$, respectively. Can the near-field wave be identified with longitudinal (P) or transverse (S) wave motion? Hint: use Cartesian coordinates to find $\boldsymbol{u}$, then project onto the $r$ and $\theta$ directions.

## 4.4 Green's functions in the ray approximation

As a brief intermezzo, let us generalize the Green's functions to the case of a smoothly varying heterogeneous medium where ray theory is valid. Note that the displacement amplitude of a P- or S-wave in a region around a single force source follows directly from (4.10) if we remain close enough to the source that the medium can be considered locally homogeneous:

$$A_s = \left| G_i^j(\boldsymbol{r}, \omega; \boldsymbol{r}') \right| = \frac{1}{4\pi \rho_s c_s^2 r},$$

where $c_s$ is the velocity for the wave of interest ($V_P$ or $V_S$) and the subscript s denotes the source. The amplitude of the particle velocity is found by time differentiation of the particle displacement, or multiplication by $\omega$ in the frequency domain. Therefore, the maximum kinetic energy (over a full cycle) within a small volume $\mathrm{d}V$ with mass $\rho_s \mathrm{d}V$ is given by $\frac{1}{2}\rho_s \omega^2 A_s^2 \mathrm{d}V$. In harmonic motions, the potential energy is zero when the kinetic energy is maximal; the total energy of $\mathrm{d}V$ at any point in time is therefore given by this expression. We find the flux by considering a volume of length $c_s \mathrm{d}t$ and cross-section $r^2 \mathrm{d}\Omega$, where $\mathrm{d}\Omega$ is the solid angle spanned by a tube of rays (Figure 4.3). The energy flowing through the tube in time $\mathrm{d}t$ is then:

$$E_s = \frac{1}{2} \rho_s \omega^2 A_s^2 c_s \mathrm{d}t \, r^2 \mathrm{d}\Omega .$$

Now consider the same ray tube – hence with the same flux because energy is conserved – at a receiver location where we denote velocity and density with a subscript r, and where the cross-sectional area for the tube is given by the geometrical spreading $\mathcal{R}_{rs}^2 \mathrm{d}\Omega$. If we denote the wave amplitude at the receiver by

$A_r$, the flux over time d$t$ is:

$$E_r = \frac{1}{2}\rho_r\omega^2 A_r^2 c_r \mathrm{d}t\, \mathcal{R}_{rs}^2 \mathrm{d}\Omega .$$

Equating $E_s = E_r$, we find for the ray-theoretical amplitude:

$$A_r = \left(\frac{\rho_s c_s}{\rho_r c_r}\right)^{\frac{1}{2}} \frac{r}{\mathcal{R}_{rs}} A_s = \frac{1}{4\pi c_s \mathcal{R}_{rs}\sqrt{\rho_s \rho_r c_s c_r}} . \tag{4.18}$$

The exponential terms in (4.10) represent time delays in the time domain. We can therefore make the transition to a heterogeneous medium by replacing:

$$\mathrm{i}kr = \mathrm{i}\omega\frac{r}{c} \rightarrow \mathrm{i}\omega\tau ,$$

where $\tau$ denotes the travel time for the P- or S-wave between source and receiver. Combining this and (4.18) we find for the single force Green's function in the ray approximation instead of (4.10):

$$\begin{aligned} G_i^j(\boldsymbol{r}, \omega; \boldsymbol{r}') = {} & \frac{\gamma_i\gamma_j}{4\pi V_P(\boldsymbol{r}_s)\mathcal{R}_{rs}\sqrt{\rho(\boldsymbol{r}_s)\rho(\boldsymbol{r}_r)V_P(\boldsymbol{r}_s)V_P(\boldsymbol{r}_r)}} \mathrm{e}^{\mathrm{i}\omega\tau_P} \\ & - \frac{\gamma_i\gamma_j - \delta_{ij}}{4\pi V_S(\boldsymbol{r}_s)\mathcal{R}_{rs}\sqrt{\rho(\boldsymbol{r}_s)\rho(\boldsymbol{r}_r)V_S(\boldsymbol{r}_s)V_S(\boldsymbol{r}_r)}} \mathrm{e}^{\mathrm{i}\omega\tau_S} . \end{aligned} \tag{4.19}$$

An important corollary of the reciprocity (2.52) of the Green's function is the reciprocity of geometrical spreading. Reciprocity of $\boldsymbol{G}$ implies that the amplitude observed at r from a source at s equals the amplitude at s if we place the source at r. Thus, interchanging subscripts in the expression for $A_r$ we find for spreading in three dimensions:

$$\frac{1}{4\pi c_s \mathcal{R}_{rs}\sqrt{\rho_s\rho_r c_s c_r}} = \frac{1}{4\pi c_r \mathcal{R}_{sr}\sqrt{\rho_r\rho_s c_r c_s}} ,$$

or

$$c_s \mathcal{R}_{rs} = c_r \mathcal{R}_{sr} . \tag{4.20}$$

Finally, we generalize the response to a moment tensor source for a smoothly heterogeneous medium. We ignore the near-field terms in (4.15) and treat the far field P-wave from a moment tensor source:

$$u_i(\boldsymbol{r}, t) = \sum_{jk} \frac{\gamma_i\gamma_j\gamma_k}{4\pi\rho V_P^3 r} \dot{M}_{jk}(t - r/V_P).$$

We can simplify the notation by assuming that all elements of the moment tensor have the same time behaviour $m(t)$, i.e. $M_{jk}(t) = M_{jk}m(t)$, and introducing the radiation pattern:

$$\mathcal{F}^{\mathrm{P}} = \sum_{jk} \gamma_j \gamma_k M_{jk}.$$

We do the analysis for the longitudinal component $u^{\mathrm{P}} = \sum_i u_i \gamma_i$:

$$u^{\mathrm{P}}(\boldsymbol{r}, t) = \frac{\mathcal{F}^{\mathrm{P}}}{4\pi \rho V_{\mathrm{P}}^3 r} \dot{m}(t - r/V_{\mathrm{P}}),$$

or, in the spectral domain:

$$u^{\mathrm{P}}(\boldsymbol{r}, \omega) = \frac{\mathcal{F}^{\mathrm{P}} \dot{m}(\omega)}{4\pi \rho V_{\mathrm{P}}^3 r} \mathrm{e}^{\mathrm{i}\omega r/V_{\mathrm{P}}}, \tag{4.21}$$

where we use (2.64) to find the phase belonging to the time delay $r/V_{\mathrm{P}}$. The spectrum $\dot{m}(\omega)$ is the Fourier transform of $\dot{m}(t)$. We preserve the 'dot' notation rather than writing $-\mathrm{i}\omega m(\omega)$ because, as we already observed in the previous section, the source time function $\dot{m}(t)$ is generally the waveform that is observed from a double couple source in the far field. As with the point force, we generalize (4.21) for a more arbitrary medium in which rays are curved, by replacing

$$\rho V_{\mathrm{P}}^3 r \to V_{\mathrm{P}}(\boldsymbol{r}_{\mathrm{s}})^2 \mathcal{R}_{\mathrm{rs}} \sqrt{\rho(\boldsymbol{r}_{\mathrm{s}})\rho(\boldsymbol{r}_{\mathrm{r}}) V_{\mathrm{P}}(\boldsymbol{r}_{\mathrm{s}}) V_{\mathrm{P}}(\boldsymbol{r}_{\mathrm{r}})}, \tag{4.22}$$

or

$$u^{\mathrm{P}}(\boldsymbol{r}_{\mathrm{r}}, \omega) = \frac{\mathcal{F}^{\mathrm{P}} \dot{m}(\omega)}{4\pi\, V_{\mathrm{P}}(\boldsymbol{r}_{\mathrm{s}})^2 \mathcal{R}_{\mathrm{rs}} \sqrt{\rho(\boldsymbol{r}_{\mathrm{s}})\rho(\boldsymbol{r}_{\mathrm{r}}) V_{\mathrm{P}}(\boldsymbol{r}_{\mathrm{s}}) V_{\mathrm{P}}(\boldsymbol{r}_{\mathrm{r}})}} \mathrm{e}^{\mathrm{i}\omega \tau_{\mathrm{P}}}. \tag{4.23}$$

## 4.5 The Born approximation

The theory in Section 4.3 is of importance to analyse the excitation of seismic waves. The fact that we assume the medium to be homogeneous is usually acceptable for a small region around the source. We can bring this one step further, and analyse the excitation of scattered waves by anomalies in an otherwise homogeneous medium. This is important for two reasons. In Chapter 5 we shall analyse the loss of energy in seismic waves because of loss to scattered waves. In Chapter 7 we shall study the interference of scattered waves with the main wavefield and its effect on the measurement of travel times and drop the assumption of homogeneity but instead use the Green's functions obtained in the previous section.

The main vehicle to study scattered elastic waves is again Born theory. As in the acoustic case, we assume that the scattered wave energy is small, such that we may ignore the fact that the scattered waves themselves may induce scattered energy ('multiple scattering'). Instead we assume that, to first order, scattered energy originates from an unperturbed incoming wavefield that strikes the scattering anomalies.

The algebraic formalism for elastic waves is quite involved, but simplification can be obtained by using the notation of Green's tensors and using an operator notation for the elastodynamic equation. Thus we write $\mathcal{L}\boldsymbol{u} = 0$ instead of (2.42), or:

$$\mathcal{L}\boldsymbol{u} \equiv -\rho\omega^2\boldsymbol{u} - \nabla(\lambda\nabla\cdot\boldsymbol{u}) - \nabla\cdot[\mu(\nabla\boldsymbol{u} + \nabla\boldsymbol{u}^T)] = 0\,,$$

where $\nabla\boldsymbol{u}$ is a second-order tensor with elements $\partial u_i/\partial x_j$, the $T$ denotes its transpose, $\partial u_j/\partial x_i$, and $(\nabla\cdot\nabla\boldsymbol{u})_i = \sum_j \partial^2 u_i/\partial x_j^2$ etc. As mentioned before, the Green's tensor $\boldsymbol{G}$ can be viewed as a $3\times 3$ matrix in which each column is the response $\boldsymbol{u}$ to a point force $\boldsymbol{f}$ in one of three coordinate directions. Arranging the point forces in the columns of a unit matrix, we can therefore write:

$$\mathcal{L}\boldsymbol{G}(\boldsymbol{r},\omega;\boldsymbol{r}') = \boldsymbol{I}\delta(\boldsymbol{r}-\boldsymbol{r}')\,.$$

If we perturb the medium with anomalies $\delta\rho$, $\delta\lambda$ and $\delta\mu$, the operator $\mathcal{L}$ will be perturbed by $\delta\mathcal{L}$ and as a consequence the wavefield will be different by $\delta\boldsymbol{u}$. We refer to $\delta\boldsymbol{u}$ as the 'scattered' wave. It is easy to see that, to first order, $\delta\boldsymbol{u}$ satisfies the same elastodynamic equation as $\boldsymbol{u}$, though with a forcing term:

$$(\mathcal{L} + \delta\mathcal{L})(\boldsymbol{u} + \delta\boldsymbol{u}) = 0\,.$$

Writing this out and using $\mathcal{L}\boldsymbol{u} = 0$:

$$\mathcal{L}\delta\boldsymbol{u} = -\delta\mathcal{L}\boldsymbol{u} + \mathcal{O}(\delta^2)\,. \tag{4.24}$$

The 'Born approximation' consists of the neglect of the second-order terms $\delta\mathcal{L}\delta\boldsymbol{u}$ in the right-hand side of (4.24). The first-order term in the right-hand side is a vector field that acts as a force, explicitly:

$$\mathcal{L}\delta\boldsymbol{u} = \omega^2\delta\rho\,\boldsymbol{u} + \nabla(\delta\lambda\nabla\cdot\boldsymbol{u}) + \nabla\cdot[\delta\mu(\nabla\boldsymbol{u} + \nabla\boldsymbol{u}^T)]\,.$$

This has the same form $\mathcal{L}\delta\boldsymbol{u} = \boldsymbol{f}$ as the original equation with a single force term,

$$f_k = \omega^2\delta\rho u_k + \sum_j \partial_k(\delta\lambda\partial_j u_j) + \partial_j[\delta\mu(\partial_k u_j + \partial_j u_k)]\,.$$

The Green's function gives the response to a point source, so we may integrate over the force field $\boldsymbol{f}$ to find the complete scattered response $\delta\boldsymbol{u}$. Since we only deal with a single force, we use only the first term on the right-hand side of (4.17) to find $\delta\boldsymbol{u}$:

$$\delta\boldsymbol{u} = \int \omega^2\delta\rho\,\boldsymbol{G}\boldsymbol{u}\mathrm{d}^3\boldsymbol{r}' + \int \boldsymbol{G}\nabla'(\delta\lambda\nabla'\cdot\boldsymbol{u})\mathrm{d}^3\boldsymbol{r}' + \int \boldsymbol{G}\nabla'\cdot[\delta\mu(\nabla'\boldsymbol{u} + \nabla'\boldsymbol{u}^T)]\mathrm{d}^3\boldsymbol{r}'.$$

Integrating by parts and application of Gauss' theorem can be used to eliminate the derivatives of $\lambda$ and $\mu$, e.g. for the term involving $\delta\lambda$ (see also Exercise 4.8):

$$\int \boldsymbol{G}\nabla'(\delta\lambda\nabla'\cdot\boldsymbol{u})\mathrm{d}^3\boldsymbol{r}' = \int \nabla'\cdot[\boldsymbol{G}(\delta\lambda\nabla'\cdot\boldsymbol{u})]\mathrm{d}^3\boldsymbol{r}' - \int(\nabla'\cdot\boldsymbol{G})\delta\lambda(\nabla'\cdot\boldsymbol{u})\mathrm{d}^3\boldsymbol{r}'$$
$$= \int_S \delta\lambda(\nabla'\cdot\boldsymbol{u})\boldsymbol{G}\hat{\boldsymbol{n}}\mathrm{d}^2\boldsymbol{r}' - \int(\nabla'\cdot\boldsymbol{G})\delta\lambda(\nabla'\cdot\boldsymbol{u})\mathrm{d}^3\boldsymbol{r}',$$

where $\hat{\boldsymbol{n}}$ is a unit vector that points outward from the surface $S$ that encloses the wavefield. There are several arguments to 'explain away' the surface integral. In exploration seismics the surface is either the free surface or practically at infinite distance, far away from the sources, where we can simply set $\delta\lambda = 0$; or argue that the wavefield $\nabla'\cdot\boldsymbol{u}$ attenuates and therefore decreases faster than $1/r$ so that the product with the perturbation field $\boldsymbol{G}$ decreases faster than the growth of the surface area and will be zero at infinity. In global seismology we can show that the stress-free boundary condition at the Earth's surface forces the contribution of the surface integral to be exactly zero. In both cases we can thus write:

$$\int \boldsymbol{G}\nabla'(\delta\lambda\nabla'\cdot\boldsymbol{u})\mathrm{d}^3\boldsymbol{r}' = -\int(\nabla'\cdot\boldsymbol{G})\delta\lambda(\nabla'\cdot\boldsymbol{u})\mathrm{d}^3\boldsymbol{r}'.$$

The notation $\nabla'\cdot\boldsymbol{G}$ can be understood as a column vector with the divergence of each row vector in $\boldsymbol{G}$ as elements, i.e. the $i$-th element is $\sum_j \partial_j G_i^j$. In the surface integral, $\hat{\boldsymbol{n}}$ is a unit vector perpendicular to the surface $S$ and here the product $\boldsymbol{G}\hat{\boldsymbol{n}}$ is a vector with elements $\sum_j G_i^j n_j$.

The term with $\delta\mu$ can be treated in the same way:

$$\int \boldsymbol{G}\nabla'\cdot[\delta\mu(\nabla'\boldsymbol{u}+\nabla'\boldsymbol{u}^T)]\mathrm{d}^3\boldsymbol{r}' = -\int[\delta\mu(\nabla'\boldsymbol{G}):(\nabla'\boldsymbol{u}+\nabla'\boldsymbol{u}^T)]\mathrm{d}^3\boldsymbol{r}',$$

so that the scattered field from an extended scatterer is given by

$$\delta\boldsymbol{u}(\boldsymbol{r},\omega) = \int \omega^2\delta\rho\boldsymbol{G}\boldsymbol{u}\,\mathrm{d}^3\boldsymbol{r}' - \int \delta\lambda(\nabla'\cdot\boldsymbol{G})(\nabla'\cdot\boldsymbol{u})]\mathrm{d}^3\boldsymbol{r}'$$
$$- \int[\delta\mu(\nabla'\boldsymbol{G}):(\nabla'\boldsymbol{u}+\nabla'\boldsymbol{u}^T)]\mathrm{d}^3\boldsymbol{r}',$$

or, writing $\partial'_k$ for $\partial/\partial x'_k$:

$$\delta u_i = \sum_k \int \omega^2\delta\rho G_i^k u_k \mathrm{d}^3\boldsymbol{r}' - \sum_{kj}\int \delta\lambda\partial'_j u_j \partial'_k G_i^k \mathrm{d}^3\boldsymbol{r}'$$
$$- \sum_{jk}\int \delta\mu\partial'_j G_i^k[\partial'_j u_k + \partial'_k u_j]\mathrm{d}^3\boldsymbol{r}'. \tag{4.25}$$

### Exercise

**Exercise 4.8** We can write the term $\delta\lambda\nabla\cdot\boldsymbol{u}$ as a scalar field $v(\boldsymbol{r})$. Write out the elements of $\boldsymbol{G}\nabla v$ under the volume integral and show that

$$\boldsymbol{G}\cdot\nabla v = \nabla\cdot(v\boldsymbol{G}) - v\nabla\cdot\boldsymbol{G}.$$

## 4.6 Scattering of a plane wave

In this section we work out the abstract notation of the last section in more useful detail, in order to study the scattering of a plane P-wave by a small heterogeneity. We dissect the problem by studying P- and S-waves separately. We denote incoming and scattered wavetypes as superscripts on $\delta\boldsymbol{u}$, in that order, e.g. $\delta\boldsymbol{u}^{\mathrm{PS}}$ is a scattered S-wave from an incoming plane P-wave. We start with (4.25), in which we insert a plane wave of unit amplitude travelling in the $z$-direction, i.e.:

$$\boldsymbol{u}(x', y', z', \omega) = \begin{pmatrix} 0 \\ 0 \\ \mathrm{e}^{\mathrm{i}k_{\mathrm{P}}z} \end{pmatrix},$$

where $k_{\mathrm{P}} = \omega/V_{\mathrm{P}}$, and the far field Green's function for a P-wave excited by a force in the $k$-direction:

$$G_i^k(\boldsymbol{r}, \omega; \boldsymbol{r}') = \frac{\gamma_i\gamma_k}{4\pi\rho V_{\mathrm{P}}^2 r}\mathrm{e}^{\mathrm{i}k_{\mathrm{P}}r},$$

where $r = |\boldsymbol{r} - \boldsymbol{r}'|$. We need the following derivatives:

$$\partial_k' r = \partial_k'\left[\sum_l (x_l - x_l')^2\right]^{\frac{1}{2}} = -(x_k - x_k')/r = -\gamma_k$$

$$(\nabla\boldsymbol{u})_{jk} = \partial_k' u_j = \delta_{jz}\delta_{kz}\,\mathrm{i}k_{\mathrm{P}}\mathrm{e}^{\mathrm{i}k_{\mathrm{P}}z'}$$

$$\nabla\cdot\boldsymbol{u} = \sum_k \partial_k' u_k = \partial_z' u_z = \mathrm{i}k_{\mathrm{P}}\mathrm{e}^{\mathrm{i}k_{\mathrm{P}}z'}$$

$$(\nabla\boldsymbol{G})_{ijk} = \partial_j' G_i^k \approx -\mathrm{i}k_{\mathrm{P}}\frac{\gamma_i\gamma_j\gamma_k}{4\pi\rho V_{\mathrm{P}}^2 r}\mathrm{e}^{\mathrm{i}k_{\mathrm{P}}r}$$

$$(\nabla\cdot\boldsymbol{G})_i = \sum_k \partial_k' G_i^k \approx -\mathrm{i}k_{\mathrm{P}}\frac{\gamma_i}{4\pi\rho V_{\mathrm{P}}^2 r}\mathrm{e}^{\mathrm{i}k_{\mathrm{P}}r},$$

where we use $\sum_j \gamma_j\gamma_j = 1$ and where the $\approx$ indicates that we have ignored the derivative of the slowly decreasing amplitude factor, since the rapidly changing phase $\mathrm{i}k_{\mathrm{P}}r$ dominates in the far field. We shall further assume that the scatterer is constant over a small volume $\mathrm{d}V$, such that, e.g. $\int\delta\rho\,\mathrm{d}V = \delta\rho\,\mathrm{d}V$. The result is

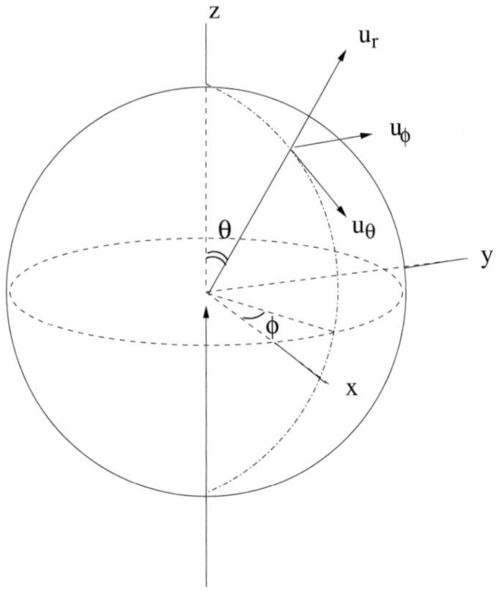


Fig. 4.4. A plane wave travelling from below in the $z$-direction hits a scatterer and generates a wave that makes an angle $\theta$ with the $z$-axis. The angle that the projection of the scattered wave makes with the $x$-axis is denoted by $\phi$. In this system $\gamma_x = \sin\theta\cos\phi$, $\gamma_y = \sin\theta\sin\phi$, $\gamma_z = \cos\theta$.

that for a scatterer in the origin where $z' = 0$ (4.25) reduces to:

$$\delta u_i^{\mathrm{PP}} = \frac{k_{\mathrm{P}}^2 \mathrm{e}^{\mathrm{i}k_{\mathrm{P}}r}}{4\pi r}\left(\frac{\delta\rho}{\rho}\gamma_i\gamma_z - \frac{\delta\lambda}{\lambda+2\mu}\gamma_i - \frac{2\delta\mu}{\lambda+2\mu}\gamma_i\gamma_z^2\right)\mathrm{d}V\,.$$

To avoid clutter, we do not use subscripts 0 as we did for $c_0$ in the acoustic case; it should be understood that $\rho$, $\lambda$, $\mu$ as well as $V_{\mathrm{P}}$ and $V_{\mathrm{S}}$ are the values for the homogeneous medium in which the scatterers are embedded. If we denote the angle that the scattered wave makes with the $z$-axis by $\theta$ (the 'scattering angle', $0 \le \theta \le \pi$) we can make the result independent of our choice of coordinate system. For the (longitudinal) component of motion of the P-wave we take the dot product with the direction vector $\boldsymbol{\gamma}$ and use $\gamma_z = \cos\theta$:

$$\begin{aligned}\delta u_r^{\mathrm{PP}} &= \sum_i \delta u_i^{\mathrm{PP}}\gamma_i \\ &= \frac{\omega^2 \mathrm{e}^{\mathrm{i}k_{\mathrm{P}}r}}{4\pi r V_{\mathrm{P}}^2}\left(\frac{\delta\rho}{\rho}\cos\theta - \frac{\delta\lambda}{\lambda+2\mu} - \frac{2\delta\mu}{\lambda+2\mu}\cos^2\theta\right)\mathrm{d}V\,,\end{aligned} \tag{4.26}$$

The treatment of the scattered S-wave is essentially the same, using the Green's function for the S-wave instead:

$$G_i^k(\boldsymbol{r}, \omega; \boldsymbol{r}') = -\frac{\gamma_i \gamma_k - \delta_{ik}}{4\pi\rho V_S^2 r} \mathrm{e}^{\mathrm{i}k_S r}$$

$$\partial'_j G_i^k = \mathrm{i}k_S - \frac{\gamma_i \gamma_k - \delta_{ik}}{4\pi\rho V_S^2 r} \gamma_j \mathrm{e}^{\mathrm{i}k_S r}$$

$$\delta u_i^{\mathrm{PS}} = \frac{k_S^2 \mathrm{e}^{\mathrm{i}k_S r}}{4\pi r} \left[ \frac{\delta\rho}{\rho}(\delta_{iz} - \gamma_i \gamma_z) + \frac{2\delta\mu V_S}{\mu V_P}(\gamma_i \gamma_z^2 - \gamma_z \delta_{iz}) \right] \mathrm{d}V \, .$$

For the component of the S-wave in the $\theta$ direction (Figure 4.4) we take the dot product with the unit vector $\hat{\boldsymbol{e}}_\theta = (\cos\theta\cos\phi, \cos\theta\sin\phi, -\sin\theta)$:

$$\begin{aligned} \delta u_\theta^{\mathrm{PS}} &= \delta u_x^{\mathrm{PS}} \cos\theta\cos\phi + \delta u_y^{\mathrm{PS}} \cos\theta\sin\phi - \delta u_z^{\mathrm{PS}} \sin\theta \\ &= \frac{\omega^2 \mathrm{e}^{\mathrm{i}k_S r}}{4\pi r V_S^2} \left( -\frac{\delta\rho}{\rho} \sin\theta - \frac{2\delta\mu V_S}{\mu V_P} \sin 2\theta \right) \mathrm{d}V \, . \end{aligned} \tag{4.27}$$

It is left to the reader to verify that the $\phi$-component of the scattered wave is zero by dotting $\delta\boldsymbol{u}^{\mathrm{PS}}$ with $\hat{\boldsymbol{e}}_\phi = \hat{\boldsymbol{e}}_r \times \hat{\boldsymbol{e}}_\theta = (-\sin\phi, \cos\phi, 0)$. If $\delta u_\phi^{\mathrm{SS}}$ is polarized in the $\phi$-direction it is often denoted as SH, and the scattered wave polarized in the $\theta$-direction as SV, for 'horizontal' and 'vertical' respectively. Thus, an incoming P results only in scattered SV energy, no SH.

The factor $\mathrm{e}^{\mathrm{i}k_P r}/4\pi r$ describes the propagation of the scattered wave away from the point scatterer. The factor $\omega^2$ gives the typical frequency dependence resulting from a small scatterer ('Rayleigh scattering'). In our first-order approximation, scattering is linear and the response of a large scattering body can be computed by decomposing it into small scatterers of size $\mathrm{d}V$ and integrating over all of them. Though we derived the expressions for a homogeneous medium, they are very useful for heterogeneous media as well, because we may consider such media locally homogeneous around the scatterer. For a small scatterer, the incoming wavefield can usually be regarded as locally plane if the source is at a distance much longer than the wavelength.

The frequency dependence changes with the size of the scatterer. The integration over a volume $V$ leads to destructive interference among high frequencies because of the rapidly varying phase, which is the sum of the phases of incoming and scattered wave. Since integration over a space coordinate involves integration over scatterers that are progressively later in time, the space integration is qualitatively equivalent to a time integration, i.e. a division by $-\mathrm{i}\omega$. Such filters act to boost low frequencies and suppress high frequencies (a 'low-pass filter' in the jargon of electronics). Imagine one integration over a curved coordinate that follows those scatterers that have the same time delay. This integration will see no destructive interference and therefore no filtering effect. But the integration over the remaining two directions divides by a factor $(-\mathrm{i}\omega)^2 = -\omega^2$. The factor $\omega^2$ in the scattered wave amplitude thus acts as a filter that restores high frequencies (a 'high-pass' filter). This we would expect to be needed, because a perfectly plane surface reflects waves with no distortion due to high-pass filtering: without the factor $\omega^2$ our face would seem very red in a mirror!

The treatment for an incoming S-wave is very similar, and left to the reader (see Exercise 4.11). Instead the next section describes a simpler, albeit more abstract, method for the computation of scattered wave amplitudes.

## *Exercises*

**Exercise 4.9** Verify that the displacement of the scattered S-wave is indeed fully in the meridian plane (i.e. $u_\phi = 0$, see Figure 4.4).

**Exercise 4.10** For scatterers $\delta\rho$, $\delta\lambda$ and $\delta\mu$ respectively, list (1) for what scattering angle(s) $\delta u_\theta^{\mathrm{PS}}$ and $\delta u_r^{\mathrm{PP}}$ are zero, and (2) where they obtain a maximum amplitude.

**Exercise 4.11** We derive the expressions for the scattering of an incoming plane S-wave in steps.

a. For an incoming S-wave with $u_k = \delta_{ky}\mathrm{e}^{\mathrm{i}k_{\mathrm{S}}z'}$ show that $\partial'_j u_k = \mathrm{i}k_{\mathrm{S}}\delta_{ky}\delta_{jz}$ in the scatterer location $z' = 0$, that

$$\partial'_j G_i^k = \mathrm{i}k_{\mathrm{S}} \frac{\gamma_i\gamma_k - \delta_{ik}}{4\pi\rho V_{\mathrm{S}}^2 r}\gamma_j \mathrm{e}^{\mathrm{i}k_{\mathrm{S}}r}\,, \quad \sum_k \partial'_k G_i^k = 0$$

for a scattered S-wave, and

$$\partial'_j G_i^k = -\mathrm{i}k_{\mathrm{P}} \frac{\gamma_i\gamma_j\gamma_k}{4\pi\rho V_{\mathrm{P}}^2 r}\mathrm{e}^{\mathrm{i}k_{\mathrm{P}}r}\,, \quad \sum_k \partial'_k G_i^k = -\mathrm{i}k_{\mathrm{P}} \frac{\gamma_i}{4\pi\rho V_{\mathrm{P}}^2 r}\mathrm{e}^{\mathrm{i}k_{\mathrm{P}}r}$$

for a scattered P-wave.

b. Insert these expressions in the term for a scatterer $\delta\rho$ in (4.25) and take the dot product with $\hat{\boldsymbol{e}}_r \equiv \boldsymbol{\gamma}$ to show that $\delta u_r^{\mathrm{SS}} = 0$ and

$$\delta u_r^{\mathrm{SP}} = \frac{\omega^2 \mathrm{e}^{\mathrm{i}k_{\mathrm{P}}r}}{4\pi r V_{\mathrm{P}}^2}\frac{\delta\rho}{\rho}\sin\theta\sin\phi\,.$$

c. For the term in (4.25) with $\delta\lambda$, show that $\delta u_i^{\mathrm{SS}}$ and $\delta u_i^{\mathrm{SP}}$ are both zero.

d. For S → S scattering by $\delta\mu$, show that:

$$\delta\boldsymbol{u}^{\mathrm{SS}} = -\frac{\omega^2\mathrm{e}^{\mathrm{i}k_{\mathrm{S}}r}}{4\pi r V_{\mathrm{S}}^2}\frac{\delta\mu}{\mu}\begin{pmatrix} 2\gamma_x\gamma_y\gamma_z \\ 2\gamma_y^2 - \gamma_z \\ 2\gamma_y\gamma_z^2 - \gamma_y \end{pmatrix} = -(\ldots)\begin{pmatrix} 2\sin^2\theta\cos\phi\sin\phi\cos\theta \\ 2\sin^2\theta\sin^2\phi\cos\theta - \cos\theta \\ 2\sin\theta\sin\phi\cos^2\theta - \sin\theta\sin\phi \end{pmatrix}$$

and verify that $\delta u_r = 0$, $\delta u_\phi \sim \cos\theta\cos\phi$ and $\delta u_\theta \sim -\sin\phi\cos 2\theta$ by taking dot products with the appropriate unit vectors.

e. Similarly, for S → P scattering, show that $u_i^{\mathrm{SP}} \sim \gamma_i\gamma_y$ in case of a $\delta\rho$ scatterer and $\sim 2\gamma_i\gamma_y\gamma_z$ in case of $\delta\mu$.

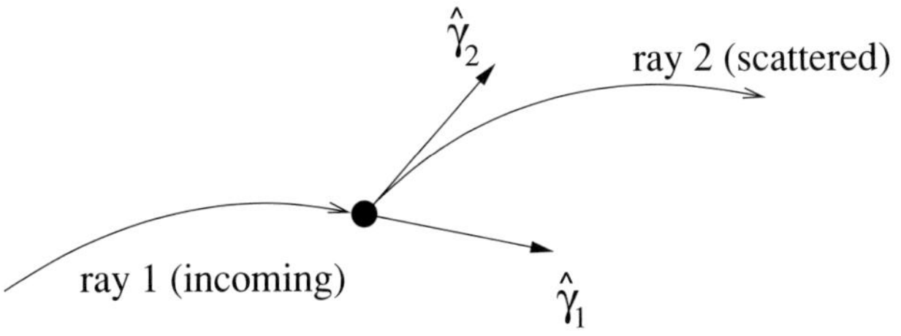


Fig. 4.5. Notation of ray directions at the scatterer with unit vectors $\hat{\boldsymbol{\gamma}}$

f. Combine the results of (b) – (e) to show that

$$\delta u_r^{\rm SP} = \frac{\omega^2 e^{ik_{\rm P}r}}{4\pi r V_{\rm P}^2}\left(\frac{\delta\rho}{\rho}\sin\theta\sin\phi - 2\frac{V_{\rm S}\delta\mu}{V_{\rm P}\mu}\sin\theta\cos\theta\sin\phi\right){\rm d}V\,, \tag{4.28}$$

$$\delta u_\theta^{\rm SS} = \frac{\omega^2 e^{ik_{\rm S}r}}{4\pi r V_{\rm S}^2}\left(\frac{\delta\rho}{\rho}\cos\theta\sin\phi - \frac{\delta\mu}{\mu}\cos 2\theta\sin\phi\right){\rm d}V\,, \tag{4.29}$$

$$\delta u_\phi^{\rm SS} = \frac{\omega^2 e^{ik_{\rm S}r}}{4\pi r V_{\rm S}^2}\left(\frac{\delta\rho}{\rho}\cos\phi - \frac{\delta\mu}{\mu}\cos\theta\cos\phi\right){\rm d}V\,, \tag{4.30}$$

where the $\phi$ component of the scattered S-wave is in a direction given by $\hat{\boldsymbol{e}}_\phi$ (Figure 4.4). Hint: use $k_{\rm P} = \omega/V_{\rm P}$, $k_{\rm S} = \omega/V_{\rm S}$ and $\rho V_{\rm S}^2 = \mu$.

## 4.7 The scattering matrix

The expressions for the scattered waves depend on the direction cosines for the incoming and outgoing rays as well as their polarizations ($\boldsymbol{u}$ and $\boldsymbol{G}$, respectively). A minor algebraic manipulation allows us to write the scattering very economically in matrix form. We deviate from Zhao and Dahlen [416] by making the factor $\rho\omega^2$ explicit in the following defining expression for the scattering matrix $\boldsymbol{S}$ for point scatterer of volume ${\rm d}V$ at location $\boldsymbol{r}'$:

$$\delta\boldsymbol{u}(\boldsymbol{r},\omega) = \omega^2\boldsymbol{G}(\boldsymbol{r},\omega;\boldsymbol{r}')(\rho\boldsymbol{S})\boldsymbol{u}(\boldsymbol{r}',\omega){\rm d}V\,. \tag{4.31}$$

The concise index notation we use in (4.25) is helpful when deriving expressions for $\boldsymbol{S}$:

$$\delta u_i = \sum_{kj}\omega^2 G_i^k(\rho S_{kj})u_j{\rm d}V\,. \tag{4.32}$$

At the location of the scatterer, the incoming and outgoing ray are defined by directions given by unit vectors $\hat{\boldsymbol{\gamma}}_1$ and $\hat{\boldsymbol{\gamma}}_2$, and P and/or S velocities $V_1$ and $V_2$, respectively (Figure 4.5). If no wavetype conversion occurs, $V_1 = V_2$. Note that

$$\partial'_j = \frac{i\omega}{V_1}\gamma_{1j} \quad \text{(incoming ray)} \tag{4.33}$$

$$\partial'_j = -\frac{i\omega}{V_2}\gamma_{2j} \quad \text{(scattered ray)}\,. \tag{4.34}$$

The full scattering matrix is a sum of contributions from heterogeneities in density and elastic parameters, and we split $\boldsymbol{S}$ into its separate contributions:

$$\boldsymbol{S} = \frac{\delta\mu}{\mu}\boldsymbol{S}_\mu + \frac{\delta\lambda}{\lambda}\boldsymbol{S}_\lambda + \frac{\delta\rho}{\rho}\boldsymbol{S}_\rho\,. \tag{4.35}$$

We find the separate contributions by substituting (4.33) and (4.34) into (4.25):

$$\begin{aligned}\delta u_i = &\sum_k \omega^2 \delta\rho G_i^k u_k \mathrm{d}V + \sum_{kj} \delta\lambda \frac{i\omega}{V_1}\gamma_{1j}u_j \frac{i\omega}{V_2}\gamma_{2k} G_i^k \mathrm{d}V \\ &+ \sum_{kj} \delta\mu \frac{i\omega}{V_2}\gamma_{2j} G_i^k \left[\frac{i\omega}{V_1}\gamma_{1j}u_k + \frac{i\omega}{V_1}\gamma_{1k}u_j\right]\mathrm{d}V\,.\end{aligned} \tag{4.36}$$

Comparing this with (4.32) we see that the scattering matrix for the density is simply $\delta_{kj}$, or the unit matrix $\boldsymbol{I}$. Similarly, the scattered signal from an inclusion $\delta\lambda$ is:

$$-G_i^k \frac{\delta\lambda\omega^2}{V_1 V_2}\gamma_{2k}\gamma_{1j}u_j \mathrm{d}V\,,$$

which shows that the scattering matrix for $\lambda$ involves the dyadic product $\hat{\boldsymbol{\gamma}}_2\hat{\boldsymbol{\gamma}}_1$. Finally, an inclusion with an anomaly $\delta\mu$ gives:

$$\begin{aligned}&-G_i^k \frac{\delta\mu\,\omega^2}{V_1 V_2}\gamma_{2j}[\gamma_{1j}u_k + \gamma_{1k}u_j]\mathrm{d}V \\ &\quad = -G_i^k \frac{\delta\mu\,\omega^2}{V_1 V_2}[(\hat{\boldsymbol{\gamma}}_1 \cdot \hat{\boldsymbol{\gamma}}_2)\delta_{jk} + \gamma_{1k}\gamma_{2j}]u_j \mathrm{d}V\,,\end{aligned}$$

which gives the following expressions for the scattering matrix:

$$\boldsymbol{S}_\mu = -\frac{V_\mathrm{S}^2}{V_1 V_2}[\hat{\boldsymbol{\gamma}}_1\hat{\boldsymbol{\gamma}}_2 + (\hat{\boldsymbol{\gamma}}_2 \cdot \hat{\boldsymbol{\gamma}}_2)\boldsymbol{I}]\,, \tag{4.37}$$

$$\boldsymbol{S}_\lambda = -\frac{\lambda}{\rho V_1 V_2}\hat{\boldsymbol{\gamma}}_2\hat{\boldsymbol{\gamma}}_1\,, \tag{4.38}$$

$$\boldsymbol{S}_\rho = \boldsymbol{I}\,. \tag{4.39}$$

In practice, we often prefer to formulate the scattering in terms of density and seismic velocities $V_\mathrm{P}$ and $V_\mathrm{S}$. Note that there is a difference between varying $\rho$ while keeping velocities constant (which forces $\mu$ and $\lambda$ to vary with $\rho$), or keeping

$\mu$ and $\lambda$ constant. The necessary expressions can be derived from $\mu = \rho V_S^2$ so that

$$\delta\mu = \left(\frac{\partial \mu}{\partial \rho}\right)_{V_S} \delta\rho + \left(\frac{\partial \mu}{\partial V_S}\right)_{\rho} \delta V_S = V_S^2 \delta\rho + 2V_S\rho\delta V_S ,$$

and similarly, since $\lambda = \rho(V_P^2 - 2V_S^2)$:

$$\delta\lambda = \left(V_P^2 - 2V_S^2\right)\delta\rho + 2V_P\rho\delta V_P - 4V_S\rho\delta V_S .$$

Substituting this gives (we keep the notation $\boldsymbol{S}_\rho$, but with a different meaning):

$$\boldsymbol{S} = \frac{\delta V_P}{V_P}\boldsymbol{S}_P + \frac{\delta V_S}{V_S}\boldsymbol{S}_S + \frac{\delta\rho}{\rho}\boldsymbol{S}_\rho , \tag{4.40}$$

with

$$\boldsymbol{S}_P = -2\frac{V_P^2}{V_1 V_2}\hat{\boldsymbol{\gamma}}_2\hat{\boldsymbol{\gamma}}_1 , \tag{4.41}$$

$$\boldsymbol{S}_S = 2\frac{V_S^2}{V_1 V_2}[2\hat{\boldsymbol{\gamma}}_2\hat{\boldsymbol{\gamma}}_1 - \hat{\boldsymbol{\gamma}}_1\hat{\boldsymbol{\gamma}}_2 - (\hat{\boldsymbol{\gamma}}_1 \cdot \hat{\boldsymbol{\gamma}}_2)\boldsymbol{I}] , \tag{4.42}$$

$$\boldsymbol{S}_\rho = \boldsymbol{I} - \frac{\lambda}{\rho V_1 V_2}\hat{\boldsymbol{\gamma}}_2\hat{\boldsymbol{\gamma}}_1 - \frac{V_S^2}{V_1 V_2}[\hat{\boldsymbol{\gamma}}_1\hat{\boldsymbol{\gamma}}_2 + (\hat{\boldsymbol{\gamma}}_1 \cdot \hat{\boldsymbol{\gamma}}_2)\boldsymbol{I}] . \tag{4.43}$$

## 4.8 Appendix B: The impulse response

This appendix assumes the reader is familiar with complex function theory, in particular with the residue theorem. The residue theorem says that the value of an integral around a pole $z_0$ of a function $f(z)$ in the complex $z$-plane is given by:

$$\int_C f(z)\mathrm{d}z = 2\pi\mathrm{i}\,\mathrm{Res}\, f(z_0) ,$$

where the residue for a simple pole $h(z_0) = 0$ of a function $f(z) = g(z)/h(z)$ is given by $g(z_0)/h'(z_0)$.

We must find the solution of (4.1):

$$c^2\nabla^2 G + \omega^2 G = \delta(\boldsymbol{r}) .$$

We shall sketch the derivation of (4.4), by making use of a common method: Fourier transforming to the wavenumber domain. In Cartesian coordinates $(x, y, z)$

we define the following transform pair:[†]

$$G(\boldsymbol{r},\omega) = \int \frac{d^3\boldsymbol{k}}{(2\pi)^3}\,\tilde{G}(\boldsymbol{k},\omega)\exp(i\boldsymbol{k}\cdot\boldsymbol{r}),$$

$$\tilde{G}(\boldsymbol{k},\omega) = \int d^3\boldsymbol{r}\,G(\boldsymbol{r},\omega)\exp(-i\boldsymbol{k}\cdot\boldsymbol{r}),$$

where $\boldsymbol{k}$ is a vector $(k_x, k_y, k_z)$ with magnitude $k$, oriented in the direction of wave propagation as can be seen from its presence in the phase factor $\boldsymbol{k}\cdot\boldsymbol{r} - \omega t$. This Fourier transformation is equivalent to a decomposition of the wave field into plane waves, since plane surfaces of constant phase can be found perpendicular to $\boldsymbol{k}$. Differentiation with respect to space and time is now reduced to a simple multiplication with $k_x$, $k_y$, $k_z$, and frequency, respectively, and the differential equation reduces to a simple algebraic equation for the Fourier transform $\tilde{G}$:

$$(-c^2k^2 + \omega^2)\tilde{G} = 1,$$

so that, inserting this $\tilde{G}$ into the Fourier transform of $G$:

$$G(\boldsymbol{r},\omega) = \frac{1}{(2\pi)^3c^2}\int \frac{\exp(i\boldsymbol{k}\cdot\boldsymbol{r})d^3\boldsymbol{k}}{\omega^2/c^2 - k^2}.$$

We could now write out the dot product $\boldsymbol{k}\cdot\boldsymbol{r}$. A simpler method consists of using spherical coordinates for the wavenumber vector $\boldsymbol{k}$. If we define the components of $\boldsymbol{k}$ as $(k,\theta,\phi)$ we have $k_x = k\sin\phi\sin\theta$, $k_y = k\cos\phi\sin\theta$, and $k_z = k\cos\theta$. We choose the $\theta$ axis in the direction of $\boldsymbol{r}$; then $\boldsymbol{k}\cdot\boldsymbol{r} = kr\cos\theta$, and $d^3\boldsymbol{k}$ becomes $k^2\sin\theta\,dk d\theta d\phi$, so that:

$$G(\boldsymbol{r},\omega) = \frac{1}{(2\pi)^3c^2}\int_0^\infty dk\int_0^\pi \sin\theta d\theta\int_0^{2\pi} d\phi\frac{k^2\exp(ikr\cos\theta)}{\omega^2/c^2 - k^2}.$$

The integration over $\theta$ and $\phi$ is elementary:

$$\int_0^{2\pi}\int_0^\pi e^{ikr\cos\theta}\sin\theta\,d\theta d\phi = 2\pi\int_0^\pi e^{ikr\cos\theta}\sin\theta d\theta = \frac{2\pi}{ikr}(e^{ikr} - e^{-ikr}),$$

and, extending the integration over the negative $k$-axis by combining the two exponents we find:

$$G(\boldsymbol{r},\omega) = \frac{2\pi}{ir(2\pi)^3c^2}\int_{-\infty}^\infty dk\frac{k\exp(ikr)}{(\omega/c - k)(\omega/c + k)}.$$

To find an analytical expression for the integral, we use the residue theorem from complex function theory. We connect the endpoints of the integration interval by

[†] The difference in sign convention with the time–frequency transform is intentional. This way, a wave with a positive wavenumber $k_x$ travels in the positive $x$ direction, etc.

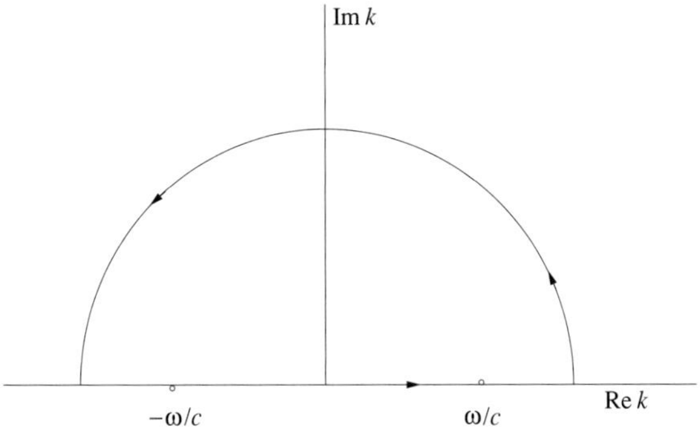


Fig. 4.6. The integration in the complex $k$-plane.

a semicircle in the positive complex $k$-plane, as shown in Figure 4.6. Because $\exp(\mathrm{i}kr) \to 0$ as $\mathrm{Im}\,k \to \infty$ this extra integration does not add to $G(\boldsymbol{r}, \omega)$, but because we now have an integration over a closed contour, we can express $G(\boldsymbol{r}, \omega)$ as the the contribution from the poles inside the contour. There are two singularities ($k = \pm\omega/c$) on the real axis. For example, the residue for $k = +\omega/c$ is:

$$\mathrm{Res} = \lim_{k \to \omega/c} (k - \omega/c) \frac{k \exp(\mathrm{i}kr)}{(\omega/c - k)(\omega/c + k)} = -\frac{\exp(\mathrm{i}\omega r/c)}{2} .$$

In a true physical situation, damping will move these poles away from the real axis and the path does not really cross the singularities. We follow a pedestrian approach, and will include the contribution from both poles. Inspection a posteriori will then tell us which we need. Closing the path of integration with a contour over the positive imaginary $k$-plane (so that the contribution of the contour at infinity disappears for $r > 0$) and including both residues yields:

$$G(\boldsymbol{r}, \omega) \;=\; 2\pi\mathrm{i} \sum \mathrm{Res} = -\frac{1}{2\pi r c^2} \left( \frac{\mathrm{e}^{\mathrm{i}\omega r/c}}{2} - \frac{\mathrm{e}^{-\mathrm{i}\omega r/c}}{2} \right) .$$

With (2.63):

$$G(\boldsymbol{r}, t) = -\frac{1}{4\pi r c^2} [\delta(t - r/c) - \delta(t + r/c)] .$$

The $\delta(t + r/c)$ solution represents an incoming wave which is, in general, unphysical. It is a consequence of the inclusion of the pole on the negative $k$-axis. In hindsight we may conclude that damping (or causality) would have moved this pole into the third quadrant, slightly below the real axis, eliminating any contribution from it.

As expected, a pulse remains a pulse. A more arbitrary waveform will therefore travel without experiencing dispersion.

# 5

# Body wave amplitudes: theory

As we have seen in Chapter 2, the energy density of a seismic wave decreases as the wave propagates because of geometrical spreading: the available energy spreads over a larger wavefront-surface, and therefore amplitudes should decrease to avoid an increase in the total energy of the system. Individual wave packets also lose energy because of scattering: part of the energy is redirected by refraction, or converted to a different kind of wave. Examples of this are P–S conversion at the Moho, or scattering at random heterogeneities.

In both cases, however, the total mechanical energy in the vibrating system (the Earth) remains the same. We know this cannot be true. At regular intervals, vibrational energy is added to the Earth, through release of potential energy (strain energy) in earthquakes, through the detonation of large explosions or simply when a train passes by. But even after a large earthquake, the activity on a short period seismograph returns to normal after a few hours. It may take days on a low frequency seismograph, but there, too, the energy eventually damps away. The mechanical energy of seismic waves is converted to other forms of energy, mostly heat. Such processes are inelastic, commonly referred to as 'intrinsic attenuation'.

In this chapter we take a closer look at the factors that determine the amplitude of a body wave.

## 5.1 Geometrical spreading

In Section 2.7 we saw that the energy of a body wave in a spherically symmetric Earth is proportional to $\partial^2 T/\partial\Delta^2 = \partial p/\partial\Delta$. Since by Snel's law the ray parameter $p$ is related to the angle at which the ray leaves the source, this can intuitively be understood as the number of rays within a bundle $\partial p$ that ends up in a stretch of epicentral distance $\partial\Delta$ – like the density of photons striking an illuminated surface. In more general 3D media we would expect, then, that the energy is somehow related to the curvature of the wavefront, since rays spread out more

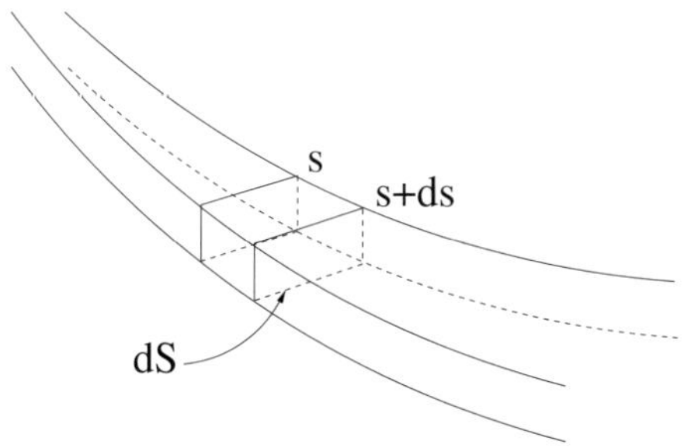


Fig. 5.1. A bundle of rays spanning a solid angle $\mathrm{d}\Omega$. By definition of geometrical spreading, the surface area $\mathrm{d}S$ perpendicular to the rays is equal to $\mathcal{R}^2\mathrm{d}\Omega$.

quickly if the wavefront is strongly curved. In Chapter 3 we saw that the curvature in ray-centred coordinates is given by:

$$\frac{\partial^2\tau}{\partial q_i \partial q_j} = H_{ij} \,.$$

The scaling factor $h = 1 - Kq_1 = 1$ on the ray itself where $q_1 = q_2 = 0$. Taking the divergence of (3.4):

$$\nabla^2\tau = \frac{1}{h^2}\frac{\partial^2\tau}{\partial s^2} + \frac{\partial^2\tau}{\partial q_1^2} + \frac{\partial^2\tau}{\partial q_2^2} = \frac{\partial}{\partial s}\left(\frac{1}{c}\right) + H_{11} + H_{22} \,,$$

where we use the fact that $\partial\tau/\partial s = 1/c$ on the ray. With the definition of divergence† we can find an expression for $\nabla^2\tau$ in terms of the geometrical spreading factor $\mathcal{R}$, introduced in Section 2.7. For this purpose we again consider a small volume $\Delta V$ enclosed by a ray tube (Figure 5.1). By definition of the geometrical spreading factor $\mathcal{R}$ the cross section of the ray tube has a surface $\mathrm{d}S = \mathcal{R}^2(\boldsymbol{r}_\mathrm{s}, \boldsymbol{r})\mathrm{d}\Omega$, where $\mathrm{d}\Omega$ is the solid angle enclosed by the ray tube at $\boldsymbol{r}_\mathrm{s}$, the source position. Since $|\nabla\tau| = \nabla\tau \cdot \hat{\boldsymbol{n}} = 1/c$ on $\mathrm{d}S$:

$$\begin{aligned}
\nabla^2\tau \equiv \nabla \cdot (\nabla\tau) &= \lim_{\Delta V \to 0} \frac{1}{\Delta V} \int_S \nabla\tau \cdot \hat{\boldsymbol{n}}\,\mathrm{d}S \\
&= \lim_{\mathrm{d}s \to 0} \frac{1}{\mathrm{d}s\mathcal{R}^2\mathrm{d}\Omega} \int \frac{\mathcal{R}^2\mathrm{d}\Omega}{c} = \lim_{\mathrm{d}s \to 0} \frac{1}{\mathcal{R}^2\mathrm{d}s}\left[\frac{\mathcal{R}^2}{c}\right]_s^{s+\mathrm{d}s} \\
&= \frac{1}{\mathcal{R}^2}\frac{\partial}{\partial s}\left(\frac{\mathcal{R}^2}{c}\right) = \frac{\partial}{\partial s}\left(\frac{1}{c}\right) + \frac{1}{c\mathcal{R}^2}\frac{\partial \mathcal{R}^2}{\partial s} \,.
\end{aligned}$$

† div $\boldsymbol{v}$ or $\nabla \cdot \boldsymbol{v} = \lim_{\Delta V \to 0} \frac{1}{\Delta V} \int_S \boldsymbol{v} \cdot \hat{\boldsymbol{n}}\,\mathrm{d}S$ for a volume $\Delta V$ bounded by a surface $S$.

Combining this with our expression of $\nabla^2\tau$ in terms of the matrix $\boldsymbol{H}$, we get the desired relationship between $\mathcal{R}$ and the curvature of the wavefront:

$$\frac{1}{\mathcal{R}^2}\frac{\partial\mathcal{R}^2}{\partial s} = \frac{\partial \ln \mathcal{R}^2}{\partial s} = c(H_{11} + H_{22}) = c\mathrm{tr}(\boldsymbol{H}), \tag{5.1}$$

where $\mathrm{tr}(\boldsymbol{H})$ denotes the trace of the matrix $\boldsymbol{H}$. The solution of this differential equation is elementary if we start from a small $s_0$ where $\mathcal{R} \approx s_0$:

$$\ln \mathcal{R}^2 = \int_{s_0}^{s} c\mathrm{tr}\boldsymbol{H}\,\mathrm{d}s + 2\ln s_0. \tag{5.2}$$

Thus, we may obtain geometrical spreading as a direct by-product of the integration of the Riccati equation (3.9). Discontinuities add jumps to (5.2).

## 5.2 The quality factor Q

Sliding on cracks and grain boundaries is probably the dominant mechanism for attenuation in rocks under physical conditions found in the upper crust. However, most cracks are closed at pressures of about 0.5 GPa (15 km depth) and here defects in the crystal structure become more important. But the exact nature of the physical processes that lead to energy loss still remains to be determined. Abundant references to the literature on attenuation can be found in Romanowicz and Mitchell [292].

In practice, the attenuation of teleseismic waves is difficult to measure. It affects both body waves and surface waves (Chapter 10) and broadens spectral peaks in the Earth's spectrum of eigenfrequencies (Chapter 9). Its effect is slow in time, and the relative contributions of wave scattering and focusing/defocusing are difficult to distinguish (see also Chapter 8). Because of this, the study of seismic wave attenuation has remained rather qualitative, at least in comparison with the more prevalent velocity studies. A decade ago, Romanowicz [290] reviewed $Q$ tomography and concluded 'in the upper mantle, agreement between body and surface wave results is variable', though some progress has been made since then. In this chapter we introduce the fundamental concepts for body waves.

In general, we speak of high quality (or high $Q$) waves when there is only a slow loss of energy with time or distance. The seismic parameter $Q$ is very similar to the $Q$ used in electronics for the decay of a current in the circuit of a damped oscillator, and we shall use the well-known physics of this to help fix some ideas. The equation for the amplitude $x$ of a damped harmonic oscillator is:

$$\ddot{x}(t) + 2\gamma\dot{x}(t) + \omega_0^2 x(t) = 0\,, \tag{5.3}$$

where the dot denotes differentiation with respect to time. The two independent solutions are:

$$x(t) = \exp(\pm i\omega_1 t - \gamma t)\,, \tag{5.4}$$

where $\omega_1^2 = \omega_0^2 - \gamma^2$. We assume that the damping $\gamma$ is small with respect to $\omega_0$ so that the wave does not damp out before having completed at least several cycles, and $\omega_1^2 > 0$. For an initial condition $x(0) = 1$ the two solutions combine to give:

$$x(t) = \cos\omega_1 t e^{-\gamma t}\,.$$

The spectrum of this signal consists of two peaks with finite width:

$$X(\omega) = \int_0^\infty x(t) e^{i\omega t}\,dt = \frac{i}{2}\left(\frac{1}{\omega + \omega_1 + i\gamma} - \frac{1}{\omega - \omega_1 + i\gamma}\right).$$

Since $\gamma \ll \omega_1$, $|X(\omega)|$ is sharply peaked near $\omega = \pm\omega_1$. We then find for positive $\omega$:

$$|X(\omega)| \approx \frac{1}{2}[(\omega - \omega_1)^2 + \gamma^2]^{-\frac{1}{2}}\,. \tag{5.5}$$

The power spectral peak $|X(\omega)|^2$ is down by a factor of 2 (−3 dB)† when $\omega - \omega_1 = \pm\gamma$. We introduce the quality factor as $Q_T = \omega_1/2\gamma$, so the width of the peak is given by $\omega_1/Q_T$. Thus, $1/Q_T$ is the relative width of the spectral peak and can in principle be measured from the spectrum of the oscillator.

An alternative definition of $Q_T$ can be formulated in terms of energy loss. If a wave has an amplitude $A = \exp(-\gamma t) = \exp(-\omega_1 t/2Q_T)$, its energy decreases like $A^2 = E = \exp(-\omega_1 t/Q_T)$. Hence $dE/dt = -\omega_1 E/Q_T$, and for $\Delta t = 2\pi/\omega_1$ (one cycle):

$$Q_T = \frac{2\pi E}{|\Delta E|}\,, \tag{5.6}$$

where $|\Delta E|$ is the energy loss in one cycle.

For a standing wave we can simply observe the attenuation of the wave in a single seismograph to measure energy loss, either with time or by measuring the width of the spectral peak in the frequency domain. This can be done for the Earth's free oscillations which we shall study in more detail in Chapter 9. A time domain approach would not be feasible for the Sun or other stars, since their oscillations are continuously excited. Very recently, reliable eigenfrequencies for variable stars have been observed. Eventually, the $Q$ of such modes will have to be determined from widths of the spectral lines.

† dB or decibel is always defined with respect to a base level. If we take the maximum power $X^2_{max}$ as a base, $|X(\omega)|^2$ can be measured in dB as $10\log_{10}|X(\omega)|^2/X^2_{max}$.

However, most types of seismic waves are propagating waves and do not have a discrete spectrum. For these waves we measure the decrease of power (or amplitude) with distance. For travelling waves we define a $Q_X$ as in (5.6), but with $\Delta E$ the energy change when the wave travels a distance of one wavelength. One easily establishes that in this case the amplitude decreases like $\exp(-kx/2Q_X)$. For body waves, where $k = \omega/c$, the two are equal ($Q_X = Q_T$). For waves with a more complicated dispersion relation, such as the surface waves we shall encounter in Chapter 10, they are different.

Just as the velocity of a seismic wave is related to the intrinsic velocity of the rock, the attenuation of the wave is caused by the attenuative properties of the rock. For a simple, monochromatic body wave, we define the intrinsic $Q$ of the rock as the $Q_X$ that is measured for the body wave passing through it. Since there are two types of body waves (assuming isotropy), a rock is characterized by $Q_P$ and $Q_S$.

### *Exercises*

**Exercise 5.1** Assume a solution $\exp(-st)$ in (5.3) to derive equation (5.4).

**Exercise 5.2** Plot $|X(\omega)|$ for various values of $\gamma$ and compare with (5.5). How good is this approximation?

**Exercise 5.3** Suppose a low-frequency wave and a high-frequency wave have the same $Q$. Which of the two damps out faster with time? What does this imply for the frequency content of seismograms?

**Exercise 5.4** What is the physical dimension of the quality factor $Q$?

## 5.3 The correspondence principle

Hooke's law, for an elastic medium in one dimension, simply states the linear relationship between stress and strain: $\sigma = G\epsilon$ . In this form, once a certain strain $\epsilon$ is imposed on the elastic medium, a constant level of stress $\sigma$ is maintained. In the real world, a number of irreversible physical mechanisms may operate to diminish the stress in course of time: for instance cracks may develop. To analyse such behaviour, we still assume a linear dependence between a strain step $\Delta\epsilon$ applied at time $\tau$ and the resulting stress increase $\Delta\sigma$, but allow the elastic modulus $G$ to be time-dependent to reflect the change of stress with time: $\Delta\sigma(t) = G(t - \tau)\Delta\epsilon(\tau)$. Physically, stress will decrease with time as microscopic structures in the rock give way, so $G$ is a decreasing function of time. This phenomenon is called *relaxation*. The time it takes $G$ to return to $1/\mathrm{e}$ of its maximum value is called the relaxation time.

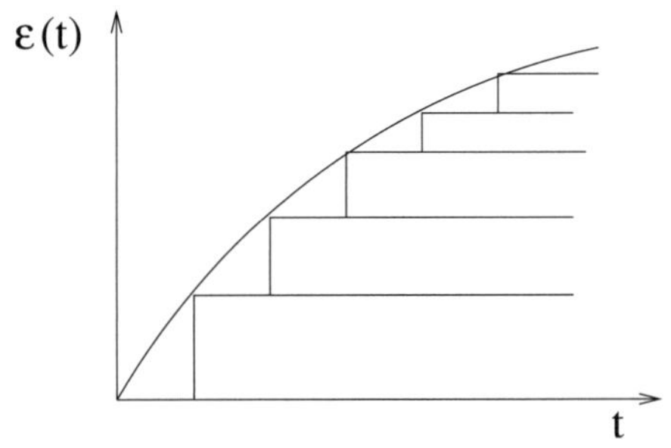


Fig. 5.2. Building a strain curve in small steps at times $\tau_1$, $\tau_2$, ...

The situation becomes slightly more complicated when we apply a strain that has an arbitrary time dependence $\epsilon(t)$. We may build this up by summing small strain steps of the form $\Delta\epsilon(\tau)H(t-\tau) = \dot{\epsilon}(\tau)\Delta\tau H(t-\tau)$, where $H(t)$ is the Heaviside function (Figure 5.2):

$$\begin{aligned} H(t) &= 1 \quad \text{if} \quad t > 0 \\ &= 0 \quad \text{otherwise.} \end{aligned}$$

The full response is obtained by integrating over all small steps:

$$\sigma(t) = \int_{-\infty}^{t} G(t-\tau)\dot{\epsilon}(\tau)\mathrm{d}\tau \,. \tag{5.7}$$

Since $G(t) = 0$ for $t < 0$, the upper limit of integration in (5.7) can be shifted to $\infty$, which makes it a convolution. The proper generalization of (5.7) to three dimensions is obvious if we look back at (2.5). The most general linear relationship between stress and strain that also incorporates a memory must clearly be:

$$\sigma_{ij}(t) = \sum_{kl} \int_{-\infty}^{t} G_{ijkl}(t-\tau)\dot{\epsilon}_{kl}(\tau)\mathrm{d}\tau \,. \tag{5.8}$$

As we found for the elasticity tensor $c_{ijkl}$ in Section 2.8, $G_{ijkl}$ has only two independent parameters, assuming that the relaxation behaviour is isotropic as well:

$$G_{ijkl}(t) = \left[\kappa(t) - \frac{2}{3}\mu(t)\right]\delta_{ij}\delta_{kl} + \mu(t)[\delta_{jl}\delta_{ik} + \delta_{jk}\delta_{il}] \,. \tag{5.9}$$

Here we write $\kappa - \frac{2}{3}\mu$ rather than the Lamé parameter $\lambda$, because the incompressibility $\kappa$ often has negligible damping associated with it, at least in non-porous media. At first sight, the complicated form of (5.8) seems to preclude a simplifying analysis of wave propagation such as we did in Chapter 4. Fortunately, things are not that bad. The clue is that the integral in (5.8) has the form of a convolution.

After Fourier transformation, it therefore reads as a simple multiplication:

$$\sigma_{ij}(\omega) = -\mathrm{i}\omega \sum_{kl} G_{ijkl}(\omega)\epsilon_{kl}(\omega)\,,$$

and substitution of this stress strain relation is as straightforward as in Chapter 2. The equations of motion are:

$$-\omega^2 \rho u_i(\omega) = \sum_j \frac{\partial \sigma_{ij}(\omega)}{\partial x_j} = \sum_{jkl} \frac{\partial}{\partial x_j}\left[-\mathrm{i}\omega G_{ijkl}(\omega)\frac{\partial u_l}{\partial x_k}\right]. \tag{5.10}$$

Because the elasticity tensor depends on frequency, we must solve (5.10) for every frequency component separately. In general we expect the elements of $G$ to be complex; therefore, the wave velocities will also be complex, e.g. in an isotropic solid we find an S-velocity $V_{\rm S}(\omega) = \sqrt{\mu(\omega)/\rho}$, which is complex because $\mu$ is complex. Unless we explicitly indicate differently, the notation $V_{\rm S}$, $V_{\rm P}$ or $c$ denotes the real part of the velocity to avoid clutter.

The fact that (5.10) has exactly the same form as (2.41) is known as the *correspondence principle* for attenuating waves. It facilitates our task of introducing attenuation into the wave propagation, because the equations remain formally the same, but the elastic constants become complex. We shall exploit this in the next section.

## 5.4 Attenuating body waves

Consider a monochromatic plane wave in a homogeneous, isotropic, attenuating medium:

$$u(x,t) = u(k,\omega)\exp[\mathrm{i}(kx - \omega t)]\,,$$

where $x$ is the distance travelled in the direction of wave propagation. This wave has a complex wavenumber $k$ since the elastic constants, and thus the velocities, are complex. In an isotropic solid, with a P-velocity $V_{\rm P}(\omega)$:

$$k = \omega/V_{\rm P}(\omega) = \mathrm{Re}(k) + \mathrm{i}\,\mathrm{Im}(k)\,.$$

We may write the exponent as:

$$\mathrm{i}kx = \mathrm{i}\,[\mathrm{Re}(k) + \mathrm{i}\,\mathrm{Im}(k)]x = \mathrm{i}\,\mathrm{Re}(k)x - \mathrm{Re}(k)x/2Q_X\,,$$

where we have introduced yet another form for the Quality factor $Q_X$:

$$Q_X = \frac{\mathrm{Re}(k)}{2\mathrm{Im}(k)}\,. \tag{5.11}$$

The imaginary component of wavenumber $k$ is directly related to the imaginary component of velocity $V_P$, hence to the elastic constants $\lambda$ or $\kappa$ and $\mu$. For S-waves we find specifically (assuming $\mathrm{Im}\mu \ll \mathrm{Re}\mu$):

$$k = \omega \left(\frac{\rho}{\mu}\right)^{\frac{1}{2}} = \omega \left(\frac{\rho}{\mathrm{Re}\,\mu + \mathrm{iIm}\,\mu}\right)^{\frac{1}{2}} = \omega \left(\frac{\rho}{\mathrm{Re}\,\mu}\right)^{\frac{1}{2}} \left(\frac{1}{1 + \mathrm{i}\frac{\mathrm{Im}\,\mu}{\mathrm{Re}\,\mu}}\right)^{\frac{1}{2}}$$

$$\approx \omega \left(\frac{\rho}{\mathrm{Re}\,\mu}\right)^{\frac{1}{2}} \left(1 - \mathrm{i}\frac{\mathrm{Im}\,\mu}{2\mathrm{Re}\,\mu}\right),$$

so that with (5.11):

$$Q_X^S = -\frac{\mathrm{Re}(\mu)}{\mathrm{Im}(\mu)}. \tag{5.12}$$

For P-waves the derivation is the same, with $\kappa + \frac{4}{3}\mu$ taking the place of the shear modulus $\mu$:

$$Q_X^P = -\frac{\mathrm{Re}(\kappa) + \frac{4}{3}\mathrm{Re}(\mu)}{\mathrm{Im}(\kappa) + \frac{4}{3}\mathrm{Im}(\mu)}. \tag{5.13}$$

For a seismic body wave propagating in a heterogeneous medium we may apply ray theory, and assume that the ray amplitude over a segment $\mathrm{d}s$ of the ray attenuates as $\exp(-\mathrm{Re}\,k\mathrm{d}s/2Q_X)$. Since the wave velocity $c$ ($V_P$ or $V_S$), depends on the location $\boldsymbol{r}$, the wavenumber $k = \omega/c(\boldsymbol{r})$ depends on $\boldsymbol{r}$ as well, and the decrease can also be written as $\exp[-\omega\mathrm{d}s/2Q_X(\boldsymbol{r})c(\boldsymbol{r})]$. Here $Q_X(\boldsymbol{r})$ is the intrinsic $Q$ of the medium at $\boldsymbol{r}$. The total attenuation is obtained by multiplying these separate attenuation factors over all segments $\mathrm{d}s$. Replacing $\sum \mathrm{d}s$ by $\int \mathrm{d}s$ in the exponent, we obtain:

$$A(\omega) = A_0(\omega) \exp\left[-\frac{\omega}{2}\int \frac{\mathrm{d}s}{Q_X(\boldsymbol{r})c(\boldsymbol{r})}\right].$$

This is usually written in the form:

$$A(\omega) = A_0(\omega) \exp(-\omega t^*/2), \tag{5.14}$$

which defines 't star':

$$t^* = \int \frac{\mathrm{d}s}{c(\boldsymbol{r})Q_X(\boldsymbol{r})}, \tag{5.15}$$

with dimension of time. The quality factor $Q$ of a particular seismic phase is defined by $Q^{-1} = t^*/t$. For a large range of teleseismic distances in the Earth, $t^* \approx 1$ s for P-waves, and roughly 4 s for S-waves, fairly constant values even though the ray length increases with distance (Figure 4.1). The explanation is that the longer rays travel deeper in the Earth and encounter less intrinsic attenuation.

An important consequence of the frequency dependence of the elastic constants is that the seismic velocities $V_P$ and $V_S$ also become frequency dependent, a phenomenon first recognized theoretically by Jeffreys [148]. The dispersion depends on the rheology of the rock, i.e. on the shape of the time response $G(t)$. One theoretical model, the standard linear solid, predicts that attenuation peaks in a narrow frequency band centred around the inverse of the relaxation time. Over this 'absorption band' the shear modulus changes from the 'relaxed' to the (higher) 'unrelaxed' value. But narrow absorption bands are not observed in natural rocks, nor visible in observed seismic spectra. For seismic frequencies below 1 Hz, $Q$ is only weakly dependent on frequency: $Q \propto \omega^{\alpha}$, with the parameter $\alpha$ somewhere between 0.1 and 0.3 in laboratory experiments on rocks. Actual observations of the frequency dependence of attenuation of teleseismic P- and S-waves are not very precise but generally give values between 0 and 0.3, with the largest $\alpha$'s found for data with frequency between about 0.1 and 1.5 Hz.

An absorption band can be modelled by introducing a series of attenuation mechanisms with increasing relaxation time, throughout the seismic band $< 1$ Hz. If this renders $Q$ sufficiently constant, the seismic velocity can be shown to change with frequency over the width of the absorption band as:

$$c(\omega) = c(\omega_0)\left[1 + \frac{1}{\pi Q} \ln\left(\frac{\omega}{\omega_0}\right)\right]. \tag{5.16}$$

## *Exercises*

**Exercise 5.5** A long-wavelength wave and one with shorter wavelength both have the same $Q_X$. Which of the two damps away fastest?

**Exercise 5.6** Show that (5.11) is indeed the relative energy loss per wavelength.

**Exercise 5.7** Give a derivation of (5.13) and (5.12). What approximations do you have to make? Are these consistent with the general condition that $|\Delta E| \ll E$?

**Exercise 5.8** A common assumption is that there is very little energy loss associated with pure compression, or $\mathrm{Im}(\kappa) \approx 0$. Show that we have then a very simple relationship between the quality factors of P and S waves:

$$Q_X^{\mathrm{P}} = \frac{3}{4}\frac{V_{\mathrm{P}}^2}{V_{\mathrm{S}}^2} Q_X^{\mathrm{S}}.$$

How large is the coefficient of proportionality for a Poisson solid, defined as a solid where $\lambda = \mu$, or, equivalently, where $V_P = \sqrt{3} V_S$ (rocks deep inside the Earth are close to Poisson solids except when partial melt occurs)?

**Exercise 5.9** Prove that

$$Q_X^{\mathrm{P}} = -\frac{\mathrm{Re}\, V_{\mathrm{P}}}{2\mathrm{Im}\, V_{\mathrm{P}}} \quad \text{and} \quad Q_X^{\mathrm{S}} = -\frac{\mathrm{Re}\, V_{\mathrm{S}}}{2\mathrm{Im}\, V_{\mathrm{S}}}. \tag{5.17}$$

## 5.5 Scattering

In addition to focusing/defocusing and energy loss by inelastic processes, the amplitude of a seismic wave is also influenced by scattering. This may take the form of organized scattering at an interface such as the core–mantle boundary or the Moho, where some of the wave energy is transmitted and some reflected, or of more random scattering off small heterogeneities encountered more or less continuously along the path. Of course, in a smoothly varying medium waves are also focused or defocused, much as light waves through a lens. But in this case the energy within the ray tube remains constant and we do not interpret the change in amplitude as energy 'loss'.

Phenomenologically, scattering removes energy from the signal. Rather than convert it into heat as an inelastic process does, it converts it into 'noise'. This is a somewhat imprecise observation, since the boundary between 'signal' and 'noise' is subjective, and depends very much on the power of our analytical toolkit. In this book, we consider scattering at frequencies equal to or higher than those that generate Mie scattering as 'noise'. We note, however, that tools to interpret scattering are improving quickly and the field is beginning to open up for more deterministic analysis: Campillo and Paul [38], and Shapiro et al. [306] have used diffusively scattered wave energy to infer a Green's function response between two stations and apply this to tomography in highly scattering surroundings. Larose et al. [170] have used this technique of 'seismic interferometry' to study the shallow subsurface of the Moon from Apollo data. Shearer and Earle [309] and Margerin and Nolet [194] used multiple scattering to make inferences about the distribution of scatterers with depth inside the Earth.

A single scattering approach such as we saw in Chapter 4 is sufficient to model the loss of energy, because we are not interested in any subsequent encounters of the scattered waves. Clearly, the regime of Rayleigh scattering is of most interest. When scatterers are small (with respect to the wavelength) the expressions (4.26)–(4.30) allow us to compute the energy loss as a function of the strength of the perturbations $\delta\rho$, $\delta\mu$ and $\delta\lambda$. The energy contained in the scattered wave $\delta\boldsymbol{u}$ is subtracted from the energy of the incident field. If $\Delta E$ is the energy lost over one wavelength, we may define a scattering $Q$ analogous to (5.6):

$$Q_X^{\mathrm{scat}} = \frac{2\pi E}{\Delta E^{\mathrm{scat}}}.$$

The total energy loss is then the sum of the inelastic and scattering losses:

$$\Delta E^{\rm tot} = \Delta E^{\rm scat} + \Delta E^{\rm inel},$$

and if we divide this expression by $2\pi E$ and define the total $Q$ factor as $2\pi E/\Delta E^{\rm tot}$ we see that:

$$\frac{1}{Q_X^{\rm tot}} = \frac{1}{Q_X^{\rm scat}} + \frac{1}{Q_X^{\rm inel}}.$$

We recall that the inelastic $Q$ factor of teleseismic waves is quite independent of frequency for frequencies below 0.1 Hz, with a significantly higher $Q$ at frequencies higher than 1 Hz. Such weak frequency dependence is not predicted for Rayleigh scattering, where we know that the scattered energy is proportional to the fourth power of the frequency if the power spectrum of the incident wave is flat. Since the loss $\Delta E$ is defined per wavelength, the energy loss per unit of length is proportional to $k^3$ or $\omega^3$ and

$$Q_X^{\rm scat} \propto \omega^{-3} \quad \text{for } ka \lesssim 0.1 \quad \text{(Rayleigh scattering)}.$$

We expect therefore that $Q^{\rm tot}$ will be dominated by scattering at the very high frequency end of the seismic spectrum, whereas below 0.1 Hz the energy loss is mostly due to the effects of inelasticity.

An 'organized' form of scattering occurs at the boundary layers in the Earth, where reflection and transmission coefficients influence the amplitude. For sharp boundaries, such coefficients are frequency independent and merely influence the amplitude, and possibly the phase if the angle of incidence is larger than the critical angle. Aki and Richards [5] list the reflection and transmission coefficients for amplitudes. For gradual boundaries Richards and Frasier [277] show that the waveshape is modified, especially when the boundary spreads out over a depth interval comparable to the wavelength. This can be modelled by frequency-dependent coefficients.

The critical quantity that plays a role in the reflection and transmission coefficients is the impedance $\rho V_{\rm P}$ or $\rho V_{\rm S}$, reflecting the fact that density heterogeneities are most effective scatterers in the backward direction.

# 6
# Travel times: observations

Historically, travel times were measured from seismograms recorded on smoked or photographic paper. An example is shown in Figure 6.1. This is a seismogram, dated July 26, 1963, from the World Wide Standardized Seismograph Network (WWSSN), the state-of-the-art at the time. Arrival times were picked from such recordings by measuring the distance to the nearest minute mark, visible as small deflections at regular intervals. One major shortcoming of such photographic recordings is that the trace becomes hard to read when the amplitude is large and the light source moves quickly, giving only a short exposure of the photographic film. Short period recordings, an example of which is shown in Figure 6.2 are to be preferred for the picking of the P-wave arrival time, but may be less suitable to correctly identify the later arrivals. Though the network is now obsolete, scanned images of seismograms for a growing number of historical earthquakes are available in the public domain.† These images can be digitized with suitable vectorising software, such as Teseo (Pintore et al., [262]).

Modern, digital instrumentation has greatly changed the practice in seismographic stations around the world. Figure 6.3 shows an example from a modern digital seismographic station. Digitized seismograms are much easier to manage, archive and analyse.

In this chapter we consider only digitized signals. Formally, a seismogram $s(t)$ is digitized by convolving it with a Dirac comb $\sum_i \delta(t - i\Delta t)$, resulting in $N$-tuple of values, e.g. $(s_1, s_2, \dots s_N)$. One can show that this does not lead to loss of information provided the sampling interval $\Delta t$ is less than half the shortest period, or: $\Delta t < \pi/\omega_{\max}$, where $\omega_{\max}$ is the highest frequency present in the signal, at least above an acceptable noise level. Any frequencies above the 'Nyquist frequency' $\omega_{\mathrm{Nyq}} = \pi/\Delta t$ will be mapped back into the spectrum at lower frequencies and contribute to the digitization noise. Modern digitizers have low digitization noise

† `http://aslwww.cr.usgs.gov/Seismic_Data/FilmScans/`.

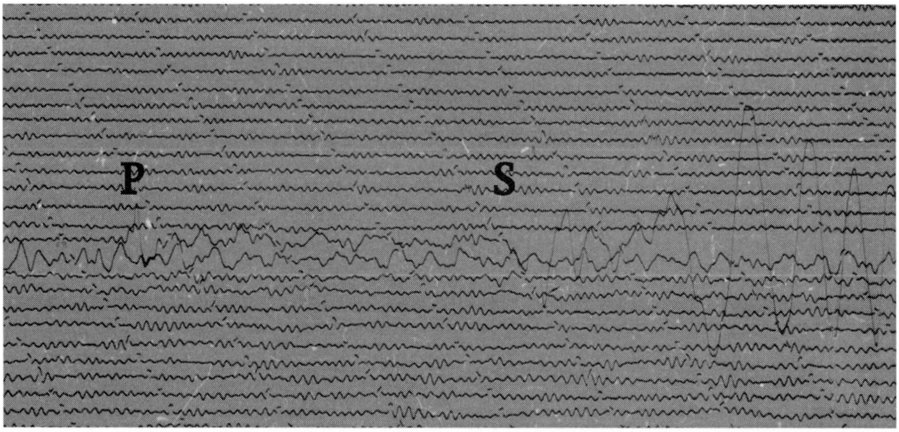


Fig. 6.1. A segment of a seismogram at WWSSN station KON (Kongsberg, Norway) of the earthquake in Skopje (Macedonia) on July 26, 1963. This is the vertical, long period component. Recording was photographic, with a radio-controlled clock deflecting the light beam every minute (small ticks). The first 13 traces of the seismogram represent microseismic noise. The epicentral distance is 19°. The wave that breaks the noise near the start of the 14th trace is the arrival of the P-wave, indicated by P. The long period wave that follows about three minutes later is the S-wave. Source USGS, reproduced with permission.

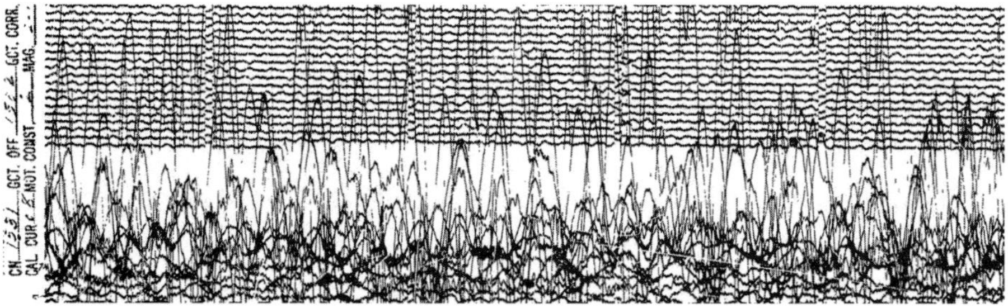

Fig. 6.2. A segment of the vertical component, short period recording of the great Alaska earthquake of March 27, 1964, recorded in WWSSN station ALQ (Albuquerque, New Mexico). Note the overlapping of the traces which makes the recognition of later phases virtually impossible. Source USGS, reproduced with permission.

(e.g., seismograms from the Global Seismograph Network or GSN are digitized with a resolution of 24 bits with close to full precision), and digital filtering and downsampling in stages keeps the effects of digitization under control. Since digitization noise from modern instruments is below that from the environment we can usually ignore it.

Digital instruments can also record seismic events over a much wider range of amplitudes. Using pen recorders, signals of very weak events may drown in the

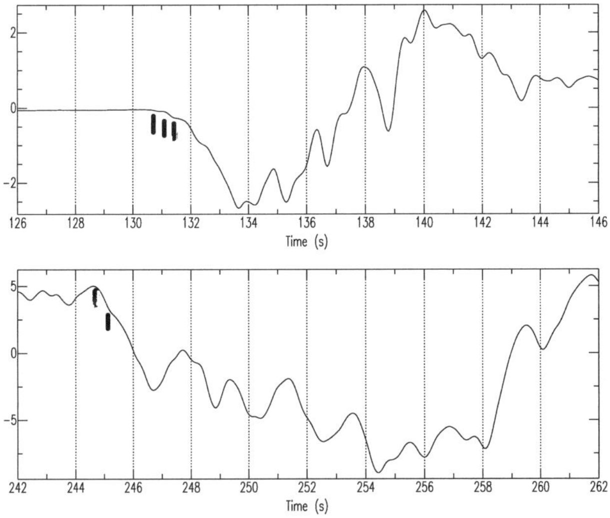


Fig. 6.3. Example of a modern seismographic recording of the arrival of a P-wave in a station of the Global Seismic Network (GSN). Shown are two segments of a recording of the vertical component of a broadband digital seismograph in Tucson (Arizona). The earthquake is in Peru (Sep 26, 2005, M = 7.5) at a distance of 50°. The time scale has an arbitrary origin. The vertical axis is in counts, roughly proportional to the ground velocity. Top: the P-wave arrival. Bottom: the (surface-reflected) PP arrival, predicted to arrive 116 s after the P-wave but about 2 s early in this case. The fat pencil marks denote possible arrival identifications and are discussed in Section 6.1.

width of the trace drawn by the pen, whereas very strong events may throw the instrument off scale. The interval between the smallest and largest recordable signal energy is called the 'dynamic range' of the instrument. It is measured in decibels (dB), a logarithmic scale defined as:

$$\text{Dynamic Range} = 20 \log_{10} \frac{A_{\text{max}}}{A_{\text{min}}} = 10 \log_{10} \frac{P_{\text{max}}}{P_{\text{min}}} \text{ dB},$$

where $A_{\text{max}}$ and $A_{\text{min}}$ are the maximum recordable and minimum discernible signal amplitudes and the power $P = A^2$. In some cases the dynamic range is not limited by the recording medium but by the accuracy of the sensor: a seismometer may respond in a nonlinear way to very large amplitude motions.

With digital recording, the dynamic range is limited by the number of bits available to describe the voltage signal from the seismometer. Analogue voltage is converted to digital numbers by an A/D converter. A commonly used low-cost A/D converter has 16 bits. One of these bits is needed to give the sign of the signal. Thus, the smallest signal unequal to zero is $2^0 = 1$, the largest in a 16-bit converter is $2^{15} - 1 \approx 2^{15}$ (or $2^{16}$ peak-to-peak) and the dynamic range is $20 \log_{10} 2^{15} = 90$ dB.

We may extend the dynamic range for a given number of bits by reserving some bits to record a gain factor. For example if we store a variable gain factor $2^n$ with $n = 0, \ldots, 7$ stored in 3 bits, and use the remaining 12 bits to record the mantissa, $A_{\min}$ is still equal to 1 (1 bit mantissa at largest gain), but this largest gain is now $2^7 = 128$ and $2^{12} \times 2^7 = 2^{19}$ giving a dynamic range of 114 dB (120 dB peak-to-peak). In this way we can accommodate larger signals, but some of the precision will be lost. If the gain is only 1 (the smallest), the minimum signal that is still visible on top of the largest amplitude signal is 11 bits (66 dB) below the amplitude of this large signal. We say that the precision is 66 dB.

The dynamic range of photographic or ink recorders is rarely more than 70 dB. Analogue tape recording has a range in the same order of magnitude. Thus, digital recording greatly enhances our capability of recording both large and small signals. A/D converters with a precision of 24 bits are good enough to eliminate the need for gain ranging.

## *Exercises*

**Exercise 6.1** What is the dynamic range of an instrument with 24-bit A/D conversion and no gain ranging? Compare this to recording on paper of 50 cm size with a pen 0.5 mm wide, defining $A_{\max}$ and $A_{\min}$, respectively. If such an instrument has a sensitivity of 5000 counts per micron at the maximum magnification frequency what is the largest signal at this frequency that can be recorded without clipping ('clipping' is a term denoting distortion of the recording, in either its mechanical, electronic or numerical form)?

**Exercise 6.2** The tides of the Earth cause the solid Earth to deflect by several decimetres (depending on latitude) with periods close to 12 and 24 hours. Microseisms on the other hand can be as small as several nanometres at very quiet sites for frequencies of a few Hertz. How many octaves (doublings in frequency) does the seismic spectrum span? What is the dynamic range (in dB) of the displacement? Is the dynamic range different for velocity and acceleration?

## 6.1 Phase picks

Arrival times of incoming phases are often estimated by 'picking' the first deflection of the seismogram trace. If the onset is sharp, one may assume that its arrival time is representative for the highest frequencies present in the signal (but see Section 7.6). If the frequency is high enough for ray theory to be valid, and the origin time $T_{\mathrm{or}}$ of the earthquake is known, the observed arrival time $T_{\mathrm{arr}}$ gives us the travel time of the ray in the form of a line integral:

$$T_{\mathrm{arr}} - T_{\mathrm{or}} = T_{\mathrm{obs}} = \int_{\mathrm{ray}} \frac{\mathrm{d}s}{c}. \tag{6.1}$$

Through (6.1), each observed arrival time represents a constraint on the velocity structure $c(\boldsymbol{r})$ of the Earth.

The origin time and location ('hypocentre') of the earthquake are determined by fitting these parameters to a large number of observed arrival times, assuming we already know the velocity in the Earth. There is a certain amount of circular reasoning in this procedure, of course, but strong sources at known locations and with known origin times – mostly tests of nuclear explosives – have reduced our uncertainty in the average, spherically symmetric, structure of the Earth to a minimum. Such radial Earth models are used very effectively in earthquake location (see Chapter 13). Nevertheless, a debate about the possibility of a small constant offset in theoretical travel times (often called the 'baseline' problem) was never completely resolved. Also, the more stations are present in the direct neighbourhood of the earthquake, the smaller the errors are in the origin time and hypocentre, but some error is inevitably propagated into the travel time when the origin time is subtracted from the observed arrival time. In Chapter 13 we shall discuss how to correct for possible biases in $T_{\text{obs}}$.

Phase picking from digital seismograms can be very precise. Leonard [179] reports a test of teleseismic P-wave picking by four analysts with different levels of experience, and finds a standard deviation in the arrival times of 0.15 s, slightly better than a standard deviation of 0.19 s obtained by the best automatic phase picker. A certain amount of subjective judgement in separating an arrival from the noise is unavoidable, and 'defining the true onset time of teleseismic events to better than 0.1 s will not normally be possible'.

For large experiments automatic phase picking may be competitive with cumbersome hand-picking by analysts. For autonomous sensors such as underwater floats proposed by Simons et al. [317] it may greatly reduce the need for digital signal transmission. The preferred algorithm for automatic phase picking models the seismogram with an autoregressive (AR) model (e.g. Leonard and Kennett [180], Morita and Hamaguchi [220], Sleeman and van Eck [322]):

$$s_t = \sum_{m=1}^{M} a_m s_{t-m} + n_t ,$$

where the $a_m$ are coefficients of the AR filter and $n_t$ is the noise digitized at discrete time $t$. One assumes that the noise has a Gaussian distribution with zero mean, i.e. the expected value $\langle n_t \rangle = 0$ and $\langle n_t^2 \rangle = \sigma^2$, and uncorrelated with the signal ($\langle n_t s_{t-m} \rangle = 0$). The quality of the filter is expressed by the likelihood function:

$$L = \left( \frac{1}{2\pi\sigma^2} \right)^{N/2} \exp \left[ -\frac{1}{2\sigma^2} \sum_{j=1}^{N} \left( s_j - \sum_{m=1}^{M} a_m s_{j-m} \right)^2 \right] ,$$

where $N$ is the length of the time window.

We find the coefficients $a_m$ by maximizing $\ln L$ and solving the resulting system of linear equations. Clearly, at the onset of a signal one expects to see a significant change in the AR coefficients. The automatic picking algorithm first determines a rough signal arrival time using the STA/LTA ratio: the ratio of energy in the seismogram averaged over a short and long time interval respectively. One then calculates the AR coefficients for a window that starts, as well as for a window that ends at a candidate arrival time $t = k$. The likelihood function for $k$ being the arrival time is given by the product of the likelihood functions for the windows before and after $t = k$. The maximum in this product (as a function of $k$) is interpreted as the arrival time.

Most – if not all – of the arrival times reported to seismological centres such as the National Earthquake Information Centre (NEIC) in the US or the International Seismological Centre (ISC) in Britain are obtained by 'picking'. Since digital recording is of recent date, many reported arrival times are picked by eye from paper records and suffer much larger observational errors and the best accuracy is probably of the order of a few tenths of a second. Paradoxically, arrival times from strong events, with a high signal-to-noise ratio are not necessarily easier to pick than those from smaller events. The reason is that a large earthquake is sometimes preceded by very small shocks, sometimes called 'subevents', in the beginning of the rupture. An example is shown in Figure 6.3. A careful observer might spot the first small deflection starting at time 130.7 s (note that these times are relative times with respect to an arbitrary zero time). However, an observer in a station with slightly more noise would only see the stronger deflection starting at time 131.2 s; if the noise is strong enough for this to be missed as well, the time pick is likely to be 131.8 s, a difference of 1.1 s with respect to the first pick.

The travel times of each ray are compared to the model predictions, leading to a delay time defined as $\Delta T = T_{\text{obs}} - T_{\text{pred}}$. The sign of $\Delta T$ tells us whether the ray has travelled through a region that is on average slower (+) or faster (−) than the Earth model used to calculate $T_{\text{pred}}$, and these deviations are the input data for seismic tomography. It is therefore important to estimate the observational errors.

Morelli and Dziewonski [219] show that it is possible to estimate the variance in P-wave travel times distributed by the ISC using the concept of 'summary rays', bundles of rays from the same receiver and source regions (of size $5^\circ \times 5^\circ$). If the bundle is narrow, one would expect all rays to have the same delay, and differences to be due to observational error. It is well known that the variance of an average of $N$ observations of the same parameter that is affected by random errors decreases as $N^{-1}$ with respect to the variance of a single observation. The reading errors are presumably random, and so would be the effect of velocity anomalies for rays following different trajectories. But the effect of the Earth's heterogeneity will be the same for rays following the same path. Thus, within a bundle of rays following

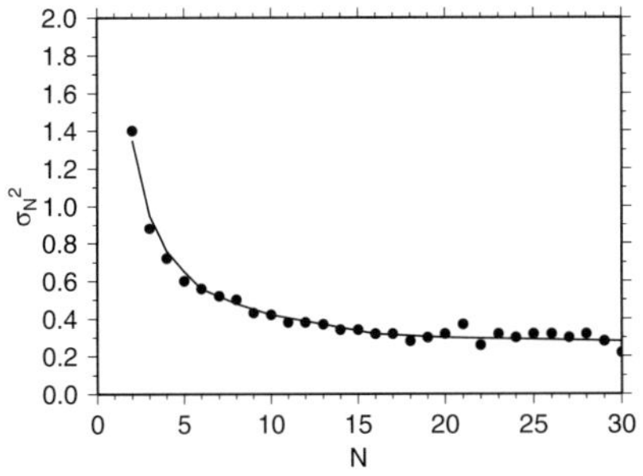


Fig. 6.4. Variance of the average delays $\sigma_N^2$ in summary P rays in s$^2$ as a function of the size of the summary ray population, $N$. From Morelli and Dziewonski [219], reproduced with permission from MacMillan Publ. (Nature ©1987).

(almost) the same path, one may model the observed variance of the summary ray averages as:

$$\sigma_N^2 = \frac{\sigma_d^2}{N} + \sigma_{\text{Earth}}^2 , \tag{6.2}$$

where $N$ is the number of rays in a summary ray, $\sigma_d$ the (unknown) standard deviation of each $\Delta T$, $\sigma_{\text{Earth}}$ the (unknown) contribution from the Earth's heterogeneity to the variability of $\Delta T$ among different bundles. Experimentally, $\sigma_N^2$ is determined by first calculating the average $\Delta T_i^{(N)}$ for a large number of summary rays with $N$ observations each. We then take the average $\Delta T_{\text{av}}^{(N)}$ of all these summary rays. If we have $M_N$ such bundles with $N$ observations the variance for them is:

$$\sigma_N^2 = \frac{1}{M_N} \sum_i^{M_N} \left[\Delta T_i^{(N)} - \Delta T_{\text{av}}^{(N)}\right]^2 .$$

We find the variances $\sigma_d^2$ and $\sigma_{\text{Earth}}^2$ by optimizing the fit of the observed $\sigma_N^2$ to (6.2). The original study found $\sigma_d$ equal to 1.4 s for P-wave data (see Figure 6.4). Later arrivals have much larger standard deviation: 2.7 s for PcP and 2.0 s for the AB branch of PKP. The contribution from the Earth's heterogeneity is very small in this study (0.50 s for the P-waves), but this may have been influenced by the rather large width of the bundles, averaging out the smallest scale heterogeneity. Gudmundsson et al. [126] did a similar analysis on the same data set, but extrapolated to bundles of zero width, and found a larger contribution from the Earth's heterogeneity, a $\sigma_{\text{Earth}}$ below 1.0 s for mantle P-waves, and a signal-to-noise ratio of about 2. Bolton and Masters [27] determined $\sigma_{\text{Earth}}$ for long-period P- and S-waves at teleseismic distances and found 1.1 and 3.2 s, respectively.

Stations with high noise are usually set to a low gain. Grand [122] and Röhm et al. [285] found a late arrival bias of as much as 0.5 s because the onset is not picked until the time when the signal arises above the noise. Thus, the signal of

ISC data is severely affected by ‘noise’. On the other hand, the centre has been operating for more than forty years, assembling millions of arrival times, enabling seismic tomographers to average over closely spaced wavepaths and reduce the observational error. Engdahl et al. [98] have winnowed the ISC data set and recomputed delays after relocating the hypocentres. This has likely reduced errors somewhat, but no new estimate of the data variance in this data set is available at this time.

The onset of later arrivals is much more difficult to pick, as can easily be seen from the PP arrival in Figure 6.3. To alleviate the severe problems associated with picking arrival times, more sophisticated and accurate techniques are known. The most widely used is the ‘matched filter’ or ‘cross-correlation’ technique.

## 6.2 Matched filters

If the assumptions of ray theory are valid, the seismogram $s(t)$ can be viewed as a succession of pulse-like arrivals $u(t)$, each with an amplitude $A_i$ and a travel time or delay $\tau_i$, plus noise $n(t)$:

$$s(t) = \sum_i A_i u(t - \tau_i) + n(t) \,. \tag{6.3}$$

Even within the framework of ray theory the pulse shape $u(t)$ may change: supercritical reflections and passage through a caustic cause phase changes (see also Section 6.4). Shallow earthquakes will be followed by ‘ghosts’, surface reflections pP and sP for example, that may change polarity depending on the epicentral distance or azimuth. The following is easily generalized to encompass these changes and we shall ignore them in the current section. Matched filters can be used to estimate the onset times $\tau_i$. The theory on matched filters is extensive, though much of it appeared outside the seismological literature (estimation of delays is important in radar research, echo sounding and ocean acoustic tomography, to name a few). In this section we briefly review the main results that are relevant. See also Turin [369].

We shall wish to send $s(t)$ through a filter that allows us to determine the $\tau_i$. In the following we further simplify the analysis by assuming that the seismogram consists of only one pulse arriving at $t = \tau$ plus noise: $s(t) = u(t - \tau) + n(t)$. Generalization to a larger number of arrivals is trivial. One might naively assume that a spiking or ‘inverse’ filter – one that transforms $u(t)$ into $\delta(t)$ – would be optimal. Since the spectrum of a delta function is 1 for all frequencies, the spiking filter involves dividing $s(\omega)$ by $u(\omega)$. The problem is that we then also divide the noise spectrum $n(\omega)$ by $u(\omega)$, and this will blow up the noise in those frequency bands where $u(t)$ has little power. The ‘matched’ filter is designed to overcome such problems and provide a stable estimate of the delay even in the presence of considerable noise.

A linear filter is characterized by its impulse response $h(t)$ (the output of the filter if the input is a delta function):

$$\gamma(t) \equiv \int s(t')h(t-t')\mathrm{d}t' = s(t) * h(t) \tag{6.4}$$

or, since convolution in the time domain is equivalent to multiplication in the frequency domain (see Appendix A):

$$\gamma(\omega) = s(\omega)h(\omega) = u(\omega)h(\omega) + n(\omega)h(\omega)\,.$$

The noise power density spectrum is assumed to be white (i.e. constant), a restriction we shall relax later. The noise power density is assumed to be $N_0^2/2$.† After passage through the filter, the noise power density is therefore $\frac{1}{2}N_0^2|h(\omega)|^2$. The expected noise power at time zero is found by setting $t = 0$ in the Fourier transform:

$$P_N = \frac{N_0^2}{4\pi}\int_{-\infty}^{\infty} h(\omega)^* h(\omega)\mathrm{d}\omega\,, \tag{6.5}$$

where the asterisk * denotes the complex conjugate. Note that (6.5) is equivalent to the total power over all frequencies. The output of the filter due to the input of $u(t)$ alone, when sampled at delay time zero in the Fourier transform is:

$$y_u(0) = \frac{1}{2\pi}\int_{-\infty}^{\infty} u(\omega)h(\omega)\mathrm{d}\omega\,. \tag{6.6}$$

We judge the filter optimal if the signal-to-noise ratio (SNR), the ratio between $y_u(0)^2$ and noise power $P_N$ is maximized:

$$\mathrm{SNR} = \frac{2\left[\int_{-\infty}^{\infty} u(\omega)h(\omega)\mathrm{d}\omega\right]^2}{N_0^2\int_{-\infty}^{\infty} h(\omega)^* h(\omega)\mathrm{d}\omega}\,. \tag{6.7}$$

Both numerator and denominator in (6.7) are real. Schwartz's inequality for real-valued integrals implies that

$$\left[\int_{-\infty}^{\infty} u(\omega)h(\omega)\mathrm{d}\omega\right]^2 \le \int_{-\infty}^{\infty} |u(\omega)|^2\mathrm{d}\omega \int_{-\infty}^{\infty} |h(\omega)|^2\mathrm{d}\omega\,,$$

with the equality only if $h(\omega)$ is a constant times $u^*(\omega)$, and one shall wish to choose the filter such that $h(\omega) = u^*(\omega)$. The complex conjugation implies that $h(t)$ is proportional to $u(-t)$. Thus, in the presence of white noise, the *matched*

† Noise power quantifications suffer an ambiguity because the total power is the sum of that at $+\omega$ and at $-\omega$; hence the factor of $1/2$, which leads to a power of $N_0^2$ if the (symmetric) negative spectrum is folded into the positive one and the spectral integral is over positive frequency only.

*filter of a signal is the signal itself, but reversed in time*. For the signal-to-noise ratio we find from (6.7):

$$\mathrm{SNR} = \frac{2}{N_0^2}\int_{-\infty}^{\infty} |u(\omega)|^2 \mathrm{d}\omega = \frac{4\pi E_u}{N_0^2},$$

where we used Parseval's theorem (2.69) to identify $E_u$ as the total energy in the signal $u(t)$. Substituting $h(t) = u(-t)$ into (6.4):

$$\gamma(t) = \int s(t')u(t'-t)\mathrm{d}t',$$

which shows that the matched filter is reduced to a simple cross-correlation between the observed time series and the known (or, more realistically, the estimated) wavelet $u(t)$.†

The maximum SNR in the cross-correlation of a delayed signal $u(t-\tau)$ is obtained for a time shift equal to $\tau$. For a perfect signal the maximum in the cross-correlation function coincides with $\tau$, but noise will cause this to deviate from the exact delay time. Carter [42] derives the following expression for the variance of the delay time estimates obtained by picking the maximum value of the cross-correlation:

$$\sigma^2_{\mathrm{CRLB}} = \frac{3}{8\pi^2}\frac{1+2\mathrm{SNR}}{\mathrm{SNR}^2}\frac{1}{\Delta f^3 T_{\mathrm{w}}}, \tag{6.8}$$

where $\Delta f = \Delta\omega/2\pi$ is the bandwidth of the filter in Hz and $T_{\mathrm{w}}$ the window length over which the cross-correlation is computed. Equation (6.8) is known as the Cramér–Rao lower bound, and does not include possible bias introduced by cutoff or tapering effects at the end of the time window. It also fails for large values $\sigma_{\mathrm{CRLB}} \gtrsim f_{\mathrm{max}}^{-1}$ for signals low-passed at $f_{\mathrm{max}}$, or when $\sigma_{\mathrm{CRLB}} \gtrsim f_{\mathrm{c}}^{-1}$ for a narrow-band signal with centre frequency $f_{\mathrm{c}}$. A more realistic estimate is the correlator performance estimate (CPE), a weighted average between $\sigma_{\mathrm{CRLB}}$ and a more empirical measure, based on the width $T_{\mathrm{c}}$ of the cross-correlation peak (Ianniello [140]):‡

$$\sigma^2_{\mathrm{CPE}} = wT_{\mathrm{c}}^2/3 + (1-w)\sigma^2_{\mathrm{CRLB}} \tag{6.9}$$

$w$ is the probability that the estimator is anomalous (e.g. the probability that a cycle skip occurs) and results in a delay time that is further than $T_{\mathrm{c}}/2$ from its true value.

A simple trick shows us how to adapt the above analysis for non-white noise. If we divide $s(\omega)$ by the noise amplitude spectrum $|n(\omega)|$ we obtain a signal,

† Following seismological practice dating from before the era of 'wavelet decompositions', we use the term 'wavelet' rather loosely for a transient signal with finite energy.

‡ $T_{\mathrm{c}}$ is known as the 'correlation time', the interval over which the cross-correlation peak is strongly positive.

say $s'(t)$, that has white noise. The optimum matched filter to apply to $s'(t)$ is then $h(\omega)/|n(\omega)|$. If we take a shortcut and apply both the prewhitening and the matched filter to $s(t)$, we see that the optimum matched filter in the presence of non-white noise is:

$$h_{nw}(\omega) = \frac{u^*(\omega)}{|n(\omega)|^2} \,.$$

We shall investigate the expected noise spectrum in the Earth in Section 6.5. In seismological practice, one almost always assumes the noise is white – at least in the spectral band of interest.

## 6.3 Wavelet estimation

Though the previous section addressed the question of finding the optimum filter for a given signal $u(t)$, it does not yet answer the question of how to find $u(t)$. There are several options, which roughly divide into 'theoretical' and 'empirical' estimators.

The simplest theoretical estimator assumes that the source time function wavelet is a very short pulse when compared to the response function of the seismograph. The estimate for $u(t)$ is then simply the impulse response of the seismograph, convolved with the effects of attenuation (see Section 13.7). This was the approach used by Woodward and Masters [404]. It does not require any knowledge of the seismic source. This approach can only be applied to long-period seismograms, for which a finite-frequency interpretation of the cross-correlation delay $\tau$ is more appropriate (see next chapter). For many earthquakes, the rupture process takes several seconds to accomplish, even more than 10 s for large ruptures. For such large sources the impulse response is not adequate. The time history of the source needs to be taken into account as soon as the rupture time is more than a fraction of the dominant period of the wave.

In order to extend the cross-correlation technique to shorter periods, and also to make full use of the power of finite-frequency theory, which allows for the interpretation of delay times and amplitude anomalies as a function of frequency, a more ambitious approach was proposed by Sigloch and Nolet [313]. They estimate the source time function moment rate function $\dot{m}(t)$ and model the source pulse as predicted by ray theory in (4.15), including the surface reflections pP, sP etc. In this approach, the seismogram $u_k(t)$ in station $k$ is modelled as a sum of pulse-like arrivals $G_k(t)$, with amplitudes and times predicted by ray theory, convolved with an unknown source time function $\dot{m}(t)$:

$$u_k(t) = a_k \int G_k(t-\tau)\dot{m}(\tau)\mathrm{d}\tau, \tag{6.10}$$

where $a_k$ is an amplitude factor close to 1 that models the amplitude anomaly due to focusing, defocusing or impedance effects, an approach first used by Ruff [297]. The pulse-like arrivals $G_k(t)$ are conveniently computed using Chapman's WKBJ algorithm [49], for which the computer code is available publicly.† The spectra of the waveforms predicted by WKBJ need to be multiplied by the attenuation response $\exp(-\omega t^*/2)$ and by the dispersive delay induced by (5.16) – this will be discussed in more detail in Section 13.7. The wavelet estimation then boils down to a correct estimation of the source time function. After discretizing all signals in (6.10) with time interval $\Delta t$:

$$\boldsymbol{u}_k = a_k \boldsymbol{G}_k \dot{\boldsymbol{m}} \,,$$

where $\boldsymbol{G}_k$ is a matrix with elements equal to the signal $G_k(t-\tau)$ for $t = i\Delta t$ in row $i$ and $\tau = j\Delta t$ in column $j$. Arranging all observed seismograms for one earthquake in a vector $\boldsymbol{u}$, and doing the same with the predicted signals, identified by a tilde:

$$\tilde{\boldsymbol{u}} = \begin{pmatrix} \tilde{\boldsymbol{u}}_1 \\ \tilde{\boldsymbol{u}}_2 \\ \vdots \\ \tilde{\boldsymbol{u}}_K \end{pmatrix} = \begin{pmatrix} a_1 \boldsymbol{G}_1 \\ a_2 \boldsymbol{G}_2 \\ \vdots \\ a_K \boldsymbol{G}_K \end{pmatrix} \dot{\boldsymbol{m}} = \boldsymbol{A}\dot{\boldsymbol{m}},$$

one finds the amplitude factors $a_k$ as well as the source time function $\dot{\boldsymbol{m}}$ by minimizing the norm of the difference vector $|\boldsymbol{u} - \tilde{\boldsymbol{u}}|^2$. This is a nonlinear problem because the amplitude factors $a_k$ and the source time function $\dot{\boldsymbol{m}}$ multiply each other. An iterative approach, in which one assumes $a_k$ is known (initially 1) and inverts for $\dot{\boldsymbol{m}}$, then estimates the $a_k$ from

$$a_k = \frac{\tilde{\boldsymbol{u}} \cdot \boldsymbol{G}_k \dot{\boldsymbol{m}}}{|\boldsymbol{G}_k \dot{\boldsymbol{m}}|^2} \,,$$

was shown to converge in practice. Sigloch's method is illustrated in Figure 6.5.

The term $\boldsymbol{G}_k$ also incorporates the source radiation pattern as predicted by the moment tensor. Source propagation effects – azimuth-dependent changes in the waveshape that occur when the length of the rupture surface cannot be neglected – can be taken into account by clustering the observed pulses and inverting for each cluster separately. Clustering techniques were introduced into seismology by Rowe et al. [295] and first applied to low frequency cross-correlations by Houser et al. [135].

† The WKBJ code is published in Chapman et al. [51], and is also embedded in the SEISAN package, available at `www.geo.uib.no/seismo/software/seisan/seisan.html`.

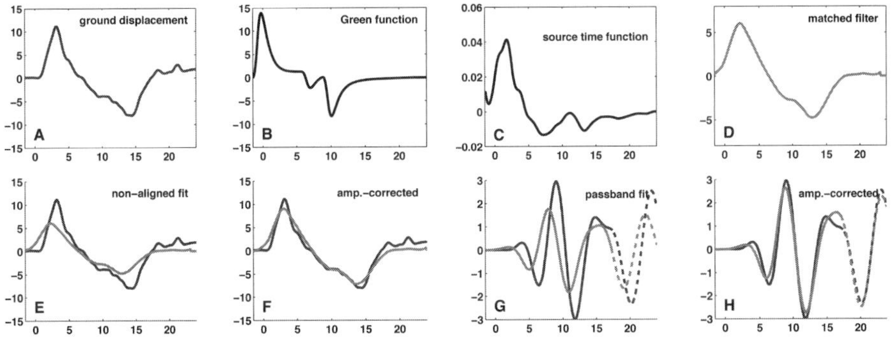


Fig. 6.5. Wavelet estimation and matched filtering. A: recorded P-wave $u(t)$, B: predicted Green's function, showing P, pP and sP arrival, C: inferred source time function $\dot{m}(t)$, D: the matched filter $\tilde{u}(t)$, E: the misfit between $u(t)$ and $\tilde{u}(t)$ (dark and light grey, respectively), F: misfit after time-shifting and multiplying $\tilde{u}(t)$ with the amplitude factor $a$, G: misfit after bandpass filtering, H: misfit after time-shifting and amplitude correcting the bandpassed signal. From Sigloch and Nolet [313], reproduced with permission from Blackwell Publishing.

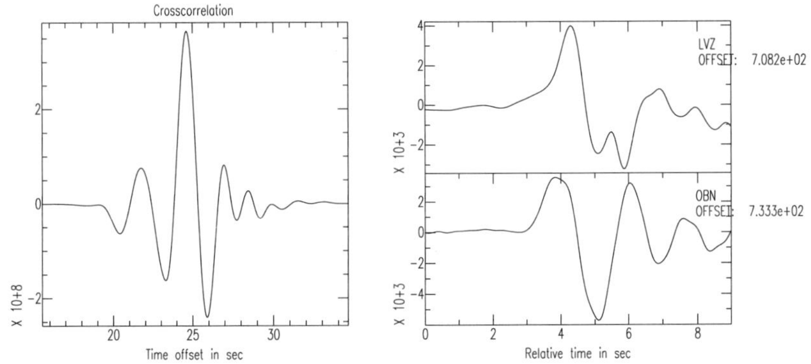


Fig. 6.6. An example of the application of cross-correlation to measure the arrival time difference between two stations. Left: the cross-correlation function showing a time offset of 24.7 s between P-waves arriving in Russian seismic stations Lovozero (LVZ, 67.9N, 34.7E) and Obninsk (OBN, 55.1N, 36.6E); Right: the individual P-waves. The time origin in the two seismograms is with respect to a predicted arrival time. Note the slow buildup of the arrival in LVZ, a possible indication of energy diffraction, which poses difficulties in picking an onset. The seismograms show ground velocity.

If it is sufficient to obtain relative delays between closely spaced instruments of a network, one may forego the estimation of $u(t)$ and instead cross-correlate pulses arriving in different sensors. An example is shown in Figure 6.6.

For a small network, an alternative to estimating the absolute arrival time for every sensor is to estimate the delay with respect to some time close to the arrival time at the centre of the array. The wavelet estimate can then be constructed by stacking individual arrivals. This gives more accurate results than simply choosing one (low noise) trace as the correlator but even more accurate is a method proposed by VanDecar and Crosson [380]. It uses the differences in arrival time between each pair of seismograms $u_i(t)$, $u_j(t)$ in a network of $N$ stations. Bagaini [15] shows that this method is superior to alternatives, in which all traces are correlated with one reference trace or a stack of traces. Thus, we measure:

$$\Delta T_{ij} = T_i - T_j \quad (i, j = 1, \ldots, N). \tag{6.11}$$

Since $T_{ij} = -T_{ji}$ we only need to cross-correlate for $j > i$. Of course, we are interested more in the absolute values $T_i$ rather than the differences. However, even though this procedure gives us $N(N-1)/2$ equations with $N$ unknowns, individual arrival times cannot be determined because any constant added to all arrival times would subtract out. In other words, if the set of absolute times $\{T_i\}$ satisfies (6.11), the set $\{T_i + c\}$ does this also. One solution is to give up on the desire to estimate absolute arrival times, but recognize that there is always a freely floating constant, which we may impose with an extra constraint:

$$\sum_{i=1}^{N} T_i = 0.$$

With this added, (6.11) is solved by:

$$T_i = \frac{1}{N} \sum_{j \neq i}^{N} \Delta T_{ij} ,$$

with a formal estimate of the variance in $T_i$:

$$\sigma_i^2 = \frac{1}{N-2} \sum_{j \neq i}^{N} [\Delta T_{ij} - (T_i - T_j)]^2 . \tag{6.12}$$

An alternative is to select the constant such that the sum of the *differences* between each $T_i$ and its value $T_i^{\mathrm{BG}}$ predicted for a background model is zero. In Section 13.8 we shall see that the tomographic inversion usually allows for a correction of the earthquake's origin time $T_0$. Since $T_0$ is, again, the same for every estimate $T_i$, this allows for different constants between different earthquakes measured at the same network, so that our choice of constraint does not really hamper the final data fit.

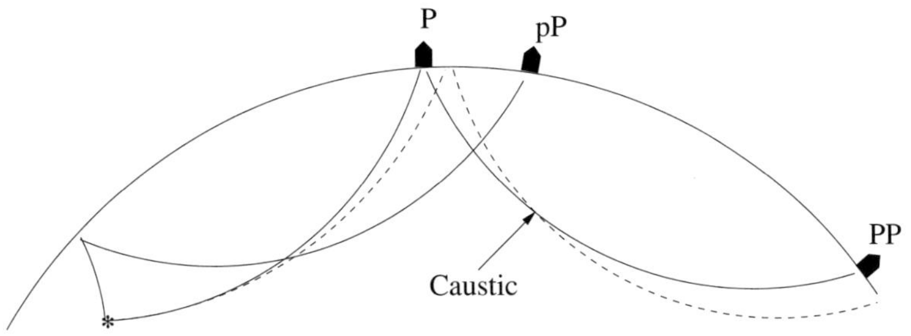


Fig. 6.7. Solid curves show the ray trajectories for the direct P-wave, the pP-wave with a surface reflection near the hypocentre (*) and the midway reflected PP-wave. The broken line is another PP-wave, giving rise to a caustic.

### *Exercises*

**Exercise 6.3** For a signal bandpassed between 0.05 and 0.15 Hz, and cross-correlated over a window of length 60 s, with a signal-to-noise ratio of 4, estimate the Cramér–Rao lower bound on the variance, then take the square root to find the standard deviation.

**Exercise 6.4** If we assume a probability of 10% that the estimate will be seriously off by one or more cycle skips, what is the correlator performance estimate (CPE) of the variance in the previous problem? Use an educated guess for $T_c$, e.g. 10 s.

**Exercise 6.5** How much does the variance reduce if we have a signal with SNR $= 20$?

**Exercise 6.6** How much does the variance reduce if we double the bandwidth?

**Exercise 6.7** In VanDecar and Crosson's method, why can we not conclude that $\Delta T_{ik} \equiv \Delta T_{ij} - \Delta T_{kj}$?

## 6.4 Differential times

We can also avoid the estimation of $u(t)$ if we cross-correlate different segments of the same seismogram. For example, the surface-reflected pP-wave (Figure 6.7) will be a close replica of the direct P-wave, apart from a possible sign change associated with the radiation pattern of the source, but for which we could easily correct. Picking the maximum of the cross-correlation of windows with the P and pP-wave, respectively, results in an estimate of the differential time:

$$T_{\text{pP-P}} = T_{\text{arr}}^{(\text{pP})} - T_{\text{arr}}^{(\text{P})} = \int_{\text{pP}} \frac{\mathrm{d}s}{V_{\text{P}}} - \int_{\text{P}} \frac{\mathrm{d}s}{V_{\text{P}}}$$

which is an integral constraint on the velocity $V_{\text{P}}$ in the Earth. A significant advantage of the observation of such 'differential' travel times is that they are insensitive to errors in the origin time, which are subtracted out when we take the difference.

Paulssen and Stutzmann [257] study the accuracy of PP-P cross-correlations. The PP waveform may be distorted by secondary arrivals, and beyond 88° the P waveform is affected by the presence of the core–mantle boundary. To avoid this, as well as interference from the upper mantle triplicated phases, the distance interval between 30° and 88° is optimal.

A variant on this is to cross-correlate body wave arrivals from closely spaced events to the same station; especially if generated by earthquakes on the same fault ('doublets') the waveforms can be practically identical. Fréchet [106] originally used this to find very precise relative earthquake locations (see also Got et al. [121], Rubin et al. [296], Waldhauser and Ellsworth [388]). Poupinet et al. [265] used the technique to monitor temporal changes in velocity in California, and Zhang and Thurber [413] inverted such differences directly ('double difference tomography').

If the earthquake is deep, a pP ray crosses the highly attenuating asthenosphere three times, whereas the P-wave, which crosses the asthenosphere only once, is far less attenuated. A correction for the difference in $t^*$ is then needed before P is cross-correlated with pP. This is most easily done by multiplying the spectrum of the P-wave with $\exp[-\omega\Delta t^*/2]$, where $\Delta t^* = t^*_{\mathrm{pP}} - t^*_{\mathrm{P}}$. Similar corrections are certainly needed for every pair of rays that is expected to experience strong differences in attenuation. If travel times are measured at more than one dominant frequency, a correction for the effects of attenuation dispersion may also be called for (see Section 13.7).

Surface reflections such as PP (but not pP) acquire a $\pi/2$ phase shift when the reflection gives rise to an arrival with minimax character: shifting the reflection point towards the source or receiver will actually result in *earlier* arrival times, whereas deflections out of the plane of propagation give the more usual positive delay of the arrival.† An intuitive argument suggested by Choy and Richards [57] can be invoked to explain this phase shift: as we saw in Chapter 2, the amplitude of a body wave is inversely proportional to the square root of the surface area of the cross section of a ray bundle, i.e. to $\mathcal{R}$. As can be seen in Figure 6.7, one of the components (the one in the plane of propagation) of the geometrical spreading $\mathcal{R}^2$ changes sign upon passage through a caustic, going from converging to diverging rays in the plane of propagation. This means that $\mathcal{R}^2 \rightarrow -\mathcal{R}^2$ or $\mathcal{R} \rightarrow \pm i\mathcal{R}$. Since the signal stays real, the phase change must have opposite sign for positive and negative frequencies. Such a sign change, which transforms a sine wave into a cosine, is known as the Hilbert transform. Of course, if we wish to determine the time difference between P and PP, we must first Hilbert transform the P-wave (or undo the Hilbert transform of the PP-wave) before we cross-correlate waveforms.

† Note that Fermat's Principle only requires the arrival time to be stationary – as opposed to a true minimum – with respect to small deflections of the path.

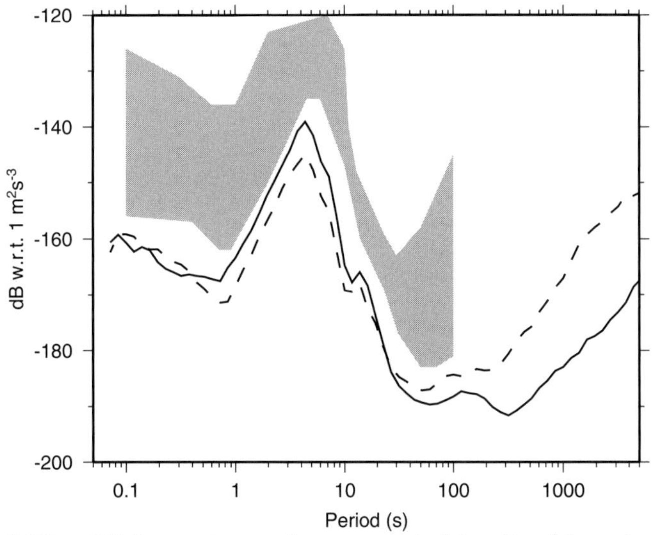

Fig. 6.8. Solid line: Minimum observed power spectral density of the noise on the vertical component in 118 stations of the Global Seismic Network (GSN), averaged over a year. Broken line: idem, horizontal component. Data for both curves come from Berger et al. [19]. The grey band denotes the noise level observed in stations in the US by McNamara and Buland [204], reproduced with permission from the AGU.

In practice, Hilbert transforms are easiest to compute in the frequency domain. If $u_{\mathrm{P}}(\omega)$ is the spectrum for the waveform of the P arrival, the spectrum of PP is given by:

$$\begin{aligned} u_{\mathrm{PP}}(\omega) &= a\, u_{\mathrm{P}}(\omega)\, \exp[-\mathrm{i}\, \mathrm{sgn}(\omega)\pi/2]\, \exp[-\omega(t^*_{\mathrm{PP}} - t^*_{\mathrm{P}})/2] \\ &= -\mathrm{i}\, \mathrm{sgn}(\omega)\, a\, u_{\mathrm{P}}(\omega)\, \exp[-\omega\Delta t^*/2]\,, \end{aligned} \tag{6.13}$$

where $\mathrm{sgn}(x) = x/|x|$ is the sign function. The amplitude factor $a$ differs from one because of differences in the source radiation pattern, the surface reflection coefficient (which for PP is negative at teleseismic distance) and different geometrical spreading.

## 6.5 Signal and noise

Even far away from cultural sources such as traffic or machinery, the Earth exhibits appreciable seismic noise, with a power density spectrum that may depend on local factors and the time of the year. A representative low-noise 'model', derived from high quality GSN stations, is given in Figure 6.8. The observed noise in most modern seismic stations has a characteristic pattern that may deviate by 20 dB or more from this low noise, as shown by the grey band in Figure 6.8. At the high frequency end of the spectrum, especially for periods below one second,

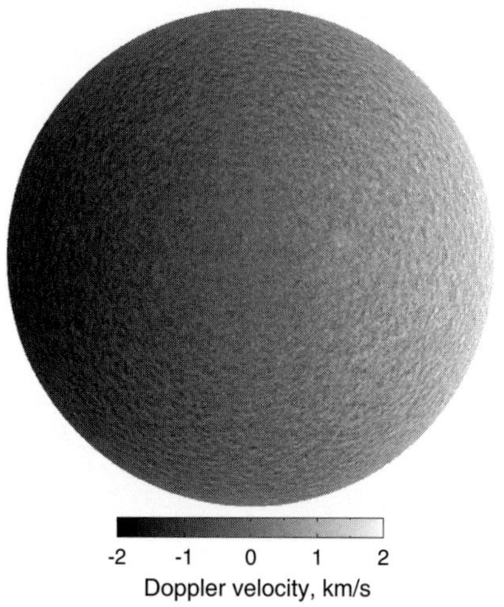


Fig. 6.9. Dopplergram of the Sun. The rotation of the Sun causes the velocity gradient from west to east, the wavefield motion is superimposed on this. Figure courtesy of Alexander Kosovichev, SOHO/MDI research group (Stanford).

cultural factors dominate the ambient noise: cars, machinery, or ship noise in the oceans. Rivers, or wind acting on buildings or trees also generate high frequency noise. Two peaks dominate the microseismic noise between periods of 1 and 20 s, generated by pressure variations on the ocean floor and waves striking the coast. The stronger one has a period of about 5 s (0.2 Hz), the second one centres around a period of 14 s (0.07 Hz). The noise peak near 5 s is especially unfortunate, since it interferes with the dominant period of P- and S-waves from earthquakes strong enough to be observable at teleseismic distances. Accurate measurements at shorter periods can only be made if the noise at the station is low, or if the noise can be suppressed by stacking signals over the closely spaced instrument of a network. If care is taken in siting and building a seismic station, ambient noise levels can be significantly reduced. Cultural noise sources vary more than noise at periods in excess of 15 s (Figure 6.8). As Figure 6.3 shows, the noise affecting later arrivals is mostly signal generated and therefore always overlapping in spectral content.

## 6.6 Time–distance analysis in helioseismology

The acoustic wavefield in the Sun is generated by motions in its convective region; this region provides a continuum of sources and the wavefield is essentially random. The situation is thus radically different from the terrestrial case in which the wavefield is generated by sources that are limited in space and time, generating a field that is largely deterministic.

The sound speed near the surface of the Sun is of the order of 10–50 km/s. The gas vibrations have velocities that are of the order of 1 km/s, which lead to only minute Doppler shifts in the spectral lines, which have a wavelength of the order of $6 \times 10^{-7}$ m and a width of about $10^{-11}$ m. For example, if the material of the Sun moves with a speed of $v_\perp$ km/s in the direction of the observing instrument, the relative wavelength shift is $v_\perp/c$ or one part per million for $v_\perp = 0.3$ km/s. At first sight, it may seem that the random nature of the field makes it impossible to obtain meaningful measurements of travel times of acoustic waves in the Sun. It turns out, however, that we can use the random nature of the field to great advantage, since the cross-correlation of the wavefield between two locations is proportional to the Green's function between these two points. This allows us to establish the travel time between each pair of pixels for which we have Doppler measurements.

The theoretical justifications for this important property of the cross-correlation are many, though each is dependent on rather strict assumptions. A proof for a reciprocity theorem 'of the correlation type' that expresses the Green's function as the cross-correlation of responses to sources distributed over a surface was given by Wapenaar and Fokkema [390], and their theorem allows us to identify the approximations needed to retrieve the Green's function by cross-correlation. The important conclusion is that though the Green's function is not recovered exactly, important properties like the phase or onset time of a wave are preserved if the sources are evenly distributed over the boundary of the domain and this property forms the basis of the method of seismic interferometry. For surface waves in the Earth, this requires an even distribution of noise sources over all azimuths. This is not usually the case, though Campillo and Paul [38] and Shapiro et al. [306] and others have obtained impressive results for diffusive noise sources that may come close, and techniques for the interpretation of arbitrary source distributions have been developed in helioseismology (see Section 7.6).

Here I give a heuristic explanation of seismic interferometry that is loosely based on Snieder's stationary phase derivation [330]. The noise cross-correlation function or $\gamma_{\rm NCF}$ is defined as:

$$\gamma_{\rm NCF}(t) = \int v_1(\tau + t) v_2(\tau) {\rm d}\tau \,,$$

for a long time window, where $v_1$ and $v_2$ are the signals from two pixels, located at $\boldsymbol{r}_1$ and $\boldsymbol{r}_2$ on the solar surface, for which we have Doppler measurements. In the frequency domain, the cross-correlation is equivalent to multiplication with the complex conjugate:

$$\gamma_{\rm NCF}(\omega) = v_1(\omega) v_2(\omega)^* = |v_1(\omega) v_2(\omega)| {\rm e}^{{\rm i}(\phi_1 - \phi_2)},$$

where $\phi_1$ and $\phi_2$ are the phases at pixel 1 and 2, respectively. Let $\boldsymbol{r}_{\rm s}$ be the location of one of the many source regions that contribute to the random wavefields at $\boldsymbol{r}_1$

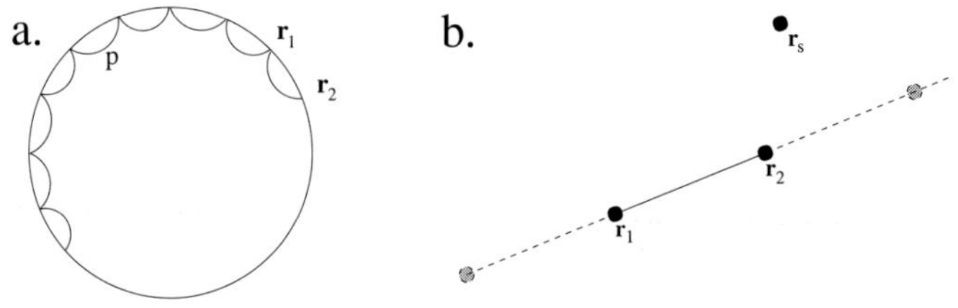


Fig. 6.10. Two pixels $\boldsymbol{r}_1$ and $\boldsymbol{r}_2$ at distance $\Delta$ on the solar surface receive signals from a noise source at $\boldsymbol{r}_s$. When the noise source is on or near the great circle at distance $n\Delta$ (grey dots) it generates one leg of a multiple P-wave between $\boldsymbol{r}_1$ and $\boldsymbol{r}_2$. All these legs sum in phase in the noise cross-correlation function.

and $\boldsymbol{r}_2$. The situation is sketched in Figure 6.10. The phase difference $\phi_1 - \phi_2$ of the wavefields in $\boldsymbol{r}_1$ and $\boldsymbol{r}_2$ depends on the source location $\boldsymbol{r}_s$ and the mode of wave propagation but is essentially random, and their contributions when summed over many different source locations will cancel out if the time window is long enough. Except for a number of sources that hit the surface at $\boldsymbol{r}_1$ at the right angle to generate a p-wave between the two (if the Sun is horizontally layered these sources would be exactly *on* the great circle through the two pixels, indicated by the dashed line in Figure 6.10, but this is not essential). To carry the argument further we must anticipate the theory of surface waves to be treated in Chapter 10. Just like PP-waves in the Earth, solar p-waves reflect off the surface. Far from the source, the wave may have undergone many such reflections and attains the character of a 'surface wave'.

The key to understanding the relationship between the noise correlation function and the Green's function is the realization that a wavefield always consists of many frequencies, and many horizontal components of the wavenumber vector $\boldsymbol{k}$. However, depending on the locations of a pair of stations, only selected frequencies and wavenumbers interfere constructively, giving rise to distinct arrivals such as p, pp etc. A theoretical justification of this statement was given by Nolet and Kennett [241] for terrestrial S-waves, and in Chapter 10 we shall see that the ray parameter $p$ is equal to $k/\omega$ if $k$ is the length of the horizontal component of $\boldsymbol{k}$. The waves that contribute to building up a pulse-like arrival such as p all travel with the same $k$. They usually interfere destructively, but at the appropriate distance $\Delta$ for a p-wave with this ray parameter they are for a brief moment in phase in a wide band of frequencies (a necessary prerequisite for constructing a pulse-like arrival is that there are many frequencies in its spectrum – see for example (2.62) for the spectrum of the delta function). At twice the distance, all phases are twice as large and therefore again in phase, and so on: this way the sequence p, pp, ppp, ... is generated (Figure 6.10).

But we can also turn the reasoning around. If some of the waves from a source near the pixel $\boldsymbol{r}_1$ at distance $\Delta$ from the pixel at $\boldsymbol{r}_2$ did generate a p-wave, the waves

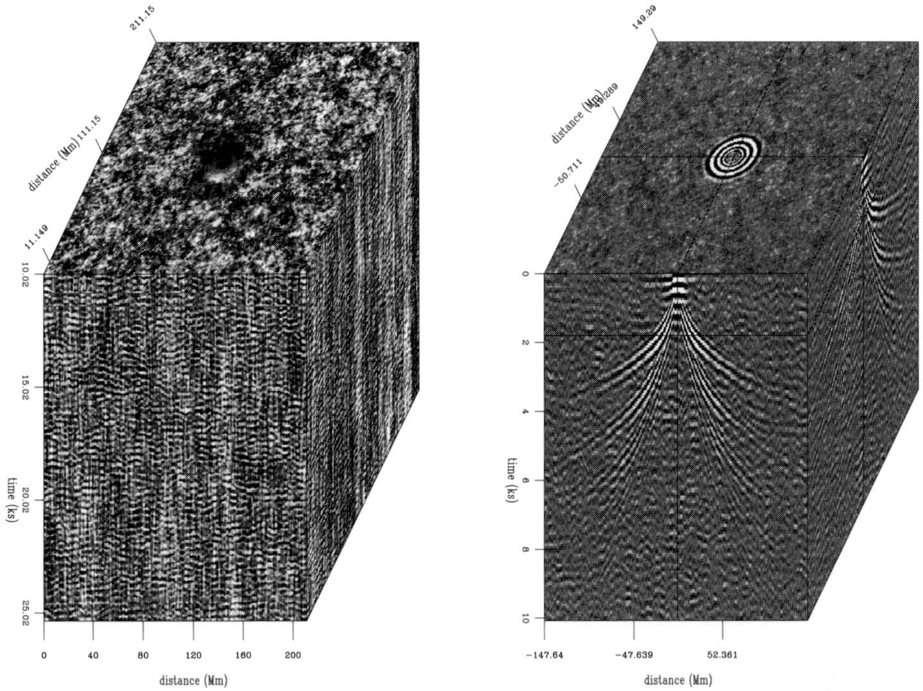


Fig. 6.11. Left: a 'data cube' for a patch of 200 × 200 Mm on the solar surface around a sunspot. The time axis runs vertically. Right: Cross-correlation of Doppler measurements shows the direct acoustic wave, followed by the pp and (weak) ppp reflections. Figure courtesy of Jon Claerbout.

with the same $\boldsymbol{k}$ from another source at $2\Delta$ will generate one leg of the pp-wave that surfaces at $\boldsymbol{r}_1$ but then again at $\boldsymbol{r}_2$. And again at $r\Delta$ etc. Now do a cross-correlation of the dopplergrams in the two pixels. The expression for the waves with horizontal wavenumber $k$ from these sources in the spectral domain is:

$$c(\omega) = u_1(\omega)u_2(\omega)^* \sim \exp[ik(\omega)(\Delta_1 - \Delta_2)]\,,$$

where $\Delta_i$ is the distance between a noise source and pixel $\boldsymbol{r}_i$. The difference $\Delta_1 - \Delta_2$ is constant for sources *on* the great circle and for those sources at a distance that is an integer multiple of $\Delta_1 - \Delta_2$, all these contributions are summed in phase. This will not be the case for waves with different wavenumber $k$ or sources located elsewhere. If we thus sum a sufficiently long record, the waves that interfere destructively should become much lower in energy than those that are summed in phase.

This was a conjecture made forty years ago by Claerbout [59] and it turns out to be good enough to be useful, as is evident from Figure 6.11. We can see this figure as a constructive interference of sources at distance $\delta\Delta = |\Delta_2 - \Delta_1|$, $2\delta\Delta$,

$3\delta\Delta, \ldots$, etc. All these sources have a p-wave surfacing at $\boldsymbol{r}_1$ that surfaces again in $\boldsymbol{r}_2$, and the constant time difference guarantees they will sum in phase in a cross-correlation over a long time window. Stations at an even number of $\delta\Delta$s will also generate multiple p-waves that surface every $\frac{1}{2}\delta\Delta$, and show up in Figure 6.11 as the second arriving branch, and so forth.

In fact, the multiple p-waves that give rise to the first branch have a phase difference equal to $\omega T_{\rm p} + \pi/2$, if $T_{\rm p}$ is the travel time of a p-wave between $\boldsymbol{r}_1$ and $\boldsymbol{r}_2$. Note the extra phase $\pi/2$ which arises from the Hilbert transform. Thus, the sum over all sources will produce two maxima, corresponding to the travel time $\pm T_{\rm p}$.

This way, we measure the travel time of a wave starting at $\boldsymbol{r}_1$ and ending at $\boldsymbol{r}_2$ by measuring the cross-correlation. A significant increase in the signal-to-noise ratio is obtained by correlating over annuli of pixels at distance $\Delta$ from a central pixel, though this obviously reduces the spatial resolution to length scales of the order of $\Delta$. The technique has led to the successful imaging of processes in the Sun's convective region.

Duvall et al. [88] describe some of the data processing steps that led to accurate travel time estimation using data from an observatory at the South Pole. Doppler shifts in the $Ca^+$ K-line were measured every minute, and images were corrected for offset and gain and interpolated onto a longitude–latitude grid, filtered and tapered. After correction for the Sun's rotation, the mean cross-correlation function between each point of the image and all points in an annulus at a fixed distance was computed and averaged over several pixels. Alternatively, one may average over a quadrant only and retain information about the flow direction that influences the travel time. Finally, travel times were measured from these cross-correlations and a map of travel time 'differences' constructed for annuli out to 15° distance. Duvall et al. [88] observe delays of a minute or more across sunspots.

The actual delay estimation can be improved by fitting one or more Gabor functions to the cross-correlations of the form:

$$G(t) = \cos[\omega_0(\tau - \tau_{\rm ph})] \exp\left(-\frac{(\tau - \tau_{\rm gr})^2}{2\sigma^2}\right),$$

where $\tau_{\rm ph}$ is the phase arrival time and $\tau_{\rm gr}$ is known as the group arrival time.† We shall wish to invert for average variations of $\tau_{\rm ph}$ over the annulus if we use ray theory. Figure 6.11 shows an example of a cross-correlation record. Gizon and Birch [117] extract a delay time from the difference between observed and

† The group arrival time is the time at which the maximum energy of a wavelet with a narrow spectral bandwidth around $\omega$ arrives; it is influenced by the dispersion of the wave.

predicted noise cross-correlation functions (see Section 7.2):

$$\tau_{\pm} = \frac{\int w(t)\dot{\gamma}_0[\gamma(t) - \gamma_0(t)]\mathrm{d}t}{\int w(t)[\dot{\gamma}_0(t)]^2\mathrm{d}t}, \tag{6.14}$$

where $\gamma$ and $\gamma_0$ are the cross-correlations as observed and predicted from a reference model, respectively, and $w(t)$ is a tapered window around the arrival. Because the cross-correlation has maxima both for positive and negative time, two travel times are obtained: $\tau_+$ for waves moving away from the starting location, and $\tau_-$ for waves moving from the averaging arc towards the centre. When summed with the reference travel time, they provide an estimate for the travel time in a non-moving medium through (2.55); the difference leads to an estimate of the flow of the medium through (2.56).

# 7

# Travel times: interpretation

Until recently, ray theory formed the backbone of all tomographic studies with body waves, but this is changing fast as seismologists push for better resolution and realize that the limitations of ray theory become tangible for long-period data and make the merging of data obtained from instruments with different passbands problematic. Ray theory is unable to model the healing of a wavefront, which depends on the ratio between wavelength and size of the heterogeneity. Delay time data observed in different frequency bands therefore contain information on the size of the heterogeneity, and one should wish to exploit this dependence by inverting for data from different frequency bands simultaneously.

Much of this chapter is therefore devoted to a 'finite-frequency' interpretation of delay times. It relies heavily on the paper by Dahlen et al. [76], which provides an efficient and comprehensive methodology for analysing teleseismic travel time data. Alternative algorithms are discussed in Sections 7.6 and 7.7.

## 7.1 The ray theoretical interpretation

The easiest way to interpret travel times of seismic or acoustic pulses is to assume that ray theory (Eq. 6.1) is valid, i.e. that the travel time of the ray is given by:

$$T = \int_P \frac{\mathrm{d}s}{c(\boldsymbol{r})},$$

where $P$ indicates the ray path, and $c$ is the intrinsic seismic velocity, $V_\mathrm{P}$ or $V_\mathrm{S}$ depending on the wave type. Since the ray path is also a function of $c$, the relationship between the travel time and the model is nonlinear. We use Fermat's Principle to linearize it. Assume that the travel time for the background (or 'starting') model is $T_0$:

$$T_0 = \int_{P_0} \frac{\mathrm{d}s}{c_0(\boldsymbol{r})}.$$

Often, the model will be spherically symmetric, i.e. $c_0(\boldsymbol{r}) \equiv c_0(r)$, but this is not necessarily the case. If lateral velocity contrasts are large, it may be advisable to use a starting model that is already laterally heterogeneous. For example, Zhao et al. [415] incorporate a slab structure that coincides with the location of deep seismicity.

We can now apply Fermat's principle to compute the true Earth integral not over the true raypath, but over the raypath computed for the background model, making only a small error. This allows us to avoid the complications of 3D ray tracing by choosing a layered background model to compute the path $P_0$:

$$T \approx \int_{P_0} \frac{\mathrm{d}s}{c(\boldsymbol{r})} .$$

Subtracting gives a linearized relationship for the travel time delay $\delta T$:

$$\delta T = T - T_0 \approx \int_{P_0} \left( \frac{1}{c} - \frac{1}{c_0} \right) \mathrm{d}s \approx - \int_{P_0} \frac{\delta c}{c^2} \mathrm{d}s \,, \tag{7.1}$$

where $\delta c(\boldsymbol{r})$ is the model velocity *anomaly*. The last linearization in (7.1) can be avoided if we define the slowness $c^{-1}$ rather than the velocity $c$ as the model variable of interest. Bijwaard and Spakman [23] have studied the effects of the raypath approximation, and conclude that, in very general terms, the approximation leads to slightly damped anomalies. Snieder and Aldridge [331] developed a raypath perturbation technique that avoids the retracing of rays in a three-dimensional model. As the amount of travel time data grows and the precision of seismic tomographic images becomes greater, it seems clear that the effects of ray bending need to be taken into account if we wish to interpret seismic velocities quantitatively – e.g. in terms of temperature or composition of the rock – and the rays are severely bent away from their 'background' model paths. There is, however, a second effect that leads to image distortion that arises from the assumption that we had to make to derive ray theory: the frequency content of P- or S-waves is often far from the 'infinite' frequency necessary to make ray theory valid (see, for example, Figure 6.6 where the maximum frequency is about 0.5 Hz).

The most severe effect caused by the frequency being finite is that of 'wavefront healing'. Figure 7.1 shows how an initial delay time diminishes after a wave with wavelength $L$ has crossed an anomaly of width $2L$. It is clear that this is a serious shortcoming of ray theory if we wish to image anomalies for which the halfwidth $L$ is of the same order as the wavelength $\lambda$. For larger anomalies the same effects will occur, only at larger distance $x$ from the heterogeneity. Hung et al. [139] conducted an extensive numerical study of wavefront healing for a wave that crosses an

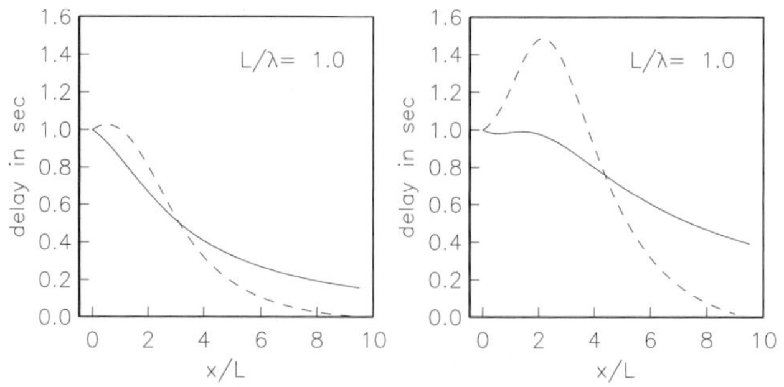


Fig. 7.1. Evolution of the absolute phase delay (left) and the delay in group arrival time (right) for a wave after passing a slow anomaly (solid line) and a fast anomaly (dashed line). To facilitate comparison, the signs of the negative delays for fast anomalies have been reversed. $L$ is the radius, or half the width of the anomaly, and is in this case equal to the wavelength $\lambda$ of the wave. From Nolet and Dahlen [239], reproduced with permission from the AGU.

isolated, spherically symmetric anomaly with a velocity perturbation

$$\frac{\delta c}{c} = \epsilon \left(1 + \cos \frac{2\pi r}{a}\right),$$

for $r < a/2$, and zero elsewhere. The numerical simulations show that the ratio between the 'observed' finite-frequency delay $\delta t_{\rm FF}$ and the ray-theoretical prediction $\delta t_{\rm RT}$ from (7.1) for a body wave arrival with a delay that has not yet completely healed is given approximately by:

$$\frac{\delta t_{\rm FF}}{\delta t_{\rm RT}} = 1 - \frac{\lambda}{a^2} \frac{SR}{S+R}, \tag{7.2}$$

where $\lambda$ is the dominant wavelength of the wave, $S$ and $R$ are the distances to the anomaly from source and receiver, respectively. For example, a P-wave at 70° has a total ray length $S + R = 7000$ km. If the dominant period is 25 s, which is typical for many broadband data, $\lambda \approx 300$ km. For an anomaly of size $a = 1000$ km, located at the turning point of the ray where $S = R$, we find that the observed delay $\delta t_{\rm FF}$ is only 48% of the ray-theoretical prediction. The dependence of healing on the square of the anomaly size $a$ makes the downfall of ray theory quite drastic: for $a = 700$ km the healing is essentially complete, i.e. the anomaly is invisible. This confirms some of the observations we made in Chapter 4.

Examples of the healing of a wavefront that has passed a fast or slow anomaly are shown in Figs. 7.2 and 7.3, respectively. Note that the healing is different in character for slow and fast anomalies. The faster wavefront in Figure 7.2 quickly loses amplitude because of the increased geometrical spreading, but the magnitude of the negative delay does not decrease: in a noise-free environment one could

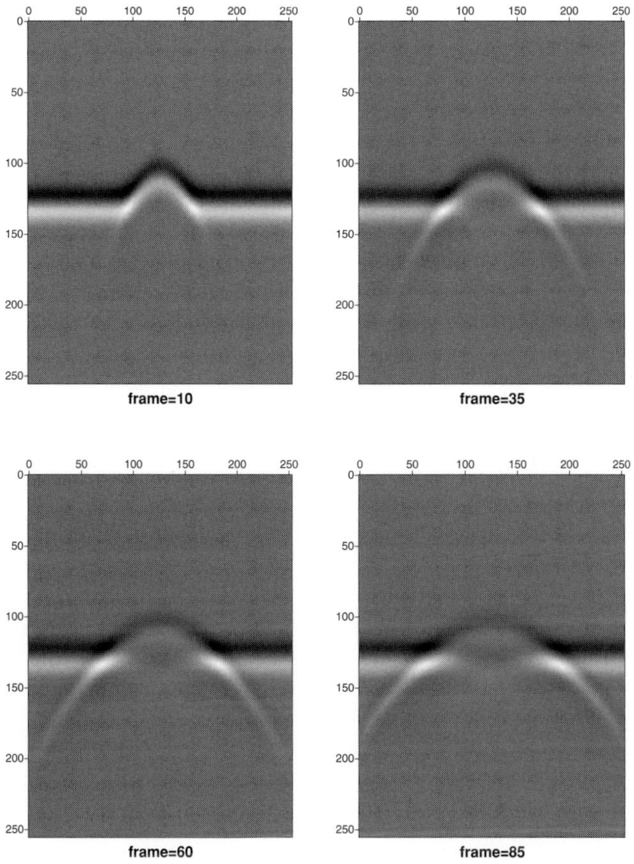


Fig. 7.2. Two-dimensional finite difference simulation of the healing of a wavefront that has passed a fast anomaly. The direction of wave propagation is upward. The visible window (frame) is shown after 10, 35, 60 and 85 time steps and also moves upward with the wavefront, such that it always appears at the centre.

continue to pick the phase close to the ray-theoretical arrival time; the slowed wavefront in Figure 7.3, on the other hand, creates a gap that, by Huygens' Principle, is slowly filled in. Thus, positive and negative anomalies show different healing behaviour, a point first made by Wielandt [394] and known as the 'Wielandt effect'. Though this seems to preclude a linear treatment of healing, the effects of noise will actually influence the time with which the fast wavefront can be 'picked' – a point we shall return to in Section 7.6. An analytical study by Nolet and Dahlen [239] and numerical studies by Hung et al. [138] show that the nonlinear effects are quite small.

The limitations of ray theory became more apparent as the quality of the seismic networks quickly improved after 1980, with broadband sensors and digital

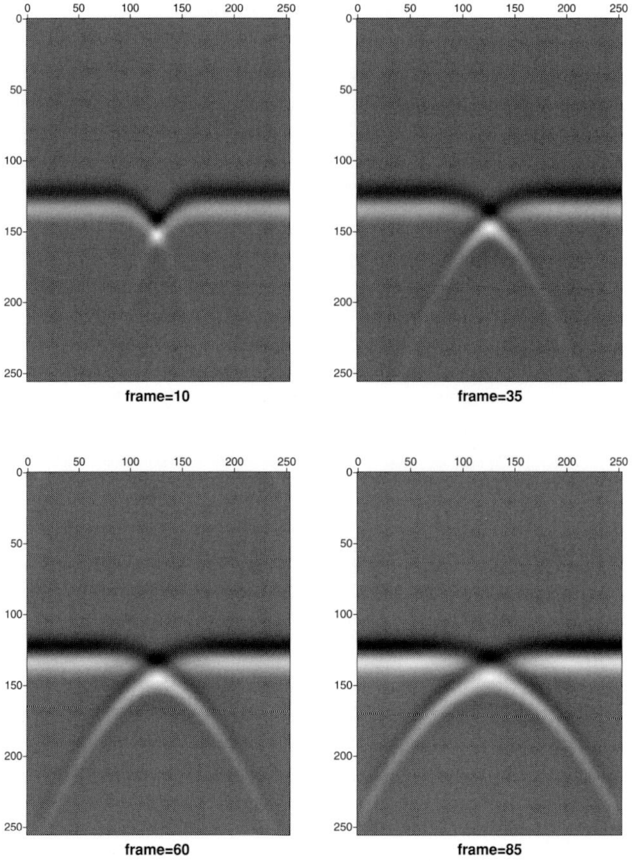


Fig. 7.3. As Figure 7.2, but for a slow anomaly.

recording offering large signal-to-noise ratios even at very low frequencies. Environmental noise has a minimum between periods of 30–100 s (Figure 6.8). For high signal-to-noise ratio (SNR) the Cramér–Rao lower bound (6.8) predicts the variance to be proportional to $\text{SNR}^{-1}$. This does not imply, however, that we obtain the most precise estimates if we bandpass between 0.01 and 0.033 Hz, because the variance decreases also as $\Delta\omega^{-3}$. Widening the passband to include high frequencies may offset the negative effects of an increase in noise. Woodward and Masters [404], who pioneered the cross-correlation technique on long-period seismograms, use a passband that includes frequencies as high as 0.06 Hz (Figure 7.4).

Such long-period data have so far mostly been used in tomographic modelling of the very long-wavelength heterogeneity (>1000 km horizontally) in the Earth, for which a ray theoretical approach is just acceptable. However, 1000 km represents a third of the thickness of the Earth's mantle, and ray theory breaks down for many smaller heterogeneities that are of considerable interest in geophysics: chunks of

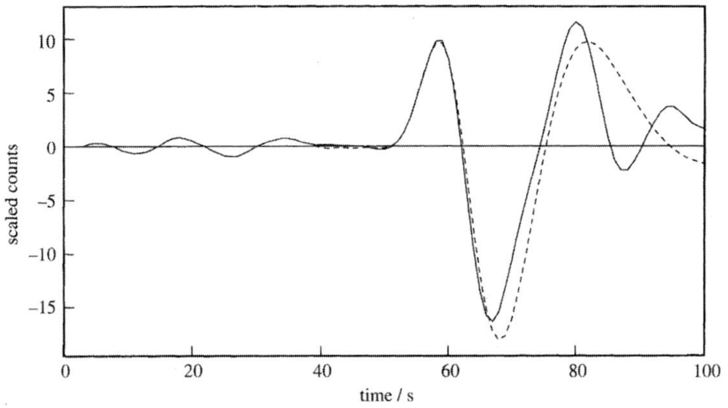


Fig. 7.4. Cross-correlation of the first half cycle of the S-wave. The cross-correlator (dashed line) is the impulse response convolved with a t* operator (5.14). In this case the cross-correlation, which puts most emphasis on the first half cycle, was done 'by eye'. The time error is subjectively estimated at 0.5 s. From Masters et al. [198], reproduced with permission from the Royal Society.

slabs sinking in the mantle, or hot rising mantle plumes, have dimensions that are likely limited in size, and progress in the imaging of such objects can only be obtained by moving away from ray theory. In this chapter we deal with the effects of diffraction by using a linear, first-order, 'finite-frequency' method.

## 7.2 Cross-correlation of seismic arrivals

Equation (6.3) gave us an expression of the seismogram as a sum of arrivals $u(t)$ with delays $\tau_i$ plus noise $n(t)$. To this we now add a contribution $\delta u_i$ of waves scattered from the wavefront around ray $i$:

$$s(t) = \sum_i A_i u(t - \tau_i) + \delta u_i(t - \tau_i) + n(t)\,.$$

If the original wave is a minimum-time phase such as P, pP, S, etc, $\delta u_i(t - \tau_i) = 0$ for $t < \tau_i$, but this is not necessarily the case for minimax phases such as PP or SS.† The scattered wavefield $\delta u(t)$ can theoretically be calculated by subdividing large scattering volumes into point scatterers of size d$V$ and integrating over the total volume $V$, using expressions (4.26)–(4.30).

As an example, consider a P-wave striking a pure density heterogeneity $\delta\rho/\rho$. Equation (4.26) tells us that the Fourier spectrum of the P → P scattered

† A phase is labelled minimax if the travel time is a minimum for raypath deflections out of the ray plane, but a maximum for deflections of the reflection point in the ray plane.

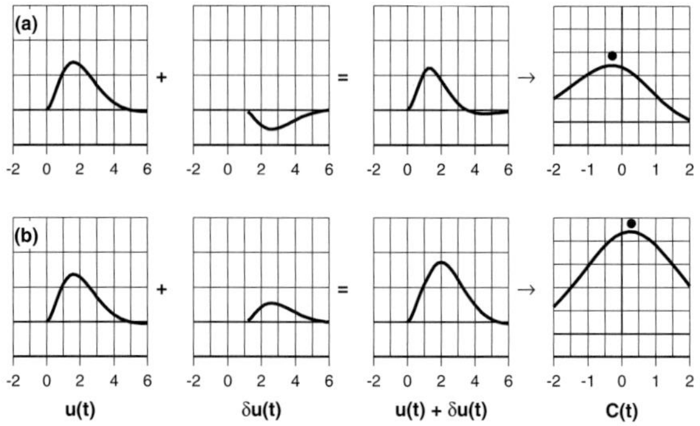


Fig. 7.5. The location of the maximum of the cross-correlation of a perturbed wave $u(t) + \delta u(t)$ with the unperturbed $u(t)$ is either advanced (a) or delayed (b), depending on the sign of the scattered wave. This occurs despite the fact that the latter is always delayed (see text for discussion). Dots denote the location of the maxima in the cross-correlation. From Nolet et al. [240], reproduced with permission from the AGU.

wave is:

$$\delta u_r^{\mathrm{PP}}(\boldsymbol{r}, \omega) = \int_V \frac{\omega^2 \mathrm{e}^{\mathrm{i}k_{\mathrm{P}}r}}{4\pi r V_{\mathrm{P}}^2} \frac{\delta\rho(\boldsymbol{r}')}{\rho(\boldsymbol{r}')} \cos\theta \, \mathrm{d}^3\boldsymbol{r}' .$$

Since the P-wave itself travels the path of minimum time, the scattered signal cannot arrive earlier than the direct wave. However, this does not mean that it always has a delaying influence on the *measured* travel time. The argument is as follows: the sign of the scattered wave may be positive or negative, depending on the sign of the density heterogeneity. In general, we deal with waves that arrive in the cross-correlation window and suffer only a small delay. This implies that $\theta$ is not large so it is reasonable to assume that the $\cos\theta$ term does not change the polarity of the scattered wave. The addition of $\delta u$ to $u$ deforms the waveshape and may have a delaying or an advancing effect depending on the sign of $\delta\rho$. Even though it seems counterintuitive to get a phase advance from a wave that arrives later than the direct wave, Figure 7.5 shows how this is possible: the effect of the addition of $\delta u$ is essentially to re-distribute energy within the cross-correlation window.

Attentive readers will have noted another deviation from ray theory: this example implies that the location of the maximum of a P-wave cross-correlation is influenced by density perturbations even if the seismic velocity remains the same. This is in fact the case. However, Hung et al. [138] show that the travel time anomalies produced by realistic variations in density are small, and perturbations of $V_{\mathrm{P}}$ and

$V_S$ are much more effective than perturbations of $\rho$ in influencing the measured travel time.

In the following we analyse only a single pulse $u(t)$ and omit the noise term $n(t)$, to enable us to analyse the contribution from scattered waves, which we regard as 'signal'. Without loss of generality we assume $A_1 = 1$. The autocorrelation of the unperturbed seismogram $u(t)$ is given by:

$$\gamma(t) = \int u(t')u(t'-t)\mathrm{d}t' \,. \tag{7.3}$$

We define the travel time delay by the maximum of the observed cross-correlation function, i.e. of the correlation of the observed signal $u + \delta u$ with the unperturbed wave $u$:

$$\gamma_{\mathrm{obs}}(t) + \delta\gamma(t) = \int [u(t') + \delta u(t')]u(t'-t)\mathrm{d}t' \,. \tag{7.4}$$

For the unperturbed wave, the cross-correlation reaches its maximum at zero lag, so:

$$\dot{\gamma}(0) = 0 \,, \tag{7.5}$$

and for the perturbed wave the maximum is reached after a delay $\delta T$:

$$\dot{\gamma}_{\mathrm{obs}}(\delta T) = \dot{\gamma}(\delta T) + \delta\dot{\gamma}(\delta T) = 0 \,, \tag{7.6}$$

where the dot denotes time differentiation. Developing $\dot{\gamma}$ to first order, we find, following Luo and Schuster [190] or Marquering et al. [195]:

$$\dot{\gamma}(\delta T) + \delta\dot{\gamma}(\delta T) = \dot{\gamma}(0) + \ddot{\gamma}(0)\delta T + \delta\dot{\gamma}(0) + \mathcal{O}(\delta^2) = 0 \,, \tag{7.7}$$

and using (7.4) and (7.5):

$$\delta T = -\frac{\delta\dot{\gamma}(0)}{\ddot{\gamma}(0)} = -\frac{\int_{-\infty}^{\infty} \dot{u}(t')\delta u(t')\mathrm{d}t'}{\int_{-\infty}^{\infty} \ddot{u}(t')u(t')\mathrm{d}t'} \,. \tag{7.8}$$

It is more convenient to express (7.8) in the frequency domain. Using $\dot{u}(\omega) = -\mathrm{i}\omega u(\omega)$ from (2.67), Parseval's theorem (2.68):

$$\int_{-\infty}^{\infty} g_1(t)g_2(t)\mathrm{d}t = \int_{-\infty}^{\infty} g_1(\omega)^* g_2(\omega)\mathrm{d}\omega \,,$$

and the spectral property of real signals $u(-\omega) = u(\omega)^*$, where an asterisk denotes the complex conjugate:

$$\begin{aligned}\delta T &= -\frac{\int_{-\infty}^{\infty}[-\mathrm{i}\omega u(\omega)]^*\delta u(\omega)\mathrm{d}\omega}{\int_{-\infty}^{\infty}[(-\mathrm{i}\omega)^2 u(\omega)]^* u(\omega)\mathrm{d}\omega} \\ &= \frac{\int_0^{\infty}\mathrm{i}\omega\{[u(\omega)^*\delta u(\omega)] - [u(\omega)^*\delta u(\omega)]^*\}\mathrm{d}\omega}{\int_0^{\infty}\omega^2\{[u(\omega)^* u(\omega)] + [u(\omega)^* u(\omega)]^*\}\mathrm{d}\omega} \\ &= -\frac{\mathrm{Re}\int_0^{\infty}\mathrm{i}\omega u(\omega)^*\delta u(\omega)\mathrm{d}\omega}{\int_0^{\infty}\omega^2 u(\omega)^* u(\omega)\mathrm{d}\omega}\,. \end{aligned} \tag{7.9}$$

Picking the maximum of $\gamma_{\mathrm{obs}}(t)$, as in (7.6) is usually accurate – in Section 6.2 we saw that it is the optimal filter judged by the signal-to-noise ratio. However, in some cases the highest absolute value of $\gamma_{\mathrm{obs}}(t)$ corresponds to a cycle skip and leads to large error. By eye, such cycle skips can often be recognized because the rest of the signal, away from the maximum, has a mismatch. Setting $w(t) = 1$ in (6.14), we simplify the estimator of Gizon and Birch [117] and minimize the integral:

$$I(t) = \int [\gamma_{\mathrm{obs}}(t') - \gamma(t'-t)]^2 \mathrm{d}t'\,. \tag{7.10}$$

Setting $\mathrm{d}I(t)/\mathrm{d}t = 0$ gives:

$$I(\delta T) = \ddot{I}(0)\delta T + \dot{I}(0) = 0 \quad\rightarrow\quad \delta T = -\frac{\dot{I}(0)}{\ddot{I}(0)}. \tag{7.11}$$

With

$$\dot{I}(0) = 2\int \dot{\gamma}(t'-t)[\gamma_{\mathrm{obs}}(t') - \gamma(t'-t)]\mathrm{d}t' = 2\int \dot{\gamma}(t')\delta\gamma(t')\mathrm{d}t'$$

and

$$\ddot{I}(0) = 2\int \ddot{\gamma}(t')\delta\gamma(t')\mathrm{d}t' + 2\int \dot{\gamma}(t')^2\mathrm{d}t' \approx 2\int \dot{\gamma}(t')^2\mathrm{d}t'\,,$$

we find

$$\delta T = \int_{-\infty}^{\infty} W(t)\delta\gamma(t)\mathrm{d}t\,, \tag{7.12}$$

where

$$W(t) = -\frac{\dot{\gamma}(t)}{\int \dot{\gamma}(t')^2\mathrm{d}t'}\,.$$

### Exercises

**Exercise 7.1** Plot (7.2) for various values of a, λ, $S$ and $R$.

**Exercise 7.2** The cartoon in Figure 7.5 assumes that the waveform of the scattered and the direct wave are the same, which in practice implies that the scatterer is large (see the discussion in Section 4.6). However, the expressions for $\delta u$ have a term $\omega^2$, so that for a very small point scatterer the scattered waveform $\delta u(t)$ is proportional to the second time derivative of $u(t)$. How would this affect the argument that scatterers can redistribute energy within the wavelet to earlier times even if they arrive late?

## 7.3 Forward scattering

The case of primary interest for transmission tomography is that of forward scattered waves. We make our observations over a rather narrow cross-correlation time window. Waves that scatter at a large angle have been scattered from anomalies distant from the direct arriving raypath, and the time lost in making the detour causes them to arrive outside of the time window of observation. It is therefore instructive to see what the scattering amplitudes become for the limiting case that $\theta \to 0$. For P→P, scattering (4.26) gives:

$$\delta\bar{u}^{\mathrm{PP}}(\omega) = \frac{\omega^2}{4\pi r V_{\mathrm{P}}^2}\left[\frac{\delta\rho}{\rho} - \frac{\delta\lambda}{\lambda+2\mu} - 2\frac{\delta\mu}{\lambda+2\mu}\right]\mathrm{e}^{ik_{\mathrm{P}}r}\,\mathrm{d}V = -\frac{\omega^2}{2\pi r V_{\mathrm{P}}^2}\frac{\delta V_{\mathrm{P}}}{V_{\mathrm{P}}}\mathrm{e}^{ik_{\mathrm{P}}r}\,\mathrm{d}V\,, \tag{7.13}$$

because to first order

$$\frac{\delta\rho}{\rho} - \frac{\delta\lambda + 2\delta\mu}{\lambda+2\mu} = -\frac{\delta V_{\mathrm{P}}^2}{V_{\mathrm{P}}^2} = -2\frac{\delta V_{\mathrm{P}}}{V_{\mathrm{P}}}. \tag{7.14}$$

The factor of −2 could have been derived as well from the expression (4.41) for the scattering matrix $\mathbf{S}_{\mathrm{P}}$. For forward scattering $\boldsymbol{\gamma}_2 = \boldsymbol{\gamma}_1$ and the polarization of the P-wave is also in the $\boldsymbol{\gamma}_1$ direction. Dotting $\mathbf{S}_{\mathrm{P}}$ left and right with the wave polarization and identifying $V_1 = V_2 = V_{\mathrm{P}}$:

$$\boldsymbol{\gamma}_1 \cdot \mathbf{S}_{\mathrm{P}}\boldsymbol{\gamma}_1 = -2(\boldsymbol{\gamma}_1\cdot\boldsymbol{\gamma}_1)(\boldsymbol{\gamma}_1\cdot\boldsymbol{\gamma}_1) = -2\,.$$

Only the heterogeneity in $V_{\mathrm{P}}$ scatters a P-wave in the forward direction.

The forward scattering for an S-wave from a heterogeneity in $V_{\mathrm{S}}$ is similar. Again $\boldsymbol{\gamma}_2 = \boldsymbol{\gamma}_2$ but the polarization of the incoming and scattered waves is different (perpendicular to the ray). To simplify the discussion it is helpful to assume that the ray propagates in the $z$-direction so that $\boldsymbol{\gamma}_1 = \hat{\boldsymbol{e}}_z$ and the wave polarization vector is $\hat{\boldsymbol{e}}_y$:

$$\hat{\boldsymbol{e}}_y \cdot \boldsymbol{S}_S \hat{\boldsymbol{e}}_y = 2\hat{\boldsymbol{e}}_y \cdot [2\hat{\boldsymbol{e}}_z\hat{\boldsymbol{e}}_z - \hat{\boldsymbol{e}}_z\hat{\boldsymbol{e}}_z - \boldsymbol{I}]\hat{\boldsymbol{e}}_y = -2\,,$$

the same result we obtained for the forward P-wave scattering.

## 7.4 Finite frequency sensitivity: a simple example

Before we tackle the algebraically more involved case of scattering under arbitrary angle $\theta$, we shall investigate the effect of purely forward scattered waves on the location of the cross-correlation maximum.

To compute the predicted delay time $\delta T$ for a given scattered field $\delta u$ we also need the incoming wavefield $u$. Our starting point is the ray-theoretical seismogram (4.23) for a P-wave with source–receiver travel time $T_{\mathrm{rs}}$:

$$u^{\mathrm{P}}(\boldsymbol{r}_{\mathrm{r}}, \omega) = \frac{\mathcal{F}^{\mathrm{P}}\dot{m}(\omega)}{4\pi\, V_{\mathrm{P}}(\boldsymbol{r}_{\mathrm{s}})^2 \mathcal{R}_{\mathrm{rs}} \sqrt{\rho(\boldsymbol{r}_{\mathrm{s}})\rho(\boldsymbol{r}_{\mathrm{r}})V_{\mathrm{P}}(\boldsymbol{r}_{\mathrm{s}})V_{\mathrm{P}}(\boldsymbol{r}_{\mathrm{r}})}} \mathrm{e}^{\mathrm{i}\omega T_{\mathrm{rs}}} \quad \text{(4.23 again).}$$

We can also use (4.22) to generalize the expression for the scattered wave (7.13) to heterogeneous media by adapting the geometrical decay and phase delay:

$$\delta u^{\mathrm{PP}}(\boldsymbol{r}_{\mathrm{x}}, \omega) = -\frac{\omega^2 \rho(\boldsymbol{r}_{\mathrm{x}})\mathrm{d}V}{2\pi \mathcal{R}_{\mathrm{rx}} V_{\mathrm{P}}(\boldsymbol{r}_{\mathrm{x}})\sqrt{V_{\mathrm{P}}(\boldsymbol{r}_{\mathrm{x}})V_{\mathrm{P}}(\boldsymbol{r}_{\mathrm{r}})\rho(\boldsymbol{r}_{\mathrm{x}})\rho(\boldsymbol{r}_{\mathrm{r}})}} \left(\frac{\delta V_{\mathrm{P}}}{V_{\mathrm{P}}}\right)_{\mathrm{x}} \mathrm{e}^{\mathrm{i}\omega T_{\mathrm{rx}}}. \tag{7.15}$$

where $T_{\mathrm{rx}}$ is the travel time of the P-wave between scatterer and receiver.

There is no need to compute the geometrical spreading $\mathcal{R}_{\mathrm{rx}}$ for each possible scattering location $x$. If we compute the geometrical spreading from the receiver to every point in the model, the amplitude reciprocity principle (4.20) gives us the spreading factor $\mathcal{R}_{\mathrm{rx}}$:

$$V_{\mathrm{P}}(\boldsymbol{r}_{\mathrm{r}})\mathcal{R}_{\mathrm{xr}} = V_{\mathrm{P}}(\boldsymbol{r}_{\mathrm{x}})\mathcal{R}_{\mathrm{rx}}. \tag{7.16}$$

If we replace the receiver r by a scatterer x, (4.23) can also be used to find the wavefield $\boldsymbol{u}^{\mathrm{P}}(\boldsymbol{r}_{\mathrm{x}}, \omega)$ that impacts on a point scatterer. Multiplying the scattered wave (7.15) for an incoming wave with unit amplitude with $\boldsymbol{u}^{\mathrm{P}}(\boldsymbol{r}_{\mathrm{x}}, \omega)$ and applying (7.16) results in:

$$\begin{aligned}\delta u^{\mathrm{PP}}&(\boldsymbol{r}_{\mathrm{x}}, \omega) \\ &= -\frac{\mathcal{F}_{\mathrm{s}}{}^{\mathrm{P}}\dot{m}(\omega)\omega^2}{8\pi^2 V_{\mathrm{P}}(\boldsymbol{r}_{\mathrm{s}})^{5/2}\rho(\boldsymbol{r}_{\mathrm{s}})^{1/2} V_{\mathrm{P}}(\boldsymbol{r}_{\mathrm{r}})^{3/2}\rho(\boldsymbol{r}_{\mathrm{r}})^{1/2}} \frac{(\delta V_{\mathrm{P}}/V_{\mathrm{P}})_{\mathrm{x}}\mathrm{d}V}{V_{\mathrm{P}}(\boldsymbol{r}_{\mathrm{x}})\mathcal{R}_{\mathrm{xs}}\mathcal{R}_{\mathrm{xr}}} \mathrm{e}^{\mathrm{i}\omega(T_{\mathrm{sx}}+T_{\mathrm{rx}})},\end{aligned} \tag{7.17}$$

where $\mathcal{F}_{\mathrm{s}}^{\mathrm{P}}$ indicates that this is the amplitude of the wave that departed in the direction of the scatterer – which may differ from the amplitude of the direct wave $\mathcal{F}^{\mathrm{P}}$. We find the time delay induced by such a point scatterer if we insert (4.23) and

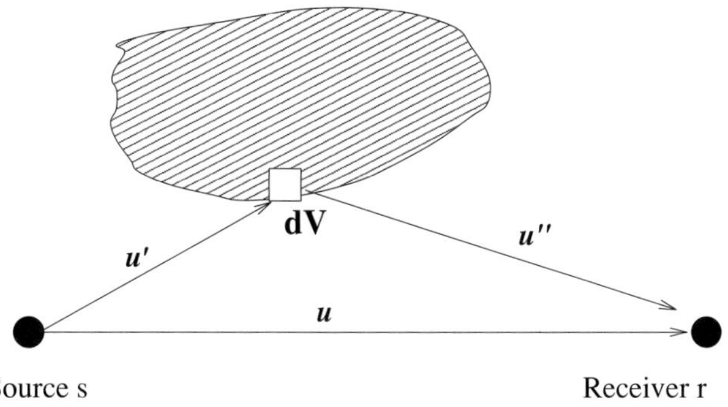


Fig. 7.6. This cartoon illustrates how Born theory works in a homogeneous medium. A wave $u'$ is scattered off a point scatterer $dV$ and generates a small perturbation $u''$ that adds to the direct wave $u$. The total contribution from the heterogeneity is found by integrating over the volume $V$ of the anomaly.

(7.17) into expression (7.9) for the cross-correlation time delay:

$$\delta T = -\frac{(\delta V_P/V_P)_x dV}{2\pi V_P(\boldsymbol{r}_r)V_P(\boldsymbol{r}_x)} \frac{\mathcal{R}_{rs}}{\mathcal{R}_{xr}\mathcal{R}_{xs}} \frac{\mathcal{F}_s^P}{\mathcal{F}^P} \frac{\int_0^\infty \omega^3 |\dot{m}(\omega)|^2 \sin[\omega \Delta T(\boldsymbol{r}_x)] d\omega}{\int_0^\infty \omega^2 |\dot{m}(\omega)|^2 d\omega}, \tag{7.18}$$

where $\Delta T$ is the extra time needed for the ray to visit the scatterer $x$ at $\boldsymbol{r}_x$ (which we shall call the 'detour time'):

$$\Delta T(\boldsymbol{r}_x) = T_{xr} + T_{xs} - T_{rs}. \tag{7.19}$$

For a more general heterogeneity, we integrate over all point scatterers (Figure 7.6):

$$\delta T = \int K_P(\boldsymbol{r}_x) \frac{\delta V_P}{V_P} d^3\boldsymbol{r}_x,$$

where $K_P(\boldsymbol{r}_x)$ is the Fréchet kernel:

$$K_P(\boldsymbol{r}_x) = -\frac{1}{2\pi V_P(\boldsymbol{r}_r)V_P(\boldsymbol{r}_x)} \frac{\mathcal{R}_{rs}}{\mathcal{R}_{xr}\mathcal{R}_{xs}} \frac{\mathcal{F}_s^P}{\mathcal{F}^P} \frac{\int_0^\infty \omega^3 |\dot{m}(\omega)|^2 \sin[\omega \Delta T(\boldsymbol{r}_x)] d\omega}{\int_0^\infty \omega^2 |\dot{m}(\omega)|^2 d\omega}. \tag{7.20}$$

An inspection of the integrals in (7.20) shows that the frequency dependence of $K_P$ enters through the scattered signal detour time $\Delta T$. Though the frequency integration is from 0 to $\infty$, the bandwidth of $\dot{m}(\omega)$ will effectively limit the integration to a finite bandwidth.

For small values of $\omega \Delta T$, i.e. for scatterers near the geometrical ray, we may approximate $\sin \omega \Delta T \approx \omega \Delta T$ and bring $\Delta T$ out of the frequency integral. In this case $K_P$ is proportional to $\Delta T$ itself. Since the ray is a minimum time path, $\partial \Delta T / \partial q = 0$ and $\Delta T$ – and thus $K_P$ – is to first order proportional to the *square* of the distance $q$ of the scatterer to the ray. This explains why the white line in

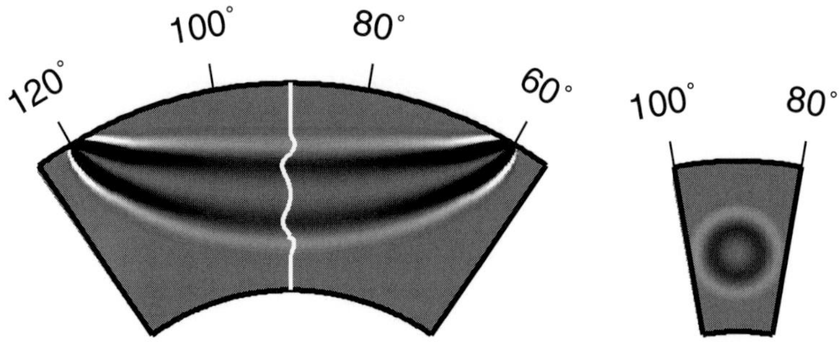


Fig. 7.7. A banana-doughnut kernel for a P-wave at a distance of 60° with a cross-section clearly showing the 'doughnut' hole, where the travel time is insensitive to variations in $V_P$. The white line is a graph of the amplitude near the midpoint of the ray. The greyscale includes negative (black) and positive (whitish) values, and shows the reversed polarity of the second Fresnel zone. Figure courtesy Tarje Nissen-Meyer.

Figure 7.7, which shows the value of $K_P$ along a cross-section, has the shape of a parabola near the location of the ray.

The most surprising characteristic of finite-frequency kernels such as (7.20) is that the cross-correlation travel time has zero sensitivity at the locations of the geometrical ray path! On the ray itself, $\Delta T = 0$, so the kernel is zero: heterogeneities on the ray do not influence $\delta T$. The right side of Figure 7.7 shows that there is a gap in the sensitivity, situated around the ray itself. Because the cross-sections shown in this figure resemble a banana and a doughnut, respectively, Marquering et al. [196] named these kernels 'banana-doughnut' kernels, though the more respectful name is 'Fréchet kernels', after the French mathematician Maurice Fréchet (1878–1972), whose thesis dealt with derivatives with respect to a *function* rather than a scalar or a vector (the function is in our case the tomographic model).

The null sensitivity on the ray is a result that is counterintuitive, because it seems to flatly contradict ray theory. In fact, the only way to remove the zero sensitivity for $\Delta T \to 0$ is to let $\omega \to \infty$. Infinite-frequency never occurs in practice, of course. Teleseismic P-waves have little energy above 1-2 Hz, and S-waves are limited to even lower frequency because of their stronger attenuation (see Exercise 5.8). The zero sensitivity can be made more intuitive if we realize that a cross-correlation measurement involves the wave energy over a time window of finite length. The energy in such a window is moved forward or backward, depending on the sign of $\delta u$, even though $\delta u$ arrives after the direct wave $u$. But if the scatterer is *on* the ray, there is no delay, so $\delta u$ can only perturb the amplitude of the wave, not its phase. The situation is sketched in Figure 7.8.

Although the term $\sin \omega \Delta T$ is modulated by the power spectrum and by a factor $\omega^3$, one may expect that the kernel has a maximum near $\omega_0 \Delta T = \pi/2$, or for

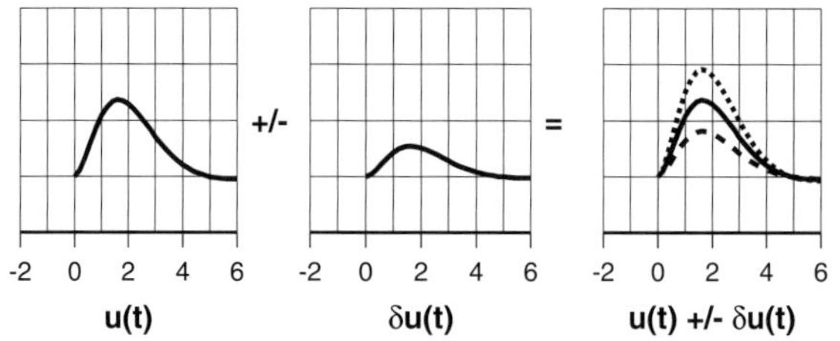


Fig. 7.8. If the scattered wave $\delta u(t)$ has no delay because the heterogeneity is located *on* the raypath, only the amplitude of $u(t)$ is affected. From Nolet et al. [240], reproduced with permission from the AGU.

$\Delta T = 1/4f_0$ if $\omega_0 = 2\pi f_0$ is the dominant frequency of the signal. This high sensitivity also tells us we shall wish to extend the length of the cross-correlation window beyond a quarter of the dominant period to capture scattered energy that arises from this part of the kernel and optimize sensitivity. Generally, we shall choose windows longer than that to retain selectivity in the frequency domain.† There are secondary maxima, corresponding to higher-order Fresnel zones. Their importance diminishes as we widen the bandwidth of the signal. For narrow band signals, however, second- or even third-order zones become important, and there will be zones with a reversed sensitivity, where a low velocity may actually lead to an advance of the energy within the cross-correlation window and lead to negative delay time observations.

### *Exercise*

**Exercise 7.3** Consider the Fresnel zone in a homogeneous medium, as in Section 2.10, for a pulse within a narrow frequency band. Show that the kernel has a maximum at $h/\sqrt{2}$, where $h$ is the radius of the first Fresnel zone.

## 7.5 Finite frequency kernels: general

We shall now drop the approximation of forward scattering and generalize these results to arbitrary rays and anomalies in both density and $V_P$ and $V_S$. It will be useful to generalize some of our previous results in a notation that is oblivious to the wavetype. Inspection of (4.13), (4.11) and (4.22) shows that we can write the

† A fundamental result of filtering theory is that the width of the frequency filter scales as the inverse of the width of the time window. The Fourier transform of a signal with a narrow spectrum thus extends far in time, and reversely a signal truncated with a short time window has a broad spectrum.

radiation pattern as a contraction of the moment tensor $\boldsymbol{M}$ with a unit vector $\hat{\boldsymbol{\gamma}}_s$ giving the ray direction and a unit vector $\hat{\boldsymbol{p}}_s$ giving the polarization of the wave at the source, yielding a source amplitude factor $\Lambda$:

$$\Lambda = (\rho_s V_s^5)^{-\frac{1}{2}} \boldsymbol{M} : \frac{1}{2}(\hat{\boldsymbol{\gamma}}_s \hat{\boldsymbol{p}}_s + \hat{\boldsymbol{p}}_s \hat{\boldsymbol{\gamma}}_s).$$

If we record the component of motion in direction $\hat{\boldsymbol{\nu}}$ at the receiver, where the ray polarization is $\hat{\boldsymbol{p}}_r$, the receiver amplitude factor is:

$$\Upsilon = (\rho_r V_r)^{-\frac{1}{2}} (\hat{\boldsymbol{\nu}} \cdot \hat{\boldsymbol{p}}_r).$$

In these expressions, $V$ denotes the appropriate velocity (P or S) at the source $s$ or receiver $r$. In addition, the amplitudes of direct and scattered wave may be affected by reflection and transmission at internal discontinuities; we denote the product of all appropriate reflection/transmission coefficients by a real factor $\Pi_{rs}$ and a possible phase advance $\Phi_{rs}$. In the absence of phase shifts acquired by supercritical reflections,

$$\Phi_{rs} = -M_{rs} \frac{\pi}{2},$$

where $M_{rs}$ is the Maslov index, a number which is usually 0 but increases by 1 upon passage of a caustic, e.g. for SS-waves.† We also note that $M_{rs} = M_{sr}$, a direct consequence of reciprocity. With that, the scalar displacement from the direct wave is:

$$u(t) = \frac{1}{4\pi} \Pi_{rs} \Lambda \Upsilon \mathcal{R}_{rs}^{-1} [\dot{m}(t - T_{rs}) \cos \Phi_{rs} + \dot{m}_H(t - T_{rs}) \sin \Phi_{rs}], \tag{7.21}$$

where $m_H(t)$ is the Hilbert transform of $m(t)$. In the frequency domain:

$$u(\omega) = \frac{1}{4\pi} \Pi_{rs} \Lambda \Upsilon \mathcal{R}_{rs}^{-1} \dot{m}(\omega) \exp i(\omega T_{rs} - \Phi_{rs}). \tag{7.22}$$

Similarly, for the perturbed seismogram:

$$\delta u(t) = -\left(\frac{1}{4\pi}\right)^2 \int \frac{\Lambda_1 \Upsilon_2 \Pi_{xs} \Pi_{xr} (\hat{\boldsymbol{p}}_2 \cdot \boldsymbol{S} \hat{\boldsymbol{p}}_1)}{V_r (V_1 V_2)^{\frac{1}{2}} \mathcal{R}_{xs} \mathcal{R}_{xr}} [\dddot{m}(t - T_{xs} - T_{xr}) \cos(\Phi_{xs} + \Phi_{xr}) + \dddot{m}_H(t - T_{xs} - T_{xr}) \sin(\Phi_{xs} + \Phi_{xr})] \mathrm{d}^3 \boldsymbol{r}_x, \tag{7.23}$$

where the triple dot indicates triple differentiation with respect to time $t$, and subscripts 1 and 2 refer to the incoming and outgoing scattered ray, respectively.

† The Maslov index is equal to the number of dimensions lost by the bundle of rays. At the caustic, the ray crossings reduce to a line. The reduction to a focal *point*, as is common in lens systems, would increase the Maslov index by 2.

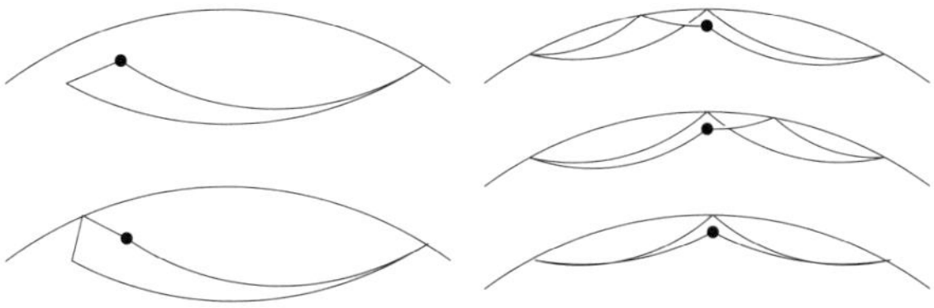

Fig. 7.9. For surface-reflected waves such as pP (left) or PP (right), raypaths for a scatterer near the surface may reflect from the surface before or after hitting the scatterer. Each of these possible rays must be included in the analysis if the detour time is short enough that the cross-correlation window is affected.

The terms $V_1$ and $V_2$ are the velocities of the medium at $\boldsymbol{r}_\mathrm{x}$. $\boldsymbol{S}$ is given by (4.40):

$$\boldsymbol{S} = \frac{\delta V_\mathrm{P}}{V_\mathrm{P}}\boldsymbol{S}_\mathrm{P} + \frac{\delta V_\mathrm{S}}{V_\mathrm{S}}\boldsymbol{S}_\mathrm{S} + \frac{\delta\rho}{\rho}\boldsymbol{S}_\rho \quad \text{(4.40 again)},$$

with $\boldsymbol{S}_\mathrm{P}$ etc. given by (4.41 - 4.43). The spectrum is:

$$\delta u(\omega) = \left(\frac{\omega}{4\pi}\right)^2 \int \frac{\Lambda_1 \Upsilon_2 \Pi_\mathrm{xs}\Pi_\mathrm{xr}(\hat{\boldsymbol{p}}_2 \cdot \boldsymbol{S}\hat{\boldsymbol{p}}_1)}{V_\mathrm{r}(V_1 V_2)^{\frac{1}{2}}\mathcal{R}_\mathrm{xs}\mathcal{R}_\mathrm{xr}}\dot{m}(\omega)\exp i[\omega(T_\mathrm{xs}+T_\mathrm{xr}) - \Phi_\mathrm{xs} - \Phi_\mathrm{xr}]\mathrm{d}^3\boldsymbol{r}_\mathrm{x}. \tag{7.24}$$

We denote the ratio of scattered to direct wave amplitude factors by $N$:

$$N = \frac{\Lambda_1 \Upsilon_2 \Pi_\mathrm{xs}\Pi_\mathrm{xr}}{\Lambda \Upsilon \Pi_\mathrm{rs}}. \tag{7.25}$$

We abbreviate the normalized scattering normalized coefficients into 'interaction coefficients' $\Omega_\mathrm{P}$, $\Omega_\mathrm{S}$ and $\Omega_\rho$ (see Table 7.1 and Figure 7.10):

$$\Omega_{\mathrm{P,S},\rho} = -\frac{1}{2}\hat{\boldsymbol{p}}_2 \cdot \boldsymbol{S}_{\mathrm{P,S},\rho}\hat{\boldsymbol{p}}_1. \tag{7.26}$$

These coefficients are normalized to 1 for forward scattering ($\hat{\boldsymbol{\gamma}}_1 = \hat{\boldsymbol{\gamma}}_2$) of unconverted waves. Putting this all together in (7.9), we find for the travel time perturbation:

$$\delta T = \int \left[ K_\mathrm{P}\left(\frac{\delta V_\mathrm{P}}{V_\mathrm{P}}\right) + K_\mathrm{S}\left(\frac{\delta V_\mathrm{S}}{V_\mathrm{S}}\right) + K_\rho\left(\frac{\delta\rho}{\rho}\right)\right]\mathrm{d}^3\boldsymbol{r}_\mathrm{x}, \tag{7.27}$$

Table 7.1. *Normalized scattering coefficients. $\hat{\boldsymbol{\gamma}}_1$ and $\hat{\boldsymbol{\gamma}}_2$ are unit vectors that give the direction of the incoming and outgoing ray, $\hat{\boldsymbol{q}}_1$ and $\hat{\boldsymbol{q}}_2$ denote the polarization of the incoming and outgoing S-wave.*

| | |
|---|---|
| $\Omega_P$ | |
| P → P | 1 |
| P → S | 0 |
| S → P | 0 |
| S → S | 0 |
| $\Omega_S$ | |
| P → P | $-2(V_S/V_P)^2[1-(\hat{\boldsymbol{\gamma}}_1\cdot\hat{\boldsymbol{\gamma}}_2)^2]$ |
| P → S | $2(V_S/V_P)(\hat{\boldsymbol{\gamma}}_1\cdot\hat{\boldsymbol{\gamma}}_2)(\hat{\boldsymbol{\gamma}}_1\cdot\hat{\boldsymbol{q}}_2)$ |
| S → P | $2(V_S/V_P)(\hat{\boldsymbol{\gamma}}_1\cdot\hat{\boldsymbol{\gamma}}_2)(\hat{\boldsymbol{\gamma}}_2\cdot\hat{\boldsymbol{q}}_1)$ |
| S → S | $(\hat{\boldsymbol{\gamma}}_1\cdot\hat{\boldsymbol{\gamma}}_2)(\hat{\boldsymbol{q}}_1\cdot\hat{\boldsymbol{q}}_2)+(\hat{\boldsymbol{\gamma}}_1\cdot\hat{\boldsymbol{q}}_2)(\hat{\boldsymbol{\gamma}}_2\cdot\hat{\boldsymbol{q}}_1)$ |
| $\Omega_\rho$ | |
| P → P | $\frac{1}{2}(1-(\hat{\boldsymbol{\gamma}}_1\cdot\hat{\boldsymbol{\gamma}}_2)-(V_S/V_P)^2[1-(\hat{\boldsymbol{\gamma}}_1\cdot\hat{\boldsymbol{\gamma}}_2)^2]$ |
| P → S | $-\frac{1}{2}(\hat{\boldsymbol{\gamma}}_1\cdot\hat{\boldsymbol{q}}_2)+(V_S/V_P)(\hat{\boldsymbol{\gamma}}_1\cdot\hat{\boldsymbol{\gamma}}_2)(\hat{\boldsymbol{\gamma}}_1\cdot\hat{\boldsymbol{q}}_2)$ |
| S → P | $(V_S/V_P)(\hat{\boldsymbol{\gamma}}_1\cdot\hat{\boldsymbol{\gamma}}_2)(\hat{\boldsymbol{\gamma}}_2\cdot\hat{\boldsymbol{q}}_1)-\frac{1}{2}(\hat{\boldsymbol{\gamma}}_2\cdot\hat{\boldsymbol{q}}_1)$ |
| S → S | $\frac{1}{2}[(\hat{\boldsymbol{\gamma}}_1\cdot\hat{\boldsymbol{\gamma}}_2)(\hat{\boldsymbol{q}}_1\cdot\hat{\boldsymbol{q}}_2)-\hat{\boldsymbol{q}}_1\cdot\hat{\boldsymbol{q}}_2+(\hat{\boldsymbol{\gamma}}_1\cdot\hat{\boldsymbol{q}}_2)(\hat{\boldsymbol{\gamma}}_2\cdot\hat{\boldsymbol{q}}_1)]$ |

where $\delta V_P/V_P$ etc. are evaluated at $\boldsymbol{r}_x$, and the Fréchet kernel for body wave delay times is:

$$K_X(\boldsymbol{r}_x) = -\frac{1}{2\pi}\sum_{\text{rays}_1}\sum_{\text{rays}_2} N(\boldsymbol{r}_x)\Omega_X\left(\frac{1}{V_1V_2}\right)^{\frac{1}{2}}\left(\frac{\mathcal{R}_{rs}}{V_r\mathcal{R}_{xr}\mathcal{R}_{xs}}\right)$$
$$\times\frac{\int_0^\infty \omega^3|\dot{m}(\omega)|^2\sin[\omega\Delta T(\boldsymbol{r}_x)-\Delta\Phi(\boldsymbol{r}_x)]\mathrm{d}\omega}{\int_0^\infty \omega^2|\dot{m}(\omega)|^2\mathrm{d}\omega}, \tag{7.28}$$

and where $\Delta\Phi = \Phi_{xs}+\Phi_{rx}-\Phi_{rs}$. As before, $V_1$ and $V_2$ are the velocities of the incoming and scattered rays at the scatterer ($V_1 = V_2$ if no wave conversion occurs). The ray summation is over all possible incoming (rays$_1$) and outgoing (rays$_2$) paths that have energy arriving in the cross-correlation window. The length of the window thus influences the effective width of the kernel. For scatterers near the surface, one should be careful to include incoming rays that hit the scatterer directly as well as those that visit the surface first. Both may have a detour time small enough to allow $\delta u$ to arrive in the cross-correlation window. Note that the Fréchet kernel should be set to zero for detour times larger than allowed by the windowing used in the cross-correlation. The same care should be taken for the outgoing rays (see Figure 7.9). Jin et al. [149] derive very similar expressions for heterogeneities expressed

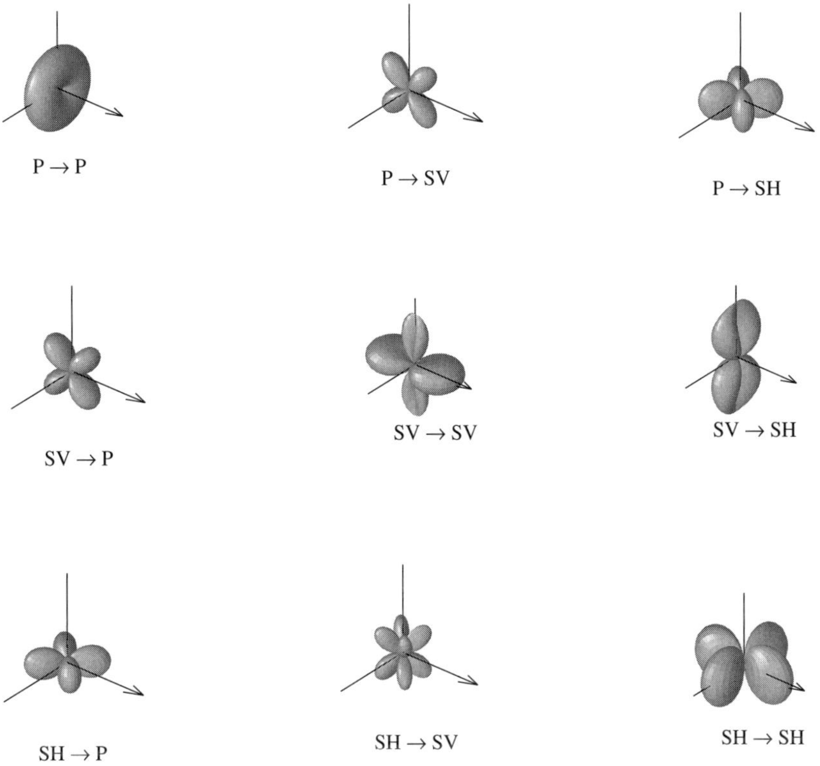


Fig. 7.10. A graphical rendition of the normalized scattering coefficient $\Omega_S$ defined by (7.26). The vector points in the direction of the incoming ray $\hat{\boldsymbol{\gamma}}_1$.

in terms of impedances $\rho V_P$ and $\rho V_S$, which may be advantageous when dealing with reflected rather than transmitted waves because it incorporates the influence of density perturbations on the amplitude in a very direct way.

The travel time is also influenced by variations in the depth to discontinuities such as the upper mantle phase transitions. We postpone a discussion of this effect to Section 13.2 where we deal with topographic corrections.

In general, we will make small errors only if we approximate $N \approx 1$ for narrow Fresnel zones, since neighbouring rays will have very similar amplitudes unless they are close to nodes in the radiation pattern. Travel times from supercritically reflected waves are rare in practice, and if they occur the shifts may be similar for direct and scattered waves. With one important exception which we have already encountered in Section 6.4: surface reflected waves such as PP or SS undergo a 90° phase shift (a Hilbert transform) so their waveform will be very different from

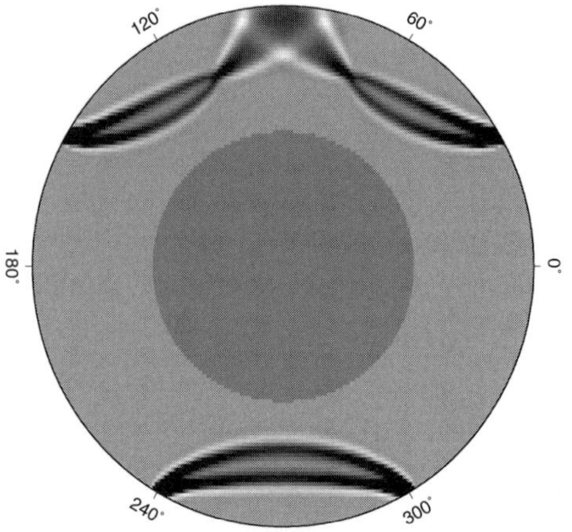


Fig. 7.11. Two examples of kernels $K_P(\boldsymbol{r}_x)$ for long-period (20 s) teleseismic P-waves in the Earth's mantle. The kernel in the Northern hemisphere is for a surface reflected PP-wave at $\Delta = 120°$, the shorter kernel in the Southern hemisphere for a P-wave at $60°$. Darker greyscale indicates more negative values of the kernel, the whitish regions have a positive value for the kernel, implying a positive delay for a positive velocity perturbation. Such 'reverse' sensitivities are located in the second Fresnel zone. Note the region of reduced sensitivity at the centre of the kernels, except near the reflection point of PP. The extra complexity of the PP kernel is caused by a $90°$ phase shift at the caustic, as well as by the fact that scattered waves may also reflect from the surface. The dark shading of the Earth's core does not indicate a sensitivity.

a scattered wave that has escaped the passage through a caustic. In this case, a scatterer on the ray will have a maximum effect on $\delta T$ and the hole in the doughnut will disappear. This is visible in Figure 7.11, where we compare $K_P$ for a PP-wave (top) with a P-wave (bottom).

If the kernel is confined to a narrow volume around the ray, which it usually is if the frequency band is not too narrow and the dominant frequency is not too low, we may equally well approximate $\Omega_X \approx 1$, which leads to savings in computation time. At any particular location of the ray, the ratio of the frequency integrals is only a function of $\Delta T$ and can be interpolated. The directional properties of $\Omega_X$ imply that the influence of scattered waves on the travel time may be different among different components of the seismogram. For example, a scattered P-wave that comes in vertically, will only influence the travel time on the vertical component. Similarly, the S-wave scattered from this same direction will only be visible on the horizontal.

The validity of the Born approximation depends on the phase change being linearizable by summing $\delta u$ to the background field $u$ – this is essentially a first-order Taylor expansion of the phase perturbation $\delta\phi$ of $u(\omega)$: $e^{i\delta\phi} \approx 1 + i\phi$. Clearly this goes completely wrong as $\phi \to \pi/2$. It seems therefore that delays must be significantly smaller than a quarter period of the wave for Born to be valid. This is, however, a pessimistic point of view. Woodward [403] compares the Born approximation with the Rytov approximation in which the phase itself rather than the displacement $\boldsymbol{u}$ is linearized, and finds very similar Fréchet kernels. In their textbook on medical tomography, Kak and Slaney [155] conclude that the validity of Born extends beyond the $\pi/2$ limit and gives good results even for phase changes as large as $\pi$. Apparently, the neglected higher-order scattering has a beneficial effect! Strictly speaking, the elastic wave equation with different velocities for P- and S-waves, cannot be treated with the Rytov approximation. However, once we assume the background medium smooth and adopt ray theory for the Green's functions, the Rytov approximation can be applied to the individual wavetypes. This gives hope that the Born approximation gives good results for delays as large as half a period and may be useful even to interpret delays twice as large.

While delays will generally be smaller than the wave period for a 20 s wave from a broadband seismometer, linearity may be a problem for times observed from short-period seismometers with a maximum response at 1 Hz, even if a 'local' one-dimensional model is adopted. The only solution in that case is to adopt a 3D model that is smooth enough to satisfy ray theory and brings predicted travel times within the allowed misfit.

Skarsoulis and Cornuelle [321] derive Fréchet kernels for ocean acoustic tomography. Their kernels are equivalent to those in the elastic case for $V_S = 0$. Comparison with forward calculations even for frequencies as high as 100 Hz shows that deviations from ray theory are large enough to warrant a finite-frequency treatment for ocean acoustic tomography.

### *Exercise*

**Exercise 7.4** Consider an SH-wave scattering from a point-like heterogeneity that is located at shallow depth to the North of a station. Which component will be most influenced by the scattered energy? Could this lead to apparent anisotropy?

## 7.6 Alternative arrival time measurements

The robust delay time estimator (7.12) leads to a different kernel than the estimator (7.6) based on the maximum in the cross-correlation. Applying Parseval's theorem to (7.12):

$$\delta T = \int_{-\infty}^{\infty} W(t)\delta\gamma(t)\mathrm{d}t = \int_{-\infty}^{\infty} W(\omega)^*\delta\gamma(\omega)\mathrm{d}\omega \tag{7.29}$$

where

$$W(\omega) = \frac{\mathrm{i}\omega\gamma(\omega)}{\int \omega^2 \gamma(\omega)^* \gamma(\omega) \mathrm{d}\omega}.$$

Again with Parseval, one easily establishes that $\gamma(\omega) = u(\omega)u(\omega)^*$ and $\delta\gamma(\omega) = \delta u(\omega)u(\omega)^*$, with $u$ and $\delta u$ given by (7.22) and (7.24); inserting all this into (7.29) yields the Fréchet kernel for robust cross-correlation estimates:

$$K_X^{\text{robust}}(\boldsymbol{r}_\text{x}) = -\frac{1}{2\pi} \sum_{\text{rays}_1} \sum_{\text{rays}_2} N(\boldsymbol{r}_\text{x})\Omega_X \left(\frac{1}{V_1 V_2}\right)^{\frac{1}{2}} \left(\frac{\mathcal{R}_\text{rs}}{V_\text{r}\mathcal{R}_\text{xr}\mathcal{R}_\text{xs}}\right)$$
$$\times \frac{\int_0^\infty \omega^3 |\dot{m}(\omega)|^4 \sin[\omega\Delta T(\boldsymbol{r}_\text{x}) - \Delta\Phi(\boldsymbol{r}_\text{x})]\mathrm{d}\omega}{\int_0^\infty \omega^2 |\dot{m}(\omega)|^4 \mathrm{d}\omega}. \tag{7.30}$$

A comparison with (7.28) shows that the only difference is in the power of $|\dot{m}(\omega)|$. The robust estimate puts more emphasis on the dominant frequency. This may lead to unwanted 'ringing' effects at the outer edge of the kernel because the higher Fresnel zones are less effectively damped out when the spectrum is narrow.

In refraction seismology, where the signal-to-noise ratio is often low, the arrival of the seismic wave is sometimes defined as the arrival time of the first maximum in the P-wavetrain, which is easier to pick by hand. We can use Born theory to develop a finite-frequency interpretation of such data as well. Again, assume that $u(t)$, which has its maximum at the 'arrival time' defined as $t = T_\text{rs}$, is perturbed by $\delta u(t)$. The shift $\delta T$ in the maximum is then found from:

$$\dot{u}(T_\text{rs} + \delta T) + \delta\dot{u}(T_\text{rs} + \delta T) = 0,$$

or to first order, using $\dot{u}(T_\text{rs}) = 0$:

$$\delta T = -\frac{\delta\dot{u}(T_\text{rs})}{\ddot{u}(T_\text{rs})}.$$

Using the expressions (7.21) and (7.23) and assuming all phase shifts $\Phi$ to be zero, we find:

$$K_X^{\text{refr}}(\boldsymbol{r}_\text{x}) = -\frac{1}{2\pi} \sum_{\text{ray1}} \sum_{\text{ray2}} N(\boldsymbol{r}_\text{x})\Omega_X \left(\frac{1}{V_1 V_2}\right)^{\frac{1}{2}} \left(\frac{\mathcal{R}_\text{rs}}{V_\text{r}\mathcal{R}_\text{xr}\mathcal{R}_\text{xs}}\right) \frac{m^{(4)}(-\Delta T)}{\dddot{m}(0)},$$

where $m^{(4)}$ denotes the fourth time derivative of $m$. The zero phase shift assumption is reasonable since picks of maxima are almost always done on the first arriving P- (and sometimes S-) waves, which have not yet passed a caustic. The theory is easily generalized to include phase shifts, though. Because the observed displacement signal is the first time derivative of $m(t)$, the time derivatives in the expression for

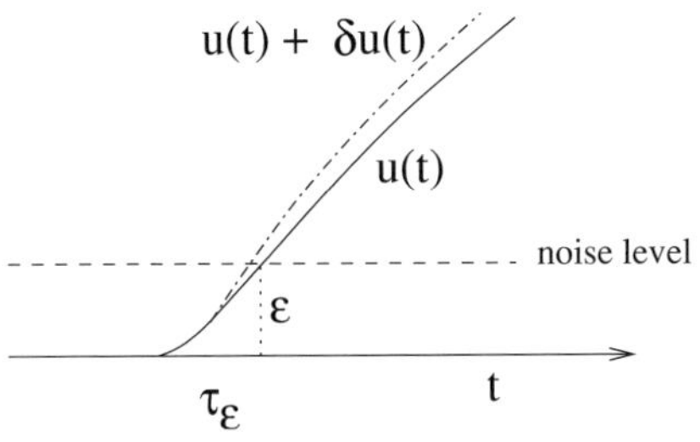


Fig. 7.12. Frequency dependent effects on a picked travel time will occur because it takes a finite time ($\tau_\epsilon$) for the signal to rise above the noise level $\epsilon$. The addition of a scattered wave perturbs the seismogram (broken line) and influences this delay in the time pick if the scattered wave arrives within $\tau_\epsilon$.

the kernel are not as high as they may seem at first sight. In fact, $\dddot{m}(0)$ is only the first derivative of the observed signal if this is recorded as ground velocity.

Note that if the arrival time is defined by the maximum value of $u(t)$, this defines zero delay as the time of the maximum of $\dot{m}(t)$. For example, if $u(t)$ has the shape of the second derivative of a Gaussian ('Ricker'-wavelet) with a width of $\alpha$ seconds:

$$\dot{m}(t) = \left(1 - \frac{2t^2}{\alpha^2}\right) \mathrm{e}^{-t^2/\alpha^2},$$

then one establishes by differentiation that:

$$\frac{m^{(4)}(-\Delta T)}{\dddot{m}(0)} = \frac{2\Delta T}{3\alpha^6}\left(15\alpha^4 - 20\alpha^2 \Delta T^2 + 4\Delta T^4\right) \mathrm{e}^{-\Delta T^2/\alpha^2}. \tag{7.31}$$

Can we use a similar reasoning to analyse the finite-frequency interpretation of a true arrival 'pick'? Stark and Nikolayev [342] point out that the 'picked' arrival is not that of the actual arrival time, but some fraction $\tau_\epsilon$ later when the incoming wavelet exceeds a (noise) threshold $\epsilon$. The shift $\tau_\epsilon$ depends on how fast the pulse rises, and thus on the dominant frequency of the signal. The situation is sketched in Figure 7.12. For the background model with wavefield $u(t)$ we would pick the arrival $\tau_\epsilon$ seconds late if $u(T_{\rm rs} + \tau_\epsilon) = \epsilon$. As can be seen in Figure 7.12, $\tau_\epsilon$ will be reduced if a positive scattered wave is added to $u(t)$. This shift is equal to $\delta\tau_\epsilon = -\delta u(\tau_\epsilon)/u(\tau_\epsilon)$. We will thus observe a finite-frequency effect even in picked data. Statistical support for the view that noise level influences the measurement of teleseismic time delays was provided by Grand [122], see also Section 6.1.

Unfortunately, we cannot apply the same treatment as in the case of picking the first maximum since the observed change is a change in the error $\tau_\epsilon$ rather than in a meaningful quantity. We also note that $\tau_\epsilon$ must be much less than the magnitude of the expected time delays, or else the picking of an arrival time would make no sense, and one would rather revert to the much more accurate cross-correlation techniques. If $\delta u$ arrives later than $\tau_\epsilon$ it will have no effect on the observed pick time. In this case there seems to be no other option than to live with the errors and simply apply ray theory. It is however remarkable that even picks of the onset are in principle affected by a finite volume around the unperturbed ray.

### *Exercise*

**Exercise 7.5** A doughnut-hole conundrum: do picked arrival times (either maxima or onsets) have sensitivity to structure on the geometrical ray in the Born approximation?

## 7.7 Alternative methods for kernel computation

Gautier et al. [109] compute finite-frequency kernels in a 3D crustal environment using the method of graph theory and ray bending outlined in Section 3.2, followed by 3D dynamic ray tracing in Cartesian coordinates as described by Virieux [385]. The ability to use 3D background models allows one to iterate and escape the limitations of Born theory. Conceivably, the existence of minor caustics in a 3D model could hamper a proper convergence – a fear raised by de Hoop and van der Hilst [79], countered by Dahlen et al. [77] – but no such problems were encountered by Gautier et al. who report convergence in five nonlinear iterations, despite the occurrence of large velocity contrasts.

Though extremely fast, the use of ray theory to compute the Fréchet kernels has some limitations. Ray theory does not handle headwaves or diffracted waves, so other techniques must be used to compute, for example, waves diffracted at the core–mantle boundary of the Earth. What is needed is an algorithm to compute the perturbed waveform $\delta u$. For waves diffracted at shallow levels, one option is to fall back on the original surface-wave mode summation kernels used by Marquering et al. [196, 195]. This has the added advantage of incorporating reverberations and wave conversions. The necessary theory for this can be found in Chapter 11, where Equation (11.5) gives an expression for $\delta u$. Substituting this into (7.8) allows one to compute Fréchet kernels for travel times of arbitrarily selected time windows. Because the modelling of P-waves by mode summation requires the computation of a very large number of higher modes with high phase velocities, this strategy is only effective for multiple S-waves that have their energy mostly confined to the upper mantle.

For core-diffracted waves, the discrete mode summation by Zhao et al. [418] has been used by Kárason and van der Hilst [156], but it is very time consuming. The reader is referred to Appendix C for more theoretical details. Since the computation of discrete normal modes tends to become unstable at frequencies above about 0.1 Hz, this method also has its limitations for modelling P-wave kernels. As an alternative, Nissen-Meyer and Dahlen [231] propose to use a 2D spectral element algorithm to compute the wavefield for a series of sources at all depth levels needed to compute $\delta u$.

## 7.8 Computational aspects

How does one efficiently compute the geometrical spreading and the detour time $\Delta T(\boldsymbol{r}_{\mathrm{x}})$? The shortest path method outlined in Chapter 3 makes this straightforward (but time consuming) in heterogeneous, 3D media. If we place the source in $\boldsymbol{r}_{\mathrm{s}}$, the travel times to all other nodes (i.e. $T_{\mathrm{xs}}$) are known, as well as the travel time $T_{\mathrm{rs}}$ to the receiver node $\boldsymbol{r}_{\mathrm{r}}$. A second application of the algorithm, but now with the source placed in the receiver location $\boldsymbol{r}_{\mathrm{r}}$, gives $T_{\mathrm{xr}}$ to every node in the model, so we have all the ingredients to compute $\Delta T$ from (7.19). To compute the geometrical spreading factors we need to use dynamic ray tracing along the (bent) rays. A software package to do this in heterogeneous media of limited extent is available from the software repository.

Note that it is imperative that the background model itself is smooth enough to allow us to compute the unperturbed signal $u(t)$ by means of ray theory. If the model is spherically symmetric, the formalism outlined in Sections 2.6 and 2.7 can be used to compute travel times, epicentral distance and geometrical spreading as a function of ray parameter for sources at any depth between centre and surface. The geometrical spreading can be obtained by finite differencing (2.38) or integrating (5.2). Calvet and Chevrot [37] set up a table of such values and interpolate.

An efficient method to compute $\mathcal{R}$ directly is provided by dynamic ray tracing, which can then also be used to estimate the detour time $\Delta T$ and the geometrical spreading factors for scatterers in the neighbourhood of the direct ray. This assumes the scatterer is reasonably close to the direct ray – an assumption that is violated for some rays such as PP near the antipode, for which rays that leave the source in all azimuths arrive with only a small delay. Again, an additional gain in efficiency can be obtained by choosing the background model layered, or spherically symmetric. Here we shall illustrate this for the case of spherical symmetry. As we have seen earlier (Equations 3.6 and 5.1), both the travel time in the neighbourhood of the ray and the geometrical spreading can be obtained by solving $\boldsymbol{H}$ from the Riccati equation (3.9). In particular, (3.6) gives:

$$\Delta T_{\mathrm{xs}} = \frac{1}{2}\boldsymbol{q} \cdot \boldsymbol{H}\boldsymbol{q} \,,$$

where $\boldsymbol{H}$ is evaluated at the location $\boldsymbol{r}_\mathrm{x}$ of the scatterer. Similarly, the Riccati equations give us $\mathcal{R}$ using:

$$\ln \mathcal{R}_{\mathrm{xs}}{}^2 = \int_0^x c\, \mathrm{tr}(\boldsymbol{H})\mathrm{d}s \qquad \text{(5.2 again).}$$

If both the source and receiver are at the surface of the Earth, the same $\boldsymbol{H}$ can be used for both $T_{\mathrm{rx}}$ and $T_{\mathrm{sx}}$, provided we reverse the sign (since the $s$-axes are in different directions). More likely is that the source is at some depth below the surface, so this would not work. Yet we may obtain a gain in efficiency by distilling the reverse $\boldsymbol{H}$ from one and the same (forward) integration. The necessary formalism was provided by Farra and Madariaga [99] and Coates and Chapman [63].

Recall the Hamiltonian system for the Hessian $\boldsymbol{H} = \boldsymbol{P}\boldsymbol{Q}^{-1}$ we encountered in Chapter 3 (Eq. 3.12): $\mathrm{d}\boldsymbol{Q}/\mathrm{d}s = c\boldsymbol{P}$, $\mathrm{d}\boldsymbol{P}/\mathrm{d}s = -c^{-2}\boldsymbol{V}\boldsymbol{Q}$. In the spherically symmetric Earth, $\boldsymbol{P}$ and $\boldsymbol{Q}$ are diagonal matrices with entries $P_1$, $P_2$, $Q_1$ and $Q_2$. Since we have four variables, there are four independent solutions. In Chapter 3 we considered only two independent initial conditions (3.15). This can be formulated as:

$$\frac{\mathrm{d}}{\mathrm{d}s}\begin{pmatrix}\boldsymbol{Q}\\ \boldsymbol{P}\end{pmatrix} = \frac{\mathrm{d}}{\mathrm{d}s}\begin{pmatrix} Q_1 & 0\\ 0 & Q_2\\ P_1 & 0\\ 0 & P_2\end{pmatrix} = \begin{pmatrix} 0 & 0 & -c^{-2}V_{11} & 0\\ 0 & 0 & 0 & -c^{-2}V_{22}\\ c & 0 & 0 & 0\\ 0 & c & 0 & 0\end{pmatrix}\begin{pmatrix} Q_1 & 0\\ 0 & Q_2\\ P_1 & 0\\ 0 & P_2\end{pmatrix}.$$

We now add two more initial conditions, which we indicate with a tilde, and which give rise to independent solutions $\tilde{\boldsymbol{Q}}$ and $\tilde{\boldsymbol{P}}$:

$$\tilde{\boldsymbol{P}} = \begin{pmatrix}\tilde{P}_1 & 0\\ 0 & \tilde{P}_2\end{pmatrix} \qquad \tilde{\boldsymbol{Q}} = \begin{pmatrix}\tilde{Q}_1 & 0\\ 0 & \tilde{Q}_2\end{pmatrix},$$

with initial conditions:

$$\tilde{P}_1(0) = \tilde{P}_2(0) = 0, \qquad \tilde{Q}_1(0) = \tilde{Q}_2(0) = 1\,,$$

and rearrange the four independent solutions in a matrix $\mathcal{P}$:

$$\mathcal{P}(s,0) = \begin{pmatrix}\tilde{Q}_1 & 0 & Q_1 & 0\\ 0 & \tilde{Q}_2 & 0 & Q_2\\ \tilde{P}_1 & 0 & P_1 & 0\\ 0 & \tilde{P}_2 & 0 & P_2\end{pmatrix}.$$

Since $\mathcal{P}(0,0) = \boldsymbol{I}$, $\mathcal{P}$ is a propagator matrix [115], i.e. when multiplied with arbitrary initial conditions $\boldsymbol{f}(0)$, the solution $\boldsymbol{f}(s) = \mathcal{P}(s,0)\boldsymbol{f}(0)$. Its inverse can be found from the following equalities (see Exercise 7.6):

$$\begin{aligned} \tilde{\boldsymbol{Q}}\boldsymbol{P} - \boldsymbol{Q}\tilde{\boldsymbol{P}} &= \boldsymbol{I} \\ \boldsymbol{P}\tilde{\boldsymbol{Q}} - \tilde{\boldsymbol{P}}\boldsymbol{Q} &= \boldsymbol{I}\,. \end{aligned} \tag{7.32}$$

The inverse of $\mathcal{P}$ is then found to be:

$$\mathcal{P}(s,0)^{-1} \equiv \mathcal{P}(0,s) = \begin{pmatrix} P_1 & 0 & -Q_1 & 0 \\ 0 & P_2 & 0 & -Q_2 \\ -\tilde{P}_1 & 0 & \tilde{Q}_1 & 0 \\ 0 & -\tilde{P}_2 & 0 & \tilde{Q}_2 \end{pmatrix}.$$

All of the elements in the inverse matrix are known if we solve the equations in the forward direction with four sets of initial conditions. Combining $\mathcal{P}$ and its inverse, we obtain the propagator that gives us the solution from $s = L$ (i.e. $\phi = \Delta$), the 'backwards solution':

$$\mathcal{P}(s,L) = \mathcal{P}(s,0)\mathcal{P}^{-1}(L,0)\,.$$

This can be used to do both the forward and the backward integration during the forward sweep along the ray, avoiding repeated evaluation of the same coefficients while obtaining $T_{\rm xs}$ and $T_{\rm rs}$ as well as $\mathcal{R}_{\rm xs}$ and $\mathcal{R}_{\rm xr}$, at the same time. In practice, using $\phi$ as variable instead of $s$, as in (3.13), may be preferable, except when the raypath is almost vertical. A more complete derivation of the propagator matrix formalism is also given by Dahlen and Baig [75].

Finally, we may avoid the integration of $\mathrm{tr}(\boldsymbol{H})$ to obtain $\mathcal{R}$. Snieder and Chapman [332] point out that the direct and reverse rays satisfy the same Riccati equations, with $\mathrm{d}/\mathrm{d}s \to -\mathrm{d}/\mathrm{d}s$ for the reversed ray:

$$\begin{aligned} \frac{\mathrm{d}\boldsymbol{H}_{\rm xs}}{\mathrm{d}s} + c\boldsymbol{H}_{\rm xs}^2 &= -\frac{1}{c^2}\boldsymbol{V} \\ -\frac{\mathrm{d}\boldsymbol{H}_{\rm xr}}{\mathrm{d}s} + c\boldsymbol{H}_{\rm xr}^2 &= -\frac{1}{c^2}\boldsymbol{V}\,. \end{aligned}$$

Subtracting these equations shows that a proper combination of the two $\boldsymbol{H}$s should be independent of the velocity derivatives in $\boldsymbol{V}$. In fact, substitution shows that:

$$\frac{\mathrm{d}}{\mathrm{d}s}\det(\boldsymbol{H}_{\rm xs}+\boldsymbol{H}_{\rm xr}) + c\,\mathrm{tr}(\boldsymbol{H}_{\rm xs}-\boldsymbol{H}_{\rm xr})\det(\boldsymbol{H}_{\rm xs}+\boldsymbol{H}_{\rm xr}) = 0\,,$$

such that

$$\int c\,\mathrm{tr}(\boldsymbol{H}_{\rm xs}-\boldsymbol{H}_{\rm xr})\mathrm{d}s = -\ln\det(\boldsymbol{H}_{\rm xs}+\boldsymbol{H}_{\rm xr})\,.$$

But we have also (see 5.1):

$$\frac{d\mathcal{R}_{xs}}{ds} = \frac{1}{2}c\ \mathrm{tr}(\boldsymbol{H}_{xs})\mathcal{R}_{xs}\,,$$
$$\frac{d\mathcal{R}_{xr}}{ds} = -\frac{1}{2}c\ \mathrm{tr}(\boldsymbol{H}_{xr})\mathcal{R}_{xr}\,,$$

such that (integrate and add):

$$\int c\ \mathrm{tr}(\boldsymbol{H}_{xs} - \boldsymbol{H}_{xr})ds = \ln(\mathcal{R}_{xs}\mathcal{R}_{xr})^2\,.$$

The two expressions imply that

$$\ln(\mathcal{R}_{xs}\mathcal{R}_{xr})^2 = -\ln\det(\boldsymbol{H}_{xs} + \boldsymbol{H}_{xr})\,,$$

or

$$(\mathcal{R}_{xs}\mathcal{R}_{xr})^{-1} = \text{constant} \times \sqrt{|\det(\boldsymbol{H}_{xs} + \boldsymbol{H}_{xr})|}\,.$$

The constant can be determined from the limiting values of $\mathcal{R}$ in a small region around the source and turns out to be $V_r/\mathcal{R}_{rs}$, so that:

$$\frac{\mathcal{R}_{rs}}{V_r\mathcal{R}_{xs}\mathcal{R}_{xr}} = \sqrt{|\det(\boldsymbol{H}_{xs} + \boldsymbol{H}_{xr})|}\,, \tag{7.33}$$

which can be used directly in the expressions for the kernels $K_X$.

When the ray passes a caustic, its Maslov index $M$ increases by 1. This is also the point in space where the curvature of the wavefront changes signs, as the rays go from a convergent to a divergent geometry or vice versa. The curvature of the wavefront is given by the eigenvalues of $\boldsymbol{H}$, and one suspects a simple relationship between the two. In fact, the following equality exists between the Maslov indices and the eigenvalue structure of the matrix $\boldsymbol{H}$:

$$\Delta M = M_{xs} + M_{xr} - M_{rs} = \frac{1}{2}[\mathrm{sig}(\boldsymbol{H}_{xs} + \boldsymbol{H}_{xr}) - 2]\,,$$

where sig[.] is the number of positive minus the number of negative eigenvalues. $\Delta M = 0$ for direct waves like P (or p in the Sun) and S, as well as for reflected waves like pP and sP.

Program `raydyntrace.f` (Tian et al. [361]), available from the software repository, can be used to compute all quantities needed for the computation of the Fréchet kernels (7.28).

### *Quadrature*

Depending on the sign of $H_{11} = \partial^2 T/\partial q_1^2$, the shape of the kernel cross-section is either elliptical $H_{11} > 0$ or hyperbolic $H_{11} < 0$ – this assumes the background model is spherically symmetric such that the horizontal wavefront curvature $H_{22} > 0$. It may be beneficial, both for purposes of accuracy and efficiency, to transform from ray coordinates $q_1$ and $q_2$ to elliptical or hyperbolic coordinates when computing the integral (7.27). Since the detour time satisfies

$$\Delta T = \frac{1}{2}(H_{11}q_1^2 + H_{22}q_2^2)\,,$$

the introduction of elliptical coordinates $\rho$ and $\theta$ such that

$$q_1 = \sqrt{2/H_{11}}\,\rho\cos\theta\,,$$

$$q_2 = \sqrt{2/H_{22}}\,\rho\sin\theta\,,$$

transforms a surface element $\mathrm{d}q_1\mathrm{d}q_2$ into a surface element $2\rho/\sqrt{H_{11}H_{22}}\,\mathrm{d}\rho\mathrm{d}\theta$. Similarly, when $H_{11} < 0$ we may use hyperbolic coordinates.

The precision of the quadrature over the frequency integral can be checked by using an analytical expression for a Gaussian power spectrum of the form (Hung et al. [139]):

$$|\dot{m}(\omega)|^2 = (\omega^2\tau^2/2\pi)\exp(-\omega^2\tau^2/4\pi^2)\,,$$

which represents a waveform in the time domain with a characteristic period $\tau$. Favier and Chevrot [100] give analytical expressions for the frequency integral ratio in the case that the time function is a Gaussian or a derivative of a Gaussian. Such filters lead to compact sensitivity kernels in which the higher Fresnel zones are suppressed. For the Gaussian spectrum above, the ratio is given by:

$$\frac{\int_0^\infty \omega^3|\dot{m}(\omega)|^2\sin\omega\Delta T\,\mathrm{d}\omega}{\int_0^\infty \omega^2|\dot{m}(\omega)|^2\mathrm{d}\omega} = \frac{8\Delta T\exp\left(-\frac{\Delta T^2\pi^2}{\tau^2}\right)\pi^6\left(\Delta T^4 - \frac{5\tau^2\Delta T^2}{\pi^2} + \frac{15\tau^4}{4\pi^4}\right)}{3\tau^6}\,,$$

and this can be used to validate code written for more general types of filters (Tian et al. [362]). Near source and receiver, as $\mathcal{R}_{\mathrm{xr}}$ or $\mathcal{R}_{\mathrm{xs}} \to 0$ the numerical value of the kernel becomes very large. Though it remains finite, the actual value of the limit depends on the direction from which one approaches this singular point. The behaviour of the kernel is very different near the ray, where $\Delta T$ is always small so it behaves smoothly, and behind the singular point, where the detour time $\Delta T$ increases rapidly with distance and the kernel oscillates heavily. Though these singularities are integrable, they can easily lead to large numerical errors. An empirical way to address this problem is to use ray theory to compute the sensitivity of the time delay to velocity perturbations of a small volume near the singularity, and to assume only a pure impedance effect for the amplitude. This is motivated by the consideration that the region near the source may considered to be homogeneous, for which a local analytical solution of the form (4.4) can be used, neglecting near-field terms for the elastic waves. Similar numerical problems may occur near caustics and here too a local assumption of homogeneity with validity of ray theory may avoid serious numerical complications.

Though the paraxial formalism is very efficient, it has some shortcomings. The first is the validity of the paraxial approximation. Hung et al. [138] tested a combination of approximations: paraxial and $N = \Omega_X = 1$ against ground truth from synthetic seismograms using a pseudospectral code and found that errors were an order of magnitude smaller than typical observational errors for kernels that were almost 1000 km wide. Discontinuities in the real Earth invariably cause a wavefront to violate the assumption of a quadratic dependence of $\Delta T$ on $\boldsymbol{q}$, but the effects of these were shown to be negligible by Tian et al. [362]. If care is taken to suppress higher Fresnel zones by a careful choice of filter – the Gaussian filter works very well – the major restriction is that PP- or SS-waves cannot be close to the antipode, where the focusing of rays from different azimuths implies that the observed wave is sensitive to heterogeneity far away from the ray and any 'paraxial' assumption breaks down completely. Tian et al. [361, 362] study the accuracy of the paraxial approximation and give diagnostics to test computer codes.

## Exercises

**Exercise 7.6** Prove (7.32) by differentiating $\tilde{\boldsymbol{Q}}\boldsymbol{P} - \boldsymbol{Q}\tilde{\boldsymbol{P}}$ and $\boldsymbol{P}\tilde{\boldsymbol{Q}} - \tilde{\boldsymbol{P}}\boldsymbol{Q}$ with respect to $s$ and applying the initial conditions. (Hint: diagonal matrices commute).

**Exercise 7.7** Consider a straight ray in a homogeneous medium with velocity $c$. Using ray coordinates $(\ell, q_1, q_2)$, show that to first order the travel time $T$ satisfies $cT = \ell + (q_1^2 + q_2^2)/2\ell$. Use this to show that for a ray of length $L$ we have

$$\sqrt{|\det(\boldsymbol{H}_{\rm xs} + \boldsymbol{H}_{\rm xr}|} = \frac{1}{c}\left|\frac{1}{\ell} + \frac{1}{L-\ell}\right|$$

(Hint: use Equation 7.33).

**Exercise 7.8** What values can $\Delta M$ take?

# 8

# Body wave amplitudes: observation and interpretation

Body wave amplitudes are influenced by three factors: the loss of energy by attenuative effects, the focusing or defocusing of rays and the local impedance (the product of density and velocity). The effect of perturbations in density or impedance can easily influence the amplitude of a wave, but it is essentially a local effect, independent of the larger-scale structure of the Earth, and we shall treat impedance variations with the method of corrections detailed in Chapter 13. Here we concentrate on focusing and attenuation.

In contrast to the numerous studies of delay times, body wave *amplitude* studies are rare. Early regional studies, resulting in 1D models for the attenuation beneath a particular province, were done by Solomon and Toksöz [336], Jordan and Sipkin [153] and Lay and Helmberger [174]. Such studies made use of the fact that the low$-Q$ asthenosphere below the source and receiver causes a major part of the body wave attenuation. Since teleseismic rays in the upper mantle travel close to the vertical, differences in amplitude observed at stations for the same event can be attributed to differences in the strength and/or thickness of the asthenosphere approximately beneath the station. This allows us to see strong differences in attenuation between different regions. Though studies of this kind continue to provide insight (e.g. Warren and Shearer [391]), this method is not really tomographic. Sanders et al. [301] and Ho-Liu et al. [134] were the first to apply the methods of tomography to attenuation data. Global studies using body wave amplitudes have been almost nonexistent, with the notable exception of tomographic upper mantle models by Bhattacharyya et al. [21] and Reid et al. [273] and a full mantle model for $Q_S$ by Lawrence and Wysession [173]. All body wave attenuation studies so far have used ray theory and have not taken focusing or defocusing into account, in spite of clear observations of the important role of focusing: Butler [36] observed a strong correlation between early arrival times and low amplitudes in the western US, the opposite of what one would expect for a hot upper mantle with both low $Q$ and low velocity. Neele et al. [228] estimate that focusing effects in this region overshadow the influence of attenuation. Ritsema et al. [279] observe focusing effects in global S/SS amplitude ratios.

Attempts to interpret focusing effects have been rare. Early efforts to attack the problem of focusing beneath seismic arrays by Haddon and Husebye [127], Moore [217] and Thomson [357] have had no serious follow-up in regional or global studies. Sensitivity kernels for amplitude focusing using the ray perturbation theory of Farra and Madariaga [99] are derived by Nowack and Lutter [245] for transmitted waves and by Nowack and Lyslo [246] for waves that are reflected or converted at an interface. Neele et al. [227] also use ray theory, whereas Thomson [357] applies perturbations to the velocity to the bending equations (3.3). Such ray-theoretical formalisms have met with only limited success in efforts to invert for local structure. Neele et al. [228]) blame the very strong sensitivity of the geometrical spreading to perturbations in velocity as a negative factor influencing the interpretation, and raise the question of the validity of ray theory for short-wavelength perturbations. One would expect wavefront healing to be effective in 'healing' strong amplitude differences along the wavefront. For example, Nolet and Dahlen [239] show that anomalies on the wavefront satisfy a diffusion equation, which causes both time and amplitude anomalies to spread out much like a thermal anomaly does.

Van der Lee et al. [376] modelled amplitudes of converted phases at the 660 km discontinuity with a finite-frequency approach but did not attempt to invert anomalies directly beyond a trial-and-error approach. Allen et al. [6] modelled fluctuations in $t^*$ for waves crossing the Iceland plume and showed that finite-frequency effects are important. Ritsema et al. [279] compare observed ratios of long period SS/S and PP/P amplitudes with synthetics computed for the 3D tomographic model S20RTS, and report variations of the order of 10% both for the observed and the modelled ratios, although crustal thickness dominates the signal for the compressional waves. No formal finite-frequency inversion of amplitudes has yet been done, although first efforts are forthcoming.

In this chapter we show that finite-frequency effects can be taken into account to first order, and can deal simultaneously with focusing and attenuation effects.

## 8.1 Amplitude observations

The traditional amplitude observation – still very much in use to quickly determine the magnitude of an event – is to measure the peak-to-peak amplitude of the largest swing of a P- or S-wave and translate this into a displacement or velocity, usually measured in microns or microns/s. This is not a very stable datum, since it is easily influenced by local noise or wave interferences. Instead we prefer to define the amplitude of a body wave pulse as a root-mean-square (RMS) average over the pulse length $T_\mathrm{p}$ as:

$$A = \sqrt{\frac{1}{T_\mathrm{p}} \int_0^{T_\mathrm{p}} u(t)^2 \mathrm{d}t}. \tag{8.1}$$

If the wavefield is perturbed from $u(t) \rightarrow u(t) + \delta u(t)$, we find, to first order:

$$\begin{aligned} A + \delta A &= \sqrt{\frac{1}{T_\mathrm{p}} \int_0^{T_\mathrm{p}} [u(t) + \delta u(t)]^2 \mathrm{d}t} \\ &\approx \sqrt{\frac{1}{T_\mathrm{p}} \int_0^{T_\mathrm{p}} u(t)^2 \mathrm{d}t} \left( 1 + \frac{\int_0^{T_\mathrm{p}} u(t)\delta u(t)\mathrm{d}t}{\int_0^{T_\mathrm{p}} u(t)^2 \mathrm{d}t} \right), \end{aligned} \tag{8.2}$$

or, interpreting $\delta A$ as the difference between an observed and a predicted amplitude from a synthetic waveform $A_0$ for the background model and applying Parseval's theorem for real signals (2.70):

$$\frac{\delta A}{A_0} = \delta \ln A = \frac{A_\mathrm{obs} - A_0}{A_0} = \frac{\mathrm{Re} \int_0^\infty u(\omega)^* \delta u(\omega) \mathrm{d}\omega}{\int_0^\infty u(\omega)^* u(\omega) \mathrm{d}\omega}. \tag{8.3}$$

For deep earthquakes, the P-wave arrives well separated from the surface ghosts pP and sP and the amplitude measurement involves a relatively simple integration of energy once the seismogram is corrected for the instrument response. Figure 8.1 shows an example of a plot of raw amplitude data, i.e. before the application of any corrections, compared to the amplitude predicted by the source radiation pattern. Figure 8.2 shows the variation of amplitudes of P-waves observed from deep earthquakes for two different durations of the P-wave itself. These amplitudes have been corrected for attenuation near the receiver and are representative for focusing and defocusing effects (see Chapter 13). In this figure, the solid line represents a subset of P-wave arrivals with the bulk of the energy below 0.1 Hz, whereas the dotted line shows all data (the study was limited to strong arrivals with a duration of at least 4.5 s). Clearly, low frequency P-wave amplitudes deviate less from predicted amplitudes, as expected if wavefront healing for amplitudes plays a role.

For shallow earthquakes, the interaction of P with pP and sP in the same time window causes the waveform to change depending on the radiation characteristics and the epicentral distance, and (8.1) cannot be used to estimate an unambiguous amplitude. One should therefore apply Sigloch's method [313] to interpret amplitudes of shallow earthquakes.

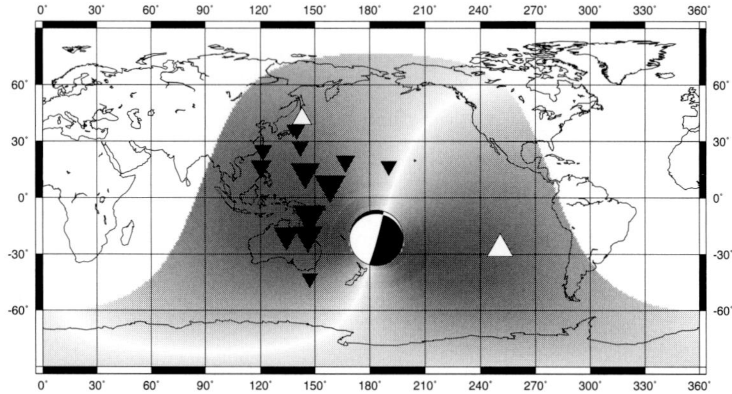


Fig. 8.1. The radiated amplitude factor for the P-wave as predicted by the Harvard Centroid Moment Tensor catalogue for the Fiji Islands Region event on April 13, 1999, mb = 6.8, 164 km depth. The amplitude is projected in greyscale onto the Earth's surface (the predicted amplitudes East of the nodal plane are positive, West are negative). White triangles represent observed dilatational (positive) arrivals, black triangles represent compressional arrivals, with the size of the symbol proportional to the P-wave amplitude. One station polarity (ERM, Japan) is reversed, likely due to an instrument error. P arrivals that surface near the edge of the shaded area graze the core–mantle boundary. The beachball shows the sign of the radiation pattern projected onto a small sphere surrounding the source (white for negative). From Tibuleac et al. [363], reproduced with permission from Blackwell Publishing.

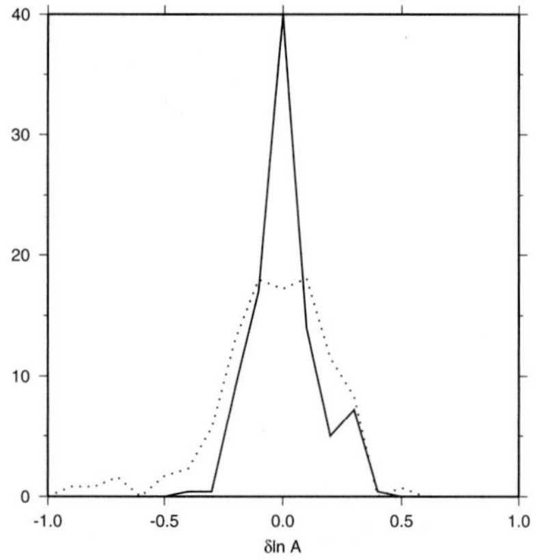


Fig. 8.2. Histogram for the misfit between the observed and predicted amplitude of P-waves from deep earthquakes, for pulse length of at least 4.5 seconds (dotted line) and more than 10 seconds (solid line). From Tibuleac et al. [363], reproduced with permission from Blackwell Publishing.

### Exercise

**Exercise 8.1** Derive (8.2) by performing two first-order Taylor approximations, first under the integral and then for the square root.

## 8.2 t* observations

For frequency-independent $Q$, we have seen in Chapter 5 that the amplitude of a body wave attenuates with frequency as $\exp(-\omega t^*/2)$, where $t^*$ is defined as:

$$t^* = \int \frac{\mathrm{d}s}{c(\mathbf{r})Q_X(\mathbf{r})} \qquad \text{(5.15 again).}$$

As in Chapter 5, we do not explicitly specify $\mathrm{Re}\, c$ but simply write $c$ for the real part of the seismic velocity. In order to determine $t^*$ from one seismogram, we would have to correct the spectrum first for the amplitude decrease with frequency because of the time behaviour of the source, forcing us to determine the source time function, as in Sigloch's method. However, we can determine the relative change in $t^*$ from two recordings and divide out the source-related spectral slope. Central to this is the assumption that the spectrum of a body wave phase is the product of four separate factors:

$$A(\omega) = A_s(\omega)A_E(\omega)A_c(\omega)A_I(\omega)\,,$$

where $A_s(\omega)$ is the source spectrum, $A_E(\omega)$ represents the influence of the Earth structure along the raypath, $A_c(\omega)$ is the crustal response – including possible reverberations that may strongly modify the spectrum – and $A_I(\omega)$ gives the instrument response. The instrument response is relatively easy to remove for modern digital instrumentation (see Chapter 13). If the spectrum is measured over a relatively short time window so that we exclude reverberations, we may assume that the crustal response is not important except for the frequency-independent impedance factor $\sqrt{\rho_\mathrm{r} c_\mathrm{r}}$ (though this assumption needs to be looked at with some suspicion in many cases). The Earth's response, for a body wave in the ray approximation, can be written as (compare Equation 4.23):

$$A_E(\omega) = \frac{\mathcal{F}}{4\pi c_\mathrm{s}^2 \mathcal{R}_\mathrm{rs}\sqrt{\rho_\mathrm{s}\rho_\mathrm{r}c_\mathrm{s}c_\mathrm{r}}}\mathrm{e}^{\mathrm{i}\omega\tau(\omega)}\mathrm{e}^{-\omega t^*/2}\,,$$

where $\mathcal{F}$ represents the source radiation term, $c_\mathrm{s}$ is the seismic velocity (P or S) at the source, $c_\mathrm{r}$ at the receiver, and similar for density $\rho$. Note that the crustal impedance factor is part of $A_E$, not $A_s$. The travel time $\tau$ is a function of frequency because of the dispersion relationship (5.16) and we have added the attenuation term.

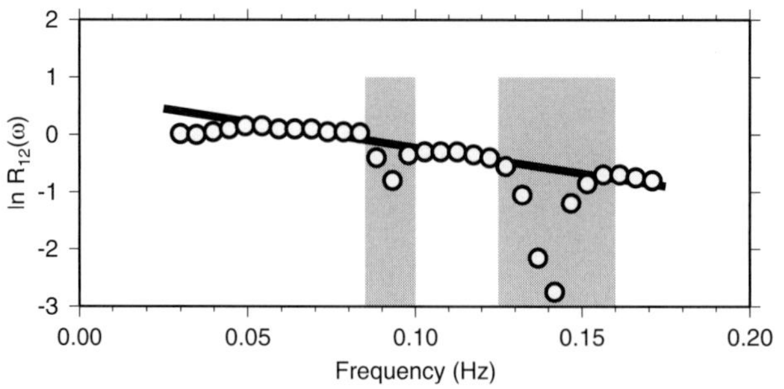


Fig. 8.3. Example of the spectral ratio of an S-wave recorded by two stations on Iceland, with the best linear fit that determines the difference in $t^*$ between the two stations. The grey areas indicate the frequency bands where microseismic noise is high. Data from Allen et al. [6], reproduced with permission from Blackwell Publishing.

Suppose now that we have two amplitude measurements, $A^{(1)}$ and $A^{(2)}$, for the same source but observed at different stations. Dividing them out removes the unknown source spectrum $A_s(\omega)$, and the ratio can be written as:

$$R_{12}(\omega) = \frac{|A_E^{(1)}(\omega)|}{|A_E^{(2)}(\omega)|} = \frac{|\mathcal{F}_1|[\mathcal{R}_{\mathrm{rs}}\rho_s c_s^{3/2}]_2}{|\mathcal{F}_2|[\mathcal{R}_{\mathrm{rs}}\rho_s c_s^{3/2}]_1} \mathrm{e}^{-\omega(t_1^*-t_2^*)}, \tag{8.4}$$

so that we can measure the difference $\delta t^* = t_1^* - t_2^*$ by estimating the slope of the spectrum plotted in a logarithmic scale (Figure 8.3). Sipkin and Jordan [320] follow a similar approach but for two phases in the same station, in which case the crustal response beneath the station is eliminated because the raypaths largely overlap.

The differential measurements for $\delta t^*$ provide linear constraints on the attenuation structure of the Earth through (5.15) and can in principle be inverted for:

$$\delta t^* = \int_{P_1} \frac{\mathrm{d}s}{c(\boldsymbol{r})Q_X(\boldsymbol{r})} - \int_{P_2} \frac{\mathrm{d}s}{c(\boldsymbol{r})Q_X(\boldsymbol{r})}, \tag{8.5}$$

where $P_1$ and $P_2$ are the raypaths to station 1 and 2, respectively. The method is strongly dependent on the assumption that the difference in spectral slope is fully determined by the attenuation. However, frequency-dependent focusing and defocusing will occur if the heterogeneities in the Earth are of the order of the wavelength of the body wave. This is demonstrated by Allen et al. [6] in a study of S-waves traversing the Iceland plume. In that case one needs to incorporate finite-frequency effects into the amplitude interpretation.

## 8.3 Amplitude healing

Just as travel time delays 'heal' after the passage of a heterogeneity that is of the order of the width of the Fresnel zone, amplitude anomalies will also diffuse over the surface of the wavefront. To analyse this phenomenon separately from the effects of attenuation we assume that the waveform $u(\omega)$ for the background model already incorporates the effects of attenuation. Like delays, amplitude variations are also subject to the effects of wavefront healing, and we can again use Born theory to linearize the problem. Dahlen and Baig [75] showed that the amplitude kernel for focusing/defocusing follows from substitution of the expressions for the scattered waves – notably (7.22) and (7.24) – into (8.3). We assume we have a prediction $A_0$ for the amplitude of the body wave in an unperturbed Earth, and that we measure:

$$\delta \ln A = \frac{A_{\rm obs} - A_0}{A_0}.$$

We then obtain:

$$\delta \ln A = \int \left[ K_{\rm P}^{\rm A} \left( \frac{\delta V_{\rm P}}{V_{\rm P}} \right) + K_{\rm S}^{\rm A} \left( \frac{\delta V_{\rm S}}{V_{\rm S}} \right) + K_{\rho}^{\rm A} \left( \frac{\delta \rho}{\rho} \right) \right] {\rm d}^3 \boldsymbol{r}_{\rm x}, \tag{8.6}$$

where $\delta V_{\rm P}/V_{\rm P}$ etc. are evaluated at $\boldsymbol{r}_{\rm x}$; writing $K_X^{\rm A}$ to represent any of the three amplitude Fréchet kernels:

$$K_X^{\rm A}(\boldsymbol{r}_{\rm x}) = -\frac{1}{2\pi} \sum_{\rm rays_1} \sum_{\rm rays_2} N(\boldsymbol{r}_{\rm x}) \Omega_X \left( \frac{1}{V_1 V_2} \right)^{\frac{1}{2}} \left( \frac{\mathcal{R}_{\rm rs}}{V_{\rm r} \mathcal{R}_{\rm xs} \mathcal{R}_{\rm xr}} \right)$$
$$\times \frac{\int_0^\infty \omega^2 |\dot{m}(\omega)|^2 \cos[\omega \Delta T(\boldsymbol{r}_{\rm x}) - \Delta\Phi(\boldsymbol{r}_{\rm x})] {\rm d}\omega}{\int_0^\infty |\dot{m}(\omega)|^2 {\rm d}\omega}, \tag{8.7}$$

with again $\Delta\Phi = \Phi_{\rm xs} + \Phi_{\rm xr} - \Phi_{\rm rs}$. A close inspection of (8.7) shows that it differs by a factor $\omega$ from the expression (7.20) for the travel time (as expected because the amplitude datum is dimensionless), and by a cosine- rather than a sine-dependence on the extra travel time $\Delta T$ for the scattered wave. Thus, a scatterer located exactly on the ray, where $\Delta T = 0$, will have a maximum effect on amplitude (see Figure 7.8).

An additional advantage of using finite-frequency theory for amplitudes is the increased stability for body wave amplitude computations, where the forward problem is often hampered by minuscule triplications of the wavefront in 3D media that may have little physical importance. From ray theory we expect the relationship between body wave amplitudes and the seismic velocity to be strongly nonlinear. However, a look at (8.7) does not confirm such strong nonlinearity, because it is in fact very similar to the travel time kernels.

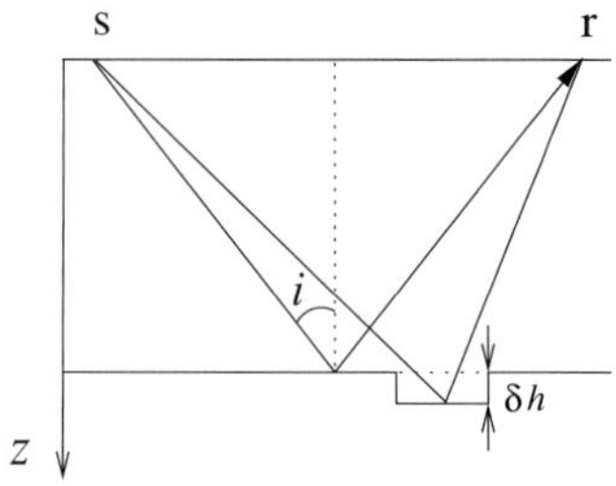


Fig. 8.4. A wave scattering off a perturbed boundary.

## 8.4 Boundary topography

Boundaries with a sharp velocity contrast, such as the Moho and the core–mantle boundary, and to a lesser degree also the upper mantle phase transitions of olivine near 410 and 660 km depth, are acting as lenses that may be very effective at focusing or defocusing seismic energy. A derivation of Fréchet kernels for the finite-frequency perturbation in amplitude due to boundary topography is given by Neele and de Regt [226]. To help understand the main ideas behind the derivation we study an example from applied geophysics in Cartesian coordinates. Consider the acoustic wave reflecting off a boundary in Figure 8.4.

Using the representation theorem (2.54), the wavefield $P(\boldsymbol{r}, \omega)$ at $\boldsymbol{r}$ can be written as the integral over the reflecting surface. To keep things simple, we assume that the interface is a perfect (rigid) reflector with a reflection coefficient 1, and the medium is homogeneous. The vertical derivative of the pressure on a rigid surface is zero, so (2.54) simplifies to:

$$P(\boldsymbol{r}, \omega) = \int_S \frac{1}{\rho} P(\boldsymbol{r}_\mathrm{x}, \omega) \nabla_\mathrm{x} G(\boldsymbol{r}, \omega; \boldsymbol{r}_\mathrm{x}) \cdot \hat{\boldsymbol{n}} \mathrm{d}^2 \boldsymbol{r}_\mathrm{x}$$
$$= \int_S \frac{1}{\rho} P(\boldsymbol{r}_\mathrm{x}, \omega) \frac{\partial G(\boldsymbol{r}, \omega; \boldsymbol{r}_\mathrm{x})}{\partial z} \mathrm{d}^2 \boldsymbol{r}_\mathrm{x} \,,$$

where the subscript x denotes the scatterer and $\nabla_\mathrm{x}$ is the gradient with respect to the scatterer coordinates. We now perturb one small area $\mathrm{d}S$ by a distance $\delta h$ in the direction given by the unit vector $\hat{z}$. Since $z$ is 'depth' its positive direction is pointing downwards. The effect of this on the pressure is found by adding the contribution from the reflecting patch at the perturbed depth and subtracting the original, unperturbed contribution to the surface integral:

$$\delta P(\boldsymbol{r}, \omega) = \frac{1}{\rho} \left[ P(\boldsymbol{r}_\mathrm{x} + \delta h \hat{z}, \omega) \frac{\partial G(\boldsymbol{r}, \omega; \boldsymbol{r}_\mathrm{x} + \delta h \hat{z})}{\partial z_\mathrm{x}} - P(\boldsymbol{r}_\mathrm{x}, \omega) \frac{\partial G(\boldsymbol{r}, \omega; \boldsymbol{r}_\mathrm{x})}{\partial z_\mathrm{x}} \right] \mathrm{d}S \,,$$

or, to first order in $\delta h$:

$$\delta P(\boldsymbol{r}, \omega) = \left( \frac{\partial P}{\partial z_\mathrm{x}} \frac{\partial G}{\partial z_\mathrm{x}} + P \frac{\partial^2 G}{\partial z_\mathrm{x}^2} \right) \delta h \mathrm{d}S \,.$$

In a homogeneous medium the acoustic wave that travels the distance $r_{xs}$ from a point source s to the boundary perturbation x and whose source spectrum is given by $\dot{m}(\omega)$ follows from (4.5):

$$P_x = P(\boldsymbol{r}_x, \omega) = -\frac{e^{i\omega r_{xs}/c}}{4\pi\rho r_{xs} c^2}\dot{m}(\omega),$$

where $r_{xs} = \sqrt{z_x^2 + x_x^2}$. The derivative with respect to $z_x$ is:

$$\frac{\partial P}{\partial z_x} \approx \frac{i\omega \cos i}{c} P_x,$$

since $\partial r_{xs}/\partial z_x = z_x/r_{xs} = \cos i$ with $i$ the angle of incidence. We neglect the change in the geometrical spreading factor $r_{xs}$, since the derivative of the phase term is expected to dominate. Physically, this means that we expect that the effects of focusing are dominant. In keeping with the paraxial ray approximation we ignore the variation in $i$ but use the incidence angle for the central ray for the off-path contributions. We also assume that the perturbation due to the change in geometrical spreading in the denominator is much smaller than that due to the phase change. In a similar fashion:

$$G_r = G(\boldsymbol{r}, \omega; \boldsymbol{r}_x) = -\frac{e^{i\omega r_{rx}/c}}{4\pi\rho r_{rx} c^2}$$
$$\frac{\partial G}{\partial z_x} \approx \frac{i\omega \cos i}{c} G_r$$
$$\frac{\partial^2 G}{\partial z_x^2} \approx -\frac{\omega^2 \cos^2 i}{c^2} G_r .$$

which gives:

$$\delta P(\boldsymbol{r}, \omega) = -\frac{2\omega^2 \cos^2 i}{\rho c^2} P_x G_r \delta h \, dS .$$

The unperturbed wave at the receiver can be written as:

$$P(\boldsymbol{r}, \omega) = -\frac{e^{i\omega T_{rs}}}{4\pi \mathcal{R}_{rs} \rho c^2}\dot{m}(\omega),$$

where $\mathcal{R}_{rs}$ stands for the total ray length between source and receiver. Substituting the expressions for $P$ and $\delta P$ for $u$ and $\delta u$ in (8.3) we find the contribution of the patch to the amplitude perturbation:

$$\delta \ln A = \frac{2\cos^2 i}{c^2} \frac{\delta h}{4\pi\rho c^2}\left(\frac{\mathcal{R}_{rs}}{\mathcal{R}_{xs}\mathcal{R}_{rx}}\right) \frac{\int_0^\infty \omega^2 |\dot{m}(\omega)|^2 \cos[\omega \Delta T] d\omega}{\int_0^\infty |\dot{m}(\omega)|^2 d\omega} dS .$$

We find the response for the full interface by integrating over the surface $S$.

Though we have adopted some drastic simplifications, the expressions for non-homogeneous media and non-rigid interfaces show the same characteristics, in particular the dependence on $\cos^2 i$ and the geometrical spreading factor. The very lengthy derivation is omitted, but the formalism adopted here is similar to that of Dahlen [74] for travel time perturbations from changes in topography. Switching from depth $\delta h$ to radius $\delta r$, which involves a change in sign, we obtain the following result for the amplitude topography kernel, which is remarkably simple:

$$\delta \ln A = \int_S K^A_{\delta r}(\boldsymbol{r}_\mathrm{x}) \delta r(\boldsymbol{r}_\mathrm{x}) \mathrm{d}^2 \boldsymbol{r}_\mathrm{x} \,,$$

$$K^A_{\delta r}(\boldsymbol{r}_\mathrm{x}) = \frac{D_{\delta r}}{2\pi V_\mathrm{r}} \left( \frac{|\cos i| \mathcal{R}_\mathrm{rs}}{\mathcal{R}_\mathrm{xs} \mathcal{R}_\mathrm{xr}} \right) \frac{\int_0^\infty \omega^2 |\dot{m}(\omega)|^2 \cos[\omega \Delta T(\boldsymbol{r}_\mathrm{x}) - \Delta\Phi(\boldsymbol{r}_\mathrm{x})] \mathrm{d}\omega}{\int_0^\infty |\dot{m}(\omega)|^2 \mathrm{d}\omega} \,, \tag{8.8}$$

where the factor $D_{\delta r}$ depends on the type of reflection:

$$D_{\delta r} = -\frac{|\cos i_1|}{V_1} - \frac{|\cos i_2|}{V_2} \quad \text{Topside reflection} \tag{8.9}$$

$$= +\frac{|\cos i_1|}{V_1} + \frac{|\cos i_2|}{V_2} \quad \text{Bottomside reflection} \tag{8.10}$$

$$= -\frac{|\cos i_1|}{V_1} + \frac{|\cos i_2|}{V_2} \quad \text{Transmission}\,. \tag{8.11}$$

Here $V$ is velocity and index 1 stands for the incident ray and index 2 for the scattered (reflected or transmitted) ray. The factor $|\cos i|\mathcal{R}_\mathrm{rs}/\mathcal{R}_\mathrm{xs}\mathcal{R}_\mathrm{xr}$ is continuous across the boundary and can be calculated on either side of the interface.

## 8.5 Finite-frequency Q tomography

The derivation of a Fréchet kernel for the quality factor $Q$ is straightforward once we recognize that the effects of attenuation can be modelled by a small imaginary component to the shear velocity as we did in (5.12). Simplifying the notation slightly by writing $Q_\mathrm{S}^{-1}$ for $(Q_X^\mathrm{S})^{-1}$:

$$\delta Q_\mathrm{S}^{-1} = -\frac{\delta \mathrm{Im}\, \mu}{\mathrm{Re}\, \mu} \,,$$

where we write a $\delta$ in front of $Q_\mathrm{S}^{-1}$ and $\mathrm{Im}\,\mu$ to indicate that we consider perturbations, assuming that the attenuation of the background model is already incorporated in $A_\mathrm{syn}$. The perturbation in the imaginary component of the shear modulus $\delta \mathrm{Im}\,\mu$ is directly related to the perturbation in $\mathrm{Im}\, V_\mathrm{S}$. To first order (recall

Exercise 5.9):

$$\frac{\delta \text{Im}\, V_S}{V_S} = -\frac{\delta \text{Im}\, \mu}{2\text{Re}\, \mu} = -\frac{1}{2}\delta Q_S^{-1}\,.$$

A similar expression can be derived for $V_P$ and $Q_P^{-1}$. However, given the large errors that still hamper the experimental investigations of $Q$, it is usually not warranted to treat Im $V_P$ and Im $V_S$ independently. We are justified in neglecting any attenuation associated with pure compression: $\text{Im}\,\kappa = 0$, since it is likely to be very small. With $V_P = \sqrt{(\kappa + \frac{4}{3}\mu)/\rho}$:

$$\frac{\delta \text{Im}\, V_P}{V_P} = \frac{\frac{2}{3}\delta \text{Im}\, \mu}{\text{Re}(\kappa + \frac{4}{3}\mu)} = -\frac{\frac{2}{3}\text{Re}\mu}{\text{Re}(\kappa + \frac{4}{3}\mu)}\delta Q_S^{-1} = -\frac{2V_S^2}{3V_P^2}\delta Q_S^{-1}\,.$$

An intuitive grasp of the finite-frequency effect of attenuation anomalies can be obtained by realizing that an imaginary component to the velocity, when inserted into the finite-frequency travel-time expression (7.27), will yield an imaginary component to the travel time anomaly $\delta T$. For a narrow band signal, in which we can assume $\delta T$ to be constant over the frequency band, the wavelet will not only be delayed by $\text{Re}\,\delta T$ seconds, it will also see its amplitude damped by the factor $\exp[\text{i}\omega(\text{i}\text{Im}\,\delta T)] = \exp[-\omega \text{Im}\,\delta T]$. Thus, we expect the Fréchet kernels for $Q$ to shape like the travel-time kernels.

To model the effects of attenuation on the amplitude we search for an expression of the form:

$$\delta \ln A = \int K_X^Q(\boldsymbol{r}_x)\left(\frac{\delta Q_X^{-1}}{Q_X^{-1}}\right)_x \text{d}^3\boldsymbol{r}_x\,,$$

while using $\delta Q_S^{-1} = -2\delta \text{Im}\, V_S/V_S$. The important term that changes with respect to purely real velocity perturbations is the scattering coefficient term in (7.24) that depends on $V_P$ and $V_S$ through $\boldsymbol{S}$ in (4.40). For its imaginary component we write:

$$\begin{aligned}\text{i}\,\text{Im}(\hat{\boldsymbol{p}}_2\cdot\boldsymbol{S}\cdot\hat{\boldsymbol{p}}_1) &= \text{i}\,\hat{\boldsymbol{p}}_2\cdot\left(\frac{\text{Im}\,\delta V_P}{V_P}\boldsymbol{S}_P + \frac{\text{Im}\,\delta V_S}{V_S}\boldsymbol{S}_S\right)\cdot\hat{\boldsymbol{p}}_1\\ &= -\frac{\text{i}}{2}\hat{\boldsymbol{p}}_2\cdot(\delta Q_P^{-1}\boldsymbol{S}_P + \delta Q_S^{-1}\boldsymbol{S}_S)\cdot\hat{\boldsymbol{p}}_1\\ &= \text{i}\left(\Omega_P\delta Q_P^{-1} + \Omega_S\delta Q_S^{-1}\right)\,.\end{aligned}$$

The expression for the unperturbed wave $u(\omega)$ remains unchanged:

$$u(\omega) = \frac{1}{4\pi}\Pi_{rs}\Lambda\Upsilon\mathcal{R}_{rs}^{-1}\dot{m}(\omega)\exp[\text{i}(\omega T_{rs} - \Phi_{rs})] \qquad \text{(7.22 again)}.$$

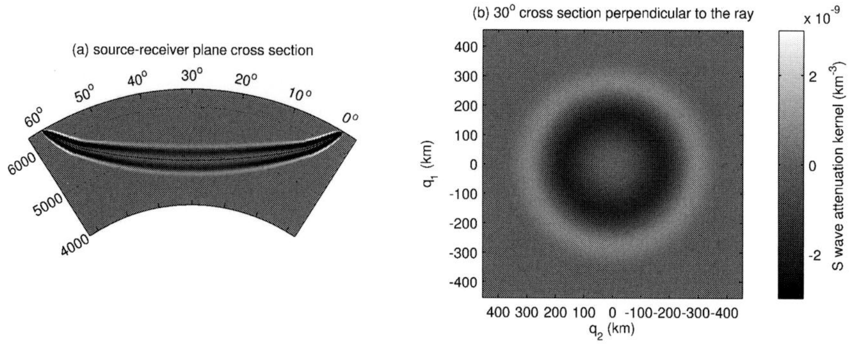


Fig. 8.5. A Fréchet kernel for attenuation ($Q_S^{-1}$). Figure courtesy Yue Tian.

The perturbation is now caused by the imaginary component of the velocity, and instead of (7.24) we get:

$$\delta u(\omega) = \left(\frac{\omega}{4\pi}\right)^2 \int \frac{\Lambda_1 \Upsilon_2 \Pi_{xs} \Pi_{xr} (\hat{\boldsymbol{p}}_2 \cdot (\mathrm{i}\, \mathrm{Im}\, \boldsymbol{S}) \cdot \hat{\boldsymbol{p}}_1)}{V_r (V_1 V_2)^{\frac{1}{2}} \mathcal{R}_{xs} \mathcal{R}_{xr}} \times \dot{m}(\omega) \exp \mathrm{i}[\omega(T_{xs} + T_{xr}) - \Phi_{xs} - \Phi_{xr}] d^3 \boldsymbol{r}_x. \tag{8.12}$$

Inserting both into (8.3) we find:

$$\delta \ln A = \int \left[ K_P^Q \left( \frac{\delta Q_P^{-1}}{Q_P^{-1}} \right) + K_S^Q \left( \frac{\delta Q_S^{-1}}{Q_S^{-1}} \right) \right] \mathrm{d}^3 \boldsymbol{r}_x, \tag{8.13}$$

with

$$K_X^Q(\boldsymbol{r}_x) = -\frac{1}{4\pi} \sum_{\mathrm{ray}_1} \sum_{\mathrm{ray}_2} N(\boldsymbol{r}_x) \Omega_X \left( \frac{1}{V_1 V_2} \right)^{\frac{1}{2}} \left( \frac{\mathcal{R}_{rs}}{V_r \mathcal{R}_{xr} \mathcal{R}_{xs}} \right) Q_X^{-1}(\boldsymbol{r}_x) \times \frac{\int_0^\infty \omega^2 |\dot{m}(\omega)|^2 \sin[\omega \Delta T(\boldsymbol{r}_x) - \Delta\Phi(\boldsymbol{r}_x)] \mathrm{d}\omega}{\int_0^\infty |\dot{m}(\omega)|^2 \mathrm{d}\omega}, \tag{8.14}$$

where, as before, $\Delta\Phi = \Phi_{xs} + \Phi_{rx} - \Phi_{rs}$, which equals $(M_{xs} + M_{rx} - M_{rs})\pi/2$ as long as there are no supercritical reflections. An expression for finite-frequency inversion of attenuation from waveforms was first derived by Tromp et al. [368], but (8.14) has the advantage that it is directly applicable to amplitude measurements. An example is shown in Figure 8.5.

At first, it may seem that the division of the kernel by $Q$, which is of the order of 100 or larger, makes inverting for $\delta Q$ a futile exercise. However, it must be realized that realistic variations in mantle velocities are of the order of 1%, whereas $Q$ can easily be doubled or halved, depending on temperature and/or the presence of volatiles or melt. As a consequence, attenuation is thus almost as important as focusing/defocusing. In areas of high attenuation ($Q < 50$) it may even be the most important effect. The importance of (8.14) is that the attenuation kernel is similar to that of velocity perturbations but quite different from that of focusing; in addition, the difference in frequency scaling for time delays and attenuation may be exploited if the body waves cover a large range of frequencies. So far, however, practical applications of (8.14) are still lacking.

### *Exercise*

**Exercise 8.2** Does the attenuation kernel have a doughnut hole? How does this compare with the focusing kernel?

# 9

# Normal modes

Since the Earth is a mechanical system with a well-defined boundary, the formal solution to the elastodynamic equations (2.3) with the boundary condition that the surface is stress-free yields a discrete spectrum of eigenfrequencies, just as is the case for a finite-length string in a violin. The reason that we have not taken this viewpoint earlier is that it is highly impractical for seismic tomography in the frequency band that is of most interest: above 10 mHz the number of eigenfrequencies in a small frequency band becomes very large. In addition, the spectral peaks are widened by the effects of attenuation and lateral heterogeneity and the discrete spectrum becomes, for all practical purposes, a continuous one because the peaks overlap. But below 10 mHz and even at higher frequency for some high $Q$ modes it becomes feasible to measure individual eigenfrequencies (Figure 9.1).

The theoretical study of terrestrial eigenfrequencies started with the historical work of Love [189]. Interest in the field really grew only after the first observations of normal modes, following the Chile earthquake of 1960. Major contributions to the development of the theory are by Pekeris et al. [259, 260], Backus and Gilbert [11], Dahlen [70, 71, 72, 73], Gilbert [112], Woodhouse and Dahlen [399], Jordan [151] and Park [252]; and of the interpretation by Backus [12], Gilbert and Dziewonski [116], Jordan [151], Masters et al. [201], Woodhouse and Girnius [401], Woodhouse and Dziewonski [400], Woodhouse and Wong [402].

Though the low frequency spectrum of the Earth's vibrations gives only limited resolution, it would be a mistake to ignore it: for a good resolution low frequencies are as essential as high frequencies. It is also likely that the Earth's lateral heterogeneity is strong for long length scales, which are adequately resolved by oscillations of similar wavelength. Finally, free oscillations, in contrast to body wave delay times, contain information on the Earth's density variations (Ishii and Tromp [142, 143]).

A complete treatment of the theory can be found in Dahlen and Tromp [78]. However, the theory is extensive, as may be clear from a page count of Dahlen

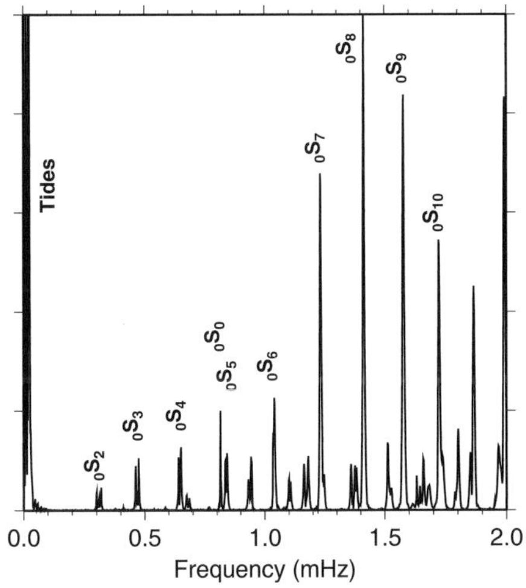


Fig. 9.1. Spectrum of the vertical ground velocity at Geoscope station UNM (Unam, Mexico) from a week-long recording directly following the large Sumatra earthquake of December 26, 2004. The first 10 fundamental spheroidal modes are indicated. Other peaks belong to the tides ($<$ 0.1 mHz), and to spheroidal overtones.

and Tromp's treatise (more than a thousand). Few observational seismologists are able to find time to educate themselves deeply on the fascinating topic of normal modes – the author is no exception to that rule. Yet tomographers must be concerned that their tomographic models satisfy both the high and low frequency ends of the seismic spectrum. In this chapter I shall therefore stay clear of a lengthy exposé of the many (and intricate) theoretical aspects and again opt for a mostly heuristic introduction. The aim is to provide enough information to tomographers to judge normal mode data sets and incorporate these into tomographic interpretations.

## 9.1 The discrete spectrum

The elastodynamic equations for an isotropic, non-rotating elastic Earth and the boundary conditions on the stress can be reformulated in spherical coordinates $(r, \theta, \phi)$. In symbolic form:

$$\rho \frac{\partial^2 \boldsymbol{u}}{\partial t^2} + \mathcal{L}\boldsymbol{u} = \boldsymbol{f} ,$$

or, in the spectral domain:

$$\mathcal{L}\boldsymbol{u} = \omega^2 \rho \boldsymbol{u} + \boldsymbol{f} . \tag{9.1}$$

The operator $\mathcal{L}$ is just a shorthand notation for the derivatives that operate on elastic constants and the displacements, to avoid lengthy expressions such as those occurring in the right-hand side of (2.41). Though compact expressions exist in Cartesian coordinates (see Exercise 9.2), spherical coordinates introduce considerable complexity. For very low frequency, the change in gravity caused by the oscillations may also play a role. We shall not delve deeply into this, but sketch the main results. Separation of variables $t, r, \theta$ and $\phi$ yields solutions that are the product of a harmonic function of time ($\mathrm{e}^{\mathrm{i}\omega t}$) as well as of longitude ($\mathrm{e}^{\mathrm{i}m\phi}$), a co-latitude dependence given by the associated Legendre function $P_\ell^m(\cos\theta)$, and a function of radius $r$. The radial dependence is not a known analytical function but needs to be found by solving a set of differential equations numerically.

For the displacement $\boldsymbol{u}$ we adopt the notation $\boldsymbol{u}(\boldsymbol{r}, t) = \exp(-\mathrm{i}_n\omega_\ell^m t)_n\boldsymbol{u}_\ell^m(\boldsymbol{r})$, where ${}_n\omega_\ell^m$ is the eigenfrequency, and the ${}_n\boldsymbol{u}_\ell^m$ have components that can be written in terms of spherical harmonic functions:

$$\begin{aligned}{}_n\boldsymbol{u}_\ell^m &= {}_nU_\ell(r)Y_\ell^m(\theta,\phi)\,\hat{\boldsymbol{r}}\\ &+\nu^{-1}[{}_nV_\ell(r)\partial_\theta Y_\ell^m(\theta,\phi) + (\sin\theta)^{-1}\,{}_nW_\ell(r)\partial_\phi Y_\ell^m(\theta,\phi)]\,\hat{\boldsymbol{\theta}}\\ &+\nu^{-1}[{}_nV_\ell(r)(\sin\theta)^{-1}\partial_\phi Y_\ell^m(\theta,\phi) - {}_nW_\ell(r)\partial_\theta Y_\ell^m(\theta,\phi)]\,\hat{\boldsymbol{\phi}},\end{aligned} \tag{9.2}$$

where

$$\nu = \sqrt{\ell(\ell+1)} \approx \ell + \frac{1}{2}, \tag{9.3}$$

and the spherical harmonics are defined as:

$$Y_\ell^m(\theta,\phi) = (-1)^m\left[\frac{(2\ell+1)(l-m)!}{4\pi(l+m)!}\right]^{\frac{1}{2}} P_\ell^m(\cos\theta)\exp(\mathrm{i}m\phi), \tag{9.4}$$

with the associated Legendre function:

$$\begin{aligned}P_\ell^m(x) &= (1-x^2)^{m/2}\frac{\mathrm{d}^m}{\mathrm{d}x^m}P_\ell(x)\\ P_\ell(x) &= \frac{1}{2^\ell\ell!}\frac{\mathrm{d}^\ell}{\mathrm{d}x^\ell}(x^2-1)^\ell.\end{aligned}$$

The first two Legendre functions are $P_0(\cos\theta) = 1$ and $P_1(\cos\theta) = \cos\theta$. The associated Legendre functions are zero for $|m| > \ell$. For $m = 0$, there are $\ell$ latitudes $\theta$ where the function is zero (nodal lines). As is evident from the factor $\mathrm{e}^{\mathrm{i}m\phi}$, the real part of $Y_\ell^m(\theta,\phi)$ has $m$ nodal lines in longitude. However, double-couple earthquake sources have only two nodal lines that come together in the epicentre, e.g. one in the direction of and one perpendicular to the strike of the fault (at least as long as the geographical extent of the fault is small with respect to the wavelength). Therefore, in a spherically symmetric Earth we do not generate modes with $m > 2$ if we choose a coordinate system such that $\theta = 0$ at the epicentre. As an example,

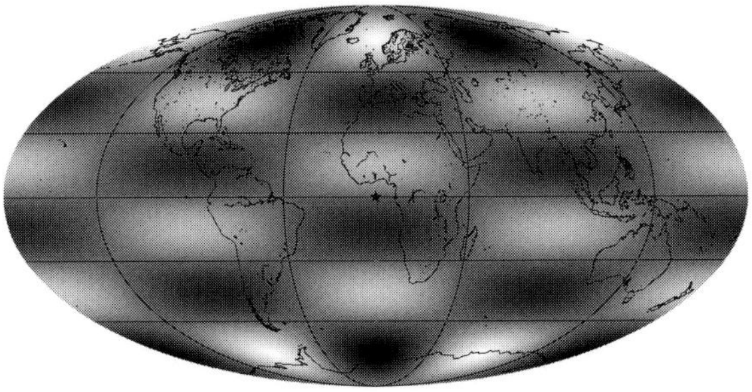

Fig. 9.2. Plot of Re $Y_7^2(\theta, \phi)$. The greyscale runs from $-0.103$ (white) to +0.103 (black).

Figure 9.2 shows the real part of $Y_7^2(\theta, \phi)$, which represents the vertical displacement pattern for ${}_n\boldsymbol{u}_7^2$ from a hypothetical earthquake located at the North Pole.

With the definitions given here the spherical harmonics $Y_\ell^m$ are orthonormal:

$$\int_0^{2\pi}\int_0^{\pi} Y_{\ell'}^{m'}(\theta,\phi)^* Y_\ell^m(\theta,\phi)\sin\theta \mathrm{d}\theta \mathrm{d}\phi = \delta_{\ell\ell'}\delta_{mm'}.$$

Different normalizations exist in the geophysical literature. In this book we adhere to the definitions used by Dahlen and Tromp [78], though we note that they prefer not to use the complex spherical harmonics, but instead combine the contributions for positive and negative $m$ to the displacement field by defining real-valued basis functions. The radial (eigen)functions ${}_nU_\ell(r)$, ${}_nV_\ell(r)$ and ${}_nW_\ell(r)$, which we shall abbreviate as $U$, $V$ and $W$ turn out to be independent of $m$ and satisfy a set of second-order differential equations. We list the equations valid for the Earth in Table 9.1. The stress-free boundary conditions are only satisfied for a discrete set of *eigenfrequencies* $\omega$. We therefore denote solutions $U$, $V$ and $W$ at such frequencies as eigenfunctions, and number them by $n$ for each given $\ell$ and $m$. For each pair $(\ell, m)$ there is a lowest eigenfrequency, the fundamental mode (indicated by index $n = 0$), and higher frequencies, the overtones, indicated by an overtone number $n > 0$. Note, however, that $m$ does not occur in (9.5 – 9.8), so that we have identical eigenfrequencies and radial eigenfunctions for each $-\ell \leq m \leq \ell$. We say that the eigenfrequency problem is degenerate in a spherically symmetric Earth. We denote such a group of eigenfrequencies as a 'multiplet'. Degeneracy implies that any linear combination of eigenvectors ${}_n\boldsymbol{u}_\ell^m$ at this same frequency ${}_n\omega_\ell$ is also an eigenvector.

Normally, we neglect the changes of the Earth's gravity field due to the oscillations, though one usually retains the static force of gravity (the Cowling approximation). The static acceleration of gravity $g$ also introduces a static displacement with respect to a non-gravitating Earth but this plays no role in the dynamic equations where only the added displacement field $\boldsymbol{u}$ is time-dependent. However, for the

Table 9.1. *Differential equations for the Earth's eigenfunctions. Dots indicate differentiation with respect to* $r$*,* $\Phi$ *represents the perturbation in the gravitational potential* $\sum_n \Phi_\ell Y_\ell^m$*,* $G$ *is the gravitational constant* ($6.67 \times 10^{-11} m^3 kg^{-1} s^{-2}$)*, and* $g(r)$ *is the acceleration due to gravity.* $U$*,* $V$*,* $W$ *and* $\Phi$ *are functions of* $r$ *and* $\nu = \sqrt{\ell(\ell+1)}$.

$$\frac{1}{r^2}\frac{\mathrm{d}}{\mathrm{d}r}[r^2(\lambda+2\mu)\dot{U} + \lambda r(2U - \nu V)] + \frac{1}{r}\left[(\lambda+2\mu)\dot{U} + \frac{\lambda}{r}(2U - \nu V)\right] - \frac{3\lambda+2\mu}{r}\left(\dot{U} + \frac{2}{r}U - \frac{\nu}{r}V\right) - \frac{\nu\mu}{r}\left(\dot{V} - \frac{V}{r} + \frac{\nu}{r}U\right) + \omega^2\rho U - \rho\left[\dot{\Phi} + \left(4\pi G\rho - \frac{4g}{r}\right)U + \frac{\nu g}{r}V\right] = 0, \quad (9.5)$$

$$\frac{1}{r^2}\frac{\mathrm{d}}{\mathrm{d}r}\left[\mu r^2\left(\dot{V} - \frac{V}{r} + \frac{\nu}{r}U\right)\right] + \frac{\mu}{r}\left(\dot{V} - \frac{V}{r} + \frac{\nu}{r}U\right) + \frac{\nu\lambda}{r}\dot{U} + \frac{\nu(\lambda+\mu)}{r^2}(2U - \nu V) + \left[\omega^2\rho - \frac{(\nu^2-2)\mu}{r^2}\right]V - \frac{\nu\rho}{r}(\Phi + gU) = 0, \quad (9.6)$$

$$\frac{1}{r^2}\frac{\mathrm{d}}{\mathrm{d}r}\left[\mu r^2\left(\dot{W} - \frac{W}{r}\right)\right] + \frac{\mu}{r}\left(\dot{W} - \frac{W}{r}\right) + \left[\omega^2\rho - \frac{(\nu^2-2)\mu}{r^2}\right]W = 0. \quad (9.7)$$

$$\frac{\mathrm{d}}{\mathrm{d}r}\dot{\Phi} + \frac{2}{r}\dot{\Phi} - \frac{\nu^2}{r^2}\Phi = -4\pi G\left[\dot{\rho}U + \rho\dot{U} + \frac{\rho}{r}(2U - \nu V)\right] \quad (9.8)$$

Boundary conditions

$$(\lambda+2\mu)\dot{U} + \frac{\lambda}{r}(2U - \nu V) = 0$$

$$\mu\left(\dot{V} - \frac{V}{r} + \frac{\nu}{r}U\right) = 0$$

$$\mu\left(\dot{W} - \frac{W}{r}\right) = 0$$

$$[\dot{\Phi} + 4\pi G\rho U]_-^+ = 0$$

Continuity of tractions and displacements at interfaces

lowest frequency eigenvibrations, the Earth's gravity field is appreciably perturbed, because the density is changed from its static value $\rho_0$ and $\mathcal{L}$ must include these changes in the gravitational potential. This gravity perturbation is also expanded into spherical harmonics. Internal gravity waves, mostly driven by variations in buoyancy, can exist in stars and are known as g-modes, in contrast to modes that are mostly acoustic (p-modes). A surface gravity mode, confined near the surface of the Sun and behaving much like water waves in the ocean, also exists and is denoted as f-mode, but for this mode the Cowling approximation is adequate.

By convention, we normalize the eigenfunctions such that

$$\int \rho[{}_nU_\ell^2 + {}_nV_\ell^2]r^2\mathrm{d}r = 1$$

$$\int \rho\, {}_nW_\ell^2 r^2\mathrm{d}r = 1\,. \tag{9.9}$$

Here, $\ell$ and $m$ are integer quantum numbers, named angular and azimuthal order, respectively, and $P_\ell^m(\cos\theta)$ is the associated Legendre function. Reliable software to solve (9.5 – 9.8) for a spherically symmetric Earth model is available in the public domain (Saito [298], Woodhouse [398]).†

In (9.2), the derivative of $Y_\ell^m$ with respect to $\theta$ introduces a factor $\ell$, so that for large $\ell \gg m$ this term dominates over the derivative with respect to $\phi$. As a result, the radial functions $V$ and $W$ become associated with dominant motion in the $\hat{\boldsymbol{\theta}}$ and $\hat{\boldsymbol{\phi}}$ direction, respectively, as $\ell$ increases. While $V$ is coupled with $U$ in (9.5–9.6) and (9.8), Equation (9.7) for $W$ is independent. From this we conclude that there are two classes of modes. Those with purely torsional motion in the $\hat{\boldsymbol{\phi}}$ direction are called 'toroidal'. Their dependence on $r$ is given by ${}_nW_\ell(r)$. Toroidal modes do not exist in gases or fluids where $\mu = 0$, and therefore not in the Sun, with the exception of a degenerate mode that can be seen as a change in the Sun's rotation. Modes with both a horizontal and a vertical component of motion are 'spheroidal'. The particle motion is elliptical, and associated with ${}_nU_\ell(r)$ and ${}_nV_\ell(r)$. Spheroidal modes correspond to p-modes in the Sun, even though the dynamic equations for stars differ from (9.5–9.7) because the equation of state that connects $\rho$ and the pressure is more complicated, and because for a higher precision one also needs to take into account the radiation pressure (Christensen-Dalsgaard [58]). For terrestrial oscillations, toroidal and spheroidal modes are usually written as ${}_nT_\ell^m$ and ${}_nS_\ell^m$, respectively. The superscript $m$ is omitted for multiplets. The special cases of spheroidal modes with $\ell = 0$ are known as radial oscillations.

The p-modes travel in the outer, convective layer of the Sun; the internal gravity modes or g-modes are trapped in the core and the radiative layer and their amplitudes decrease exponentially towards the surface, which makes them hard to measure from Doppler observations. Early observations of g-modes in the Sun were never confirmed, but the Global Oscillations at Low Frequencies (GOLF) instrument

† One battle-tested program that has been around for a long time is Guy Masters' MINEOS, available from `http://geodynamics.org/cig/software/packages/seismo/mineos/`; Ruedi Widmer at the University of Karlsruhe wrote a servlet that lets one compute and plot one mode at a time for a variety of Earth models, see `www-gpi.physik.uni-karlsruhe.de/pub/widmer/Modes/modes.html`

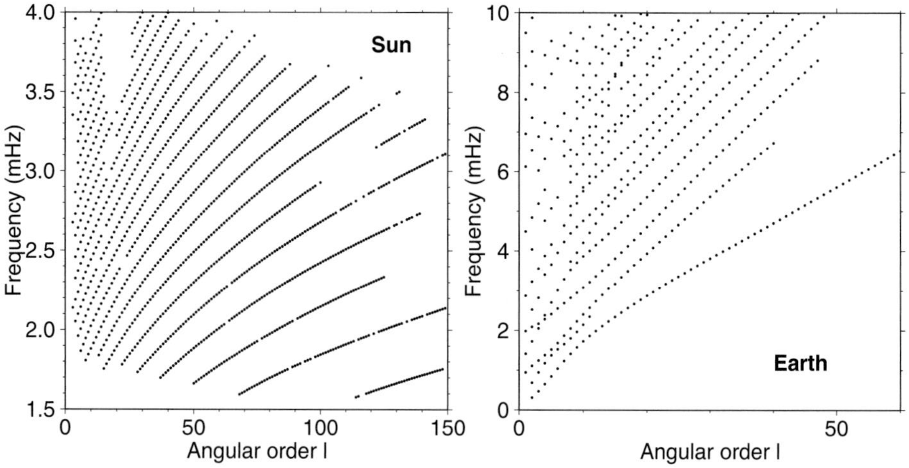


Fig. 9.3. Observed eigenfrequencies for p-modes in the Sun (left) and for spheroidal oscillations in the Earth (right). Modes with equal radial order $n$ align along branches.

on board the SOHO satellite has produced weak but persistent oscillations near the predicted frequency of 220 $\mu$Hz (Cox and Guzik [66], García et al. [108]). Figure 9.3 shows the observed eigenfrequencies for p-modes in the Sun and for spheroidal modes in the Earth.

The theory of free oscillations extends even beyond the domain of our solar system. Very long-period changes in the magnitude or luminosity of variable stars such as $\beta$ Cepheids are known to be due to radial oscillations; isolated spectral peaks with the expected line spacing are now being observed for a number of stars, opening up the possibility to obtain independent information on their mass and radii. This new field of 'astroseismology' is very exciting. Stellar oscillations manifest themselves in very slight periodic changes in magnitude as well as by periodic Dopplershifts in spectral lines. Although the Dopplershift averages to zero over the full surface of the star except for the radial modes with $\ell = 0$, the fact that we observe only half of the surface, and under different angles, allows us to see a few lowest-order modes, and even to obtain some information on $\ell$ and $m$. No fundamental limitation to the quantum number observability exists for changes in luminosity, but this technique requires the measurement of very small changes in magnitude (of order $10^{-4}$), which can only be done reliably from space.

Theoretically, internal gravity waves exist also in the fluid core of the Earth, with periods of several hours, but even if such 'undertones' can be excited, their signal would be completely overshadowed by tidal waves in the oceans with similar periods and no reliable identifications of terrestrial gravity modes exist.

In the following we simplify the notation and rank the eigenvalues with a single index $k = (n, \ell, m)$ and write $\boldsymbol{u}_k$ instead of ${}_n\boldsymbol{u}_\ell^m$. The differential operator $\mathcal{L}$ is Hermitian in an elastic body with homogeneous boundary conditions. It is convenient to define an inner product by

$$(\boldsymbol{u}, \boldsymbol{v}) = \int \boldsymbol{u} \cdot \boldsymbol{v}^* \, \mathrm{d}^3\boldsymbol{r}, \tag{9.10}$$

where, as usual, $\boldsymbol{u} \cdot \boldsymbol{v} = \sum_i u_i v_i$, the asterisk $*$ denotes complex conjugation, and the integration is over the volume of the Earth or Sun. The Hermiticity implies:

$$(\boldsymbol{u}_p, \mathcal{L}\boldsymbol{u}_q) \equiv \int \boldsymbol{u}_p \cdot \mathcal{L}\boldsymbol{u}_q^* \, \mathrm{d}^3\boldsymbol{r} = \int \mathcal{L}\boldsymbol{u}_p \cdot \boldsymbol{u}_q^* \, \mathrm{d}^3\boldsymbol{r} = (\mathcal{L}\boldsymbol{u}_p, \boldsymbol{u}_q). \tag{9.11}$$

Since $\mathcal{L}\boldsymbol{u} = \omega^2 \rho \boldsymbol{u}$ in the absence of a force term, $\omega^2$ takes on the function of an eigenvalue of $\mathcal{L}$. Equation (9.11) implies that such eigenvalues are real and that the eigenvectors are orthogonal with a norm that involves the density $\rho$ (see Exercise 9.2):

$$(\rho \boldsymbol{u}_p, \boldsymbol{u}_q) = \int \rho \boldsymbol{u}_p \cdot \boldsymbol{u}_q^* \, \mathrm{d}^3\boldsymbol{r} = \delta_{pq} \, . \tag{9.12}$$

If we add anelasticity, the eigenvalues become complex. It is advantageous to neglect anelasticity initially and retain the Hermitian nature of the problem. Effects of damping can then be dealt with using perturbation theory, as we did in Chapter 5, by adding a small imaginary component to the seismic velocities. As a consequence, the degenerate eigenvalue perturbs:

$$\omega_k \to \omega_k \left(1 + \frac{\mathrm{i}}{2Q_k}\right) \tag{9.13}$$

where $Q_k$ is yet another quality factor, this one belonging to mode $k$, giving decay of energy with time as in (5.6). Each mode thus attenuates at its own rate. Postponing the incorporation of damping effects to a perturbation approach greatly simplifies the perturbation theory for lateral heterogeneity in elastic constants and density.

## Exercises

**Exercise 9.1** In Cartesian coordinates, $\mathcal{L}$ represents the divergence of the stress tensor (see Equation 2.3) or, with Hooke's law, $(\mathcal{L}\boldsymbol{u})_i = \sum_{jkl} \partial_j (c_{ijkl} \partial_k u_l)$. The boundary of the medium with normal $\hat{\boldsymbol{n}}$ is stress-free, so that

$$\sum_j \sigma_{ij} n_j = \sum_{jkl} \partial_j (c_{ijkl} \partial_k u_l) n_j = 0.$$

Use Gauss's theorem to show that

$$(\mathcal{L}\boldsymbol{u}, \boldsymbol{v}) = \int \sum_i (\mathcal{L}\boldsymbol{u})_i v_i^* \, \mathrm{d}^3\boldsymbol{r} = -\sum_{ijkl} \int c_{ijkl}(\partial_l u_k)(\partial_j v_i^*) \, \mathrm{d}^3\boldsymbol{r} \, .$$

Use the symmetry properties of the elastic tensor ($c_{ijkl} = c_{jikl} = c_{ijlk} = c_{lkij}$) to show also that:

$$(\boldsymbol{u}, \mathcal{L}\boldsymbol{v}) = -\sum_{ijkl} \int c_{ijkl}(\partial_l u_k)(\partial_j v_i^*) \, \mathrm{d}^3\boldsymbol{r} \, ,$$

by which we establish the Hermitian nature of the operator $\mathcal{L}$.

**Exercise 9.2** a. Use the Hermitian property of $\mathcal{L}$ to show that:

$$\omega_p^2(\rho \boldsymbol{u}_p, \boldsymbol{u}_q) = (\omega_q^2)^*(\boldsymbol{u}_p, \rho \boldsymbol{u}_q) \, .$$

b. Use this result to show that the squared eigenfrequencies are real, and that eigenvectors belonging to different eigenfrequencies are orthogonal.

## 9.2 Rayleigh's Principle

An important variational principle, which goes back to Rayleigh [272], can be used to relate variations in the eigenfrequency $\omega_k$ to spherically symmetric perturbations in the Earth model, i.e. in $\rho$ and $c_{ijkl}$ (or $\mathcal{L}$). After a short time the source rupture has been completed and we can ignore the force term. We transform the elastodynamic equations to the frequency domain:

$$\rho \omega_k^2 \boldsymbol{u}_k = \mathcal{L}\boldsymbol{u}_k . \tag{9.14}$$

Perturbing this, we find:

$$(\rho + \delta\rho)(\omega_k^2 + \delta\omega^2)(\boldsymbol{u}_k + \delta\boldsymbol{u}_k) = (\mathcal{L} + \delta\mathcal{L})(\boldsymbol{u}_k + \delta\boldsymbol{u}_k) \, ,$$

and taking the inner product with $\boldsymbol{u}_k$ and using $(\mathcal{L}\delta\boldsymbol{u}_k, \boldsymbol{u}_k) = (\delta\boldsymbol{u}_k, \mathcal{L}\boldsymbol{u}_k)$:

$$(\delta\boldsymbol{u}_k, \rho\omega_k^2\boldsymbol{u}_k - \mathcal{L}\boldsymbol{u}_k) + \delta\omega^2(\rho\boldsymbol{u}_k, \boldsymbol{u}_k) = (\delta\mathcal{L}\boldsymbol{u}_k, \boldsymbol{u}_k) - \omega_k^2(\delta\rho\boldsymbol{u}_k, \boldsymbol{u}_k) \, .$$

The first term is zero because of (9.14):

$$\delta\omega^2 = 2\omega_k\delta\omega = \frac{(\delta\mathcal{L}\boldsymbol{u}_k, \boldsymbol{u}_k) - \omega_k^2(\delta\rho\boldsymbol{u}_k, \boldsymbol{u}_k)}{(\rho\boldsymbol{u}_k, \boldsymbol{u}_k)} \, . \tag{9.15}$$

Note that the eigenvector perturbation $\delta\boldsymbol{u}_k$ is absent in (9.15). This insensitivity to eigenvector perturbations is reminiscent of the absence of the raypath perturbation in Equation (7.1) for delay times. Thus, we need only to compute the eigenvectors for a background model, just as we appealed to Fermat's principle to compute ray trajectories only in the background model to find travel time anomalies.

Table 9.2. *Partial derivatives for eigenfrequencies. U, V and W are normalized using (9.9). Dots indicate differentiation with respect to r.*

Toroidal oscillations

$$2\omega K_\mu = (r\dot{W} - W)^2 + (\nu^2 - 2)W^2$$
$$2\omega K'_\rho = -\omega^2 r^2 W^2$$

Spheroidal oscillations

$$2\omega K_\kappa = (r\dot{U} + 2U - \nu V)^2$$
$$2\omega K_\mu = \frac{1}{3}\left(2r\dot{U} - 2U + \nu V\right)^2 + (r\dot{V} - V + \nu U)^2 + (\nu^2 - 2)V^2$$
$$2\omega K'_\rho = -\omega^2 r^2 (U^2 + V^2) + 8\pi G\rho r^2 U^2 + 2r^2\left(U\dot{\Phi} + \frac{\nu V\Phi}{r}\right) - 2grU(2U - \nu V) - 8\pi G r^2 \int_r^a \rho U \frac{2U - \nu V}{r'} \mathrm{d}r'$$

Conversions

$$K_\mathrm{P} = 2\rho V_\mathrm{P} K_\kappa$$
$$K_\mathrm{S} = 2\rho V_\mathrm{S}\left(K_\mu - \frac{4}{3}K_\kappa\right)$$
$$K_\rho = \frac{\kappa}{\rho}K_\kappa + V_\mathrm{S}^2 K_\mu + K'_\rho$$

Equation (9.15) then provides a linear relationship between $\delta\omega^2$ and perturbations $\delta\rho$ in density, as well as perturbations in Lamé's constants $\delta\mu$ and $\delta\lambda$ that are implicit in the operator perturbations $\delta\mathcal{L}$. For spherically symmetric perturbations in $\rho$, $\mu$ and $\kappa = \lambda + \frac{2}{3}\mu$, the integration over the two angle parameters yields a constant that divides out in (9.15), leaving only an integration over radius $r$:

$$\delta\omega = \int [K_\kappa \delta\kappa + K_\mu \delta\mu + K'_\rho \delta\rho]\, \mathrm{d}r. \tag{9.16}$$

Table 9.2 gives expressions of the partial derivatives $K_X$ of the eigenfrequencies, scaled by a factor $2\omega$. Though $\kappa$, $\mu$ and $\rho$ are usually the preferred independent parameters when inverting eigenfrequencies, a parametrization directly in terms of the seismic velocities $V_\mathrm{P}$ and $V_\mathrm{S}$ is needed when eigenfrequency data are combined with travel times:

$$\delta\omega = \int [K_\mathrm{P} \delta V_\mathrm{P} + K_\mathrm{S} \delta V_\mathrm{S} + K_\rho \delta\rho]\, \mathrm{d}r. \tag{9.17}$$

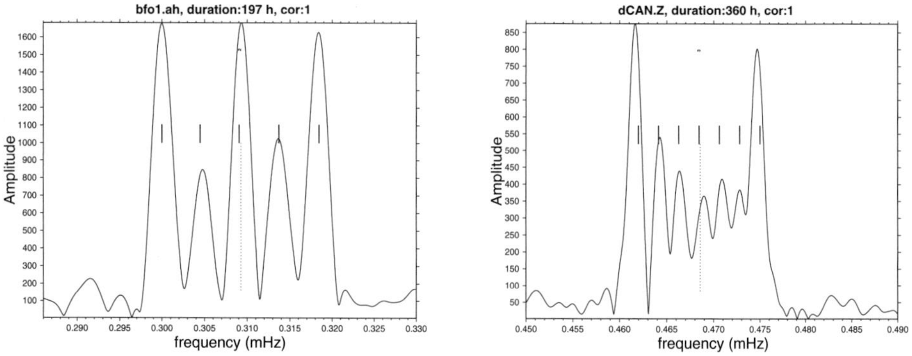


Fig. 9.4. The very strong earthquake of Dec. 24, 2006 that caused a disastrous tsunami in the Indian Ocean allowed the signal/noise ratio to remain high for more than a week. Such very long time records allow for the resolving of line splitting of the Earth's lowest eigenvibrations, such as ${}_0S_2$ (left) and ${}_0S_3$ (right) from the recording in only one station. The vertical lines gives the predicted splitting due to the Coriolis force. Figure courtesy Geneviève Roult.

We use a $'$ on $K_\rho$ if the density is varied while keeping $\kappa$ and $\mu$ constant. The conversion rules for the kernels are given in Table 9.2 (see also Section 4.7 and Exercise 9.3). Rayleigh's Principle has been widely applied to adjust the spherically symmetric Earth models to satisfy observed normal mode frequencies. To study the effect of lateral variations on the wavefield $\boldsymbol{u}$, we need to find the eigenvectors belonging to the asymmetric Earth; alternatively, we may decompose the eigenvector in travelling waves ('surface waves') and apply ray theory along a segment of the surface wave path. Here we shall take a brief look at the first option. Surface waves are treated in Chapter 10.

*Exercise*

**Exercise 9.3** Derive the conversions in Table 9.2 by expressing $\delta\kappa = \delta[\rho(V_P^2 - \frac{4}{3}V_S^2)]$ and $\delta\mu = \delta(\rho V_S^2)$ in terms of perturbations in $\rho$, $V_P$ and $V_S$.

## 9.3 Mode splitting

Three mechanisms break the symmetry – and with that the degeneracy of the eigenvalue problem: the rotation, the ellipticity and the lateral heterogeneity of the Earth (Figure 9.4). Park [252] showed that coupling of toroidal and spheroidal modes by the Coriolis force due to the Earth's rotation only affects the fundamental ($n = 0$) modes at frequencies lower than about 4 mHz (angular orders 9–23), though occasional coupling effects can be noted for higher modes. The heterogeneity is the most important effect in tomographic interpretations.

We may use the fact that the set of eigenfunctions for the background model span a basis in which the new ('perturbed') eigenfunctions can be expanded. Only eigenfunctions with almost equal eigenfrequency interact; we can therefore limit the problem by envisaging a seismogram band-filtered around eigenfrequency $\omega_k$. Normally, we try to keep the band so narrow that it contains only one multiplet, in which case the index $k$ is shorthand for just one pair of quantum numbers $(n, \ell)$ and the mode type (toroidal, spheroidal). For closely spaced multiplets, coupling across branches with different $n$ or along the mode branch with the same $n$ but for different $\ell$ or mode type may force us to extend this to a larger set, as shown by Deuss and Woodhouse [86, 87]. In the following, we ignore the minor complications introduced by such coupled multiplets (see Masters et al. [201] for a particularly clear discussion of this issue). In the unperturbed Earth, there are then $2\ell + 1$ eigenfunctions $(-\ell \leq m \leq \ell)$ with the same eigenfrequency $\omega_k$ that satisfy (9.14), whereas each such spectral line splits in $2\ell + 1$ singlets when the Earth is perturbed away from spherical symmetry. The wavefield has three components and is a function of location $\boldsymbol{r} = (r, \theta, \phi)$, although we omit the spatial dependence in the notation.

We now again analyse the equations of motion for a perturbed Earth around the centre frequency $\omega_k$ of a multiplet, but recognize that the eigenvector perturbations may not be small.

$$(\rho + \delta\rho)(\omega_k^2 + \delta\omega^2)\boldsymbol{u} = (\mathcal{L} + \delta\mathcal{L})\boldsymbol{u} \, .$$

As long as we can ignore the coupling with other multiplets, the space spanned by the eigenvectors of the perturbed Earth is the same as that spanned by the original eigenvectors $\boldsymbol{u}_k$. Thus, the wavefield $\boldsymbol{u}$ can be expressed in terms of the unperturbed eigenvectors:

$$\boldsymbol{u} = \sum_m \zeta_m \boldsymbol{u}_m \, . \tag{9.18}$$

Inserting this, applying (9.14) and retaining first order only:

$$\sum_m \omega_k^2 \delta\rho \zeta_m \boldsymbol{u}_m + \delta\omega^2 \rho \zeta_m \boldsymbol{u}_m = \sum_m \zeta_m \delta\mathcal{L} \boldsymbol{u}_m \, .$$

We multiply by $\boldsymbol{u}_n^*$, integrate, and use the orthogonality (9.12):

$$\sum_m [(\delta\mathcal{L}\boldsymbol{u}_m, \boldsymbol{u}_n) - \omega_k^2(\delta\rho\boldsymbol{u}_m, \boldsymbol{u}_n)]\zeta_m = \delta\omega^2\zeta_n = 2\omega_k\delta\omega\zeta_n \, ,$$

to get:

$$\sum_m H_{nm}\zeta_m = \delta\omega\zeta_n \, , \tag{9.19}$$

where we define the elements of the real, symmetric splitting matrix $\boldsymbol{H}$ for multiplet $k$ due to lateral heterogeneity:

$$2\omega_k H_{nm} = (\delta\mathcal{L}\boldsymbol{u}_m, \boldsymbol{u}_n) - \omega_k^2(\delta\rho\boldsymbol{u}_m, \boldsymbol{u}_n). \tag{9.20}$$

Equation (9.19) is an eigenvalue problem for the vector of coefficients $(\zeta_1, \zeta_2, ...)$ with eigenvalue $\delta\omega$. Arranging the coefficients for each eigenvector as columns in a matrix $\boldsymbol{Z}$, such that $Z_{ij}$ is the coefficient $\zeta_i^{(j)}$ for eigenvector $\boldsymbol{s}_j$, and writing $\boldsymbol{\Omega}$ for diag$\{\delta\omega_1, \delta\omega_2, ...\}$ we get (Woodhouse and Dahlen [399]):

$$\boldsymbol{H}\boldsymbol{Z} = \boldsymbol{Z}\boldsymbol{\Omega}\,. \tag{9.21}$$

Since $\mathcal{L}$, and thus also $\delta\mathcal{L}$, is Hermitian, the matrix $\boldsymbol{H}$ is Hermitian and $\boldsymbol{Z}$ is consequently unitary, i.e. the complex conjugate transpose of $\boldsymbol{Z}$ is its inverse: $\boldsymbol{Z}^{-1} = (\boldsymbol{Z}^*)^T$, or:

$$\sum_m Z_{mi}^* Z_{mj} = \sum_m Z_{im}^* Z_{jm} = \delta_{ij}\,. \tag{9.22}$$

The eigenvectors of the perturbed Earth are, explicitly:

$$\boldsymbol{s}_j = \sum_m \zeta_m^{(j)}\boldsymbol{u}_m = \sum_m Z_{mj}\boldsymbol{u}_m\,, \tag{9.23}$$

where index $m$ sums over all the modes that participate in the multiplet. In a more complete analysis, the splitting matrix $\boldsymbol{H}$ is a complicated amalgam of terms relating to elastic perturbations, perturbations in the depths of discontinuities, initial stress, anisotropy, ellipticity and (for the lowest eigenfrequencies) Coriolis force terms induced by the Earth's rotation.

The splitting matrix $\boldsymbol{H}$ defines the evolution of the wavefield, as we show in the following. Suppose the initial amplitudes of the singlets are given by $a_j$ in a symmetric Earth and by $b_j$ in the perturbed Earth. The wavefield $\boldsymbol{u}_0$ in the unperturbed Earth is then:

$$\boldsymbol{u}_0(t) = \sum_m a_m\boldsymbol{u}_m \mathrm{e}^{-\mathrm{i}\omega_k t}\,,$$

but in the perturbed Earth:

$$\boldsymbol{u}(t) = \sum_j b_j\boldsymbol{s}_j \mathrm{e}^{-\mathrm{i}(\omega_k+\delta\omega_j)t}\,.$$

Although one commonly determines the coefficients $a_i$ by representing the seismic source by torques with 'equivalent' forces (a moment tensor), in the present context it is actually more transparent to follow Vlaar [387] and see the Earth's motion as more naturally arising from an initially stressed situation, i.e. as an initial value problem with a nonzero deflection $\boldsymbol{u}(0)$. At $t = 0$ this displacement field does not yet 'know' if it is in the perturbed or unperturbed Earth so $\boldsymbol{u}_0(0) = \boldsymbol{u}(0)$, from which we deduce that

$$\sum_j b_j \boldsymbol{s}_j = \sum_{jm} b_j Z_{mj} \boldsymbol{u}_m = \sum_m a_m \boldsymbol{u}_m \quad \rightarrow \quad \sum_j b_m Z_{mj} = a_m ,$$

or, reversely:

$$b_j = \sum_m Z^*_{mj} a_m .$$

The wavefield in the perturbed Earth can be written as

$$\begin{aligned} \boldsymbol{u}(t) &= \sum_j b_j \boldsymbol{s}_j \mathrm{e}^{-\mathrm{i}(\omega_k + \delta\omega_j)t} \\ &= \sum_j \left( \sum_m Z^*_{mj} a_m \right) \left( \sum_{m'} Z_{m'j} \boldsymbol{u}_{m'} \right) \mathrm{e}^{-\mathrm{i}(\omega_k + \delta\omega_j)t} . \end{aligned} \tag{9.24}$$

To understand the following step it helps to assume that time is short, i.e. the heterogeneities have not had sufficient time to perturb the wavefield much because $\delta\omega_j t \ll 1$, so that $\exp(-\mathrm{i}\delta\omega_j t) \approx 1 - \mathrm{i}\delta\omega_j t$. Using the eigenvector properties of $\boldsymbol{Z}$: $\sum_j Z_{m'j} \delta\omega_j Z^*_{mj} = H_{m'm}$ and $\sum_j Z_{m'j} Z^*_{mj} = \delta_{m'm}$:

$$\boldsymbol{u}(t) = \sum_{m'm} (\delta_{m'm} - \mathrm{i}t H_{m'm}) a_m \boldsymbol{u}_{m'} \mathrm{e}^{-\mathrm{i}\omega_k t} . \tag{9.25}$$

The result of Exercise 9.5 shows that we could have continued the Taylor expansion, and used the fact that $\boldsymbol{Z}$ is unitary to drop the assumption of a short time span and write this as:

$$\boldsymbol{u}(t) = \sum_{m'm} (\mathrm{e}^{-\mathrm{i}t\boldsymbol{H}})_{m'm} a_m \boldsymbol{u}_{m'} \mathrm{e}^{-\mathrm{i}\omega_k t} , \tag{9.26}$$

where $\mathrm{e}^{-\mathrm{i}t\boldsymbol{H}}$ is defined by its Taylor expansion:

$$\mathrm{e}^{-\mathrm{i}t\boldsymbol{H}} = \boldsymbol{I} - \mathrm{i}t\boldsymbol{H} - \frac{t^2}{2!}\boldsymbol{H}^2 + \dots$$

We developed this result in the time domain, but the practical use is in the frequency domain, where the spectra of $\boldsymbol{u}$ are a function of location $\boldsymbol{r}$ through the eigenvectors $\boldsymbol{u}_{m'}$. The differences in the spectrum of an isolated multiplet therefore contain information about $\boldsymbol{H}$. Though the forward problem that leads to (9.26) is straightforward, the inverse problem – to find $\boldsymbol{H}$ from observed spectra – is significantly more complicated. But if we can extract $\boldsymbol{H}$ from the multiplet observations, (9.20) gives us a set of linear constraints on the Earth's heterogeneity.

To find $\boldsymbol{H}$ requires ingenuity. Fitting multiplet spectra using (9.26) – after introducing a complex component to $\omega_k$ as in (9.13) to account for the widening of spectral peaks by attenuation – is nonlinear. Nonlinear optimization problems such as this one become increasingly difficult as the number of model parameters increases and the process may get stuck in local minima of the misfit function. Early efforts, notably by Giardini et al. [111] and Ritzwoller et al. [282], reduced the number of model parameters significantly by developing $\delta\rho$ and the elastic perturbations in $\delta\mathcal{L}$ also in spherical harmonics, for example:

$$\delta\rho(\boldsymbol{r}) = \sum_{s,t} \delta\rho_{st}(r) Y_s^t(\theta, \phi),$$

and similar equations for $\mu$ and $\lambda$ or the incompressibility $\kappa$. In this case the computation of $\boldsymbol{H}$ using (9.20) involves triple products of spherical harmonics. Such integral products are well known in quantum mechanics:

$$\int_0^{2\pi} \int_0^{\pi} Y_\ell^{m*}(\theta, \phi) Y_s^t(\theta, \phi) Y_\ell^{m'}(\theta, \phi) \sin\theta \, \mathrm{d}\theta \mathrm{d}\phi = \gamma_s^{mm't}, \tag{9.27}$$

where $\gamma_s^{mm't}$ can be expressed in terms of Wigner 3j symbols that can be computed by recursion (Luscombe and Luban [191]). The $\gamma_s^{mm't}$ are nonzero only if $s$ is even, $0 \le s \le 2\ell$ and $t = m - m'$. For an isolated multiplet, in which only one angular order $\ell$ is present, these 'selection rules' help to reduce the number of parameters that we need to vary to optimize the data fit. The use of expansion coefficients in spherical harmonics leads to a compact expression for the splitting matrix in terms of the Earth's heterogeneity:

$$H_{mm'} = (a + bm + cm^2)\delta_{mm'} + \sum_{st} \gamma_s^{mm't} c_{st}, \tag{9.28}$$

where

$$\begin{aligned} c_{st} = & \int \left[ C_s^\rho(r)\delta\rho_{st}(r) + C_s^\mu(r)\delta\mu_{st}(r) + C_s^\kappa(r)\delta\kappa_{st}(r) \right] r^2 \mathrm{d}r \\ & - \sum_d [D_s^\rho(r_d)\rho(r_d) + D_s^\mu \mu(r_d) + D_s^\kappa \kappa(r_d)]_-^+ r_d^2 \delta r_{st}(r_d), \end{aligned} \tag{9.29}$$

and where for completeness we add the effects of boundary perturbations $\delta r(\boldsymbol{r}) = \sum_{st} \delta r_{st}(r_d) Y_s^t(\theta, \phi)$ at radii $r_d$, as well as the effects of rotation and ellipticity, described by the known coefficients $a$, $b$ and $c$ listed in Table 14.1 of Dahlen and Tromp [78]. The integral kernels $C_s$ and the discontinuity kernels $D_s$ are listed in Tables 9.3 and 9.4. Thus, if we can estimate $\boldsymbol{H}$ from the data, (9.28) and (9.29) give us linear constraints on the lateral heterogeneity of the Earth. Giardini et al. [111] and Ritzwoller et al. [282] determine the $c_{st}$ such that spectra predicted

Table 9.3. *Integral kernels $C_s^X(r)$. $U$, $V$ and $W$ are normalized as in (9.9).*

Toroidal oscillations

$$2\omega C_s^{\mu} = \left[\nu^2 - \frac{1}{2}s(s+1)\right]\frac{1}{\nu^2}\left(\dot{W} - \frac{W}{r}\right)^2 + \left[\nu^2(\nu^2-2) - \frac{1}{2}s(s+1)(4\nu^2 - s(s+1) - 2)\right]\frac{W^2}{\nu^2 r^2}$$

$$2\omega C_s^{\rho} = -\left[\nu^2 - \frac{1}{2}s(s+1)\right]\frac{\omega^2 W^2}{\nu^2}$$

Spheroidal oscillations

$$F = \frac{2U - \nu V}{r}, \qquad 2\omega G_s^{(2)} = \rho\left[-2UF + \frac{1}{2}s(s+1)\frac{UV}{\nu r}\right]$$

$$2\omega G_s^{(1)} = \rho\left[\frac{s(s+1)}{r^2}U^2 + \frac{1}{2\nu r}s(s+1)\left[U\dot{V} - V(\dot{U} + 2F - \frac{U}{r})\right]\right]$$

$$2\omega C_s^{\kappa} = (\dot{U} + F)^2$$

$$2\omega C_s^{\mu} = \frac{1}{3}\left(2\dot{U} - F\right)^2 + \left[\nu^2 - \frac{1}{2}s(s+1)\right]\left(\frac{r\dot{V} - V + \nu U}{\nu r}\right)^2 + \left[\nu^2(\nu^2-2) - \frac{1}{2}s(s+1)(4\nu^2 - s(s+1) - 2)\right]\frac{V^2}{\nu^2 r^2}$$

$$\begin{aligned} 2\omega C_s^{\rho} &= 8\pi G\rho U^2 - gU\left(F + \frac{2U}{r}\right) + 2U\dot{\Phi} - \omega^2 U^2 \\ &+ \left[\nu^2 - \frac{1}{2}s(s+1)\right]\left(-\omega^2\frac{V^2}{\nu^2} + \frac{2V\Phi}{\nu r} + \frac{gUV}{\nu r}\right) \\ &+ \frac{4\pi G}{2s+1}\left[r^s\int_r^a \frac{1}{(r')^s}[(s+1)G_s^{(2)} - r'G_s^{(1)}]\mathrm{d}r' \right. \\ &\left. - \frac{1}{r^{s+1}}\int_0^r (r')^{s+1}\left[sG_s^{(2)} + r'G_s^{(1)}\right]\mathrm{d}r\right] \end{aligned}$$

with the Fourier transform of (9.26) give an optimal fit to the observed spectra. In this approach one makes no attempt to find the individual peaks of a multiplet, but models the variations in multiplet shape as a function of station location. An iterative gradient search for the optimal $c_{st}$ leads to an acceptable (though not always perfect) fit of predicted to observed multiplets. However, because of the strong nonlinearity, the method is time-consuming and there may be tradeoffs between the excitation at the source and the anelastic structure assumed (or simultaneously inverted for). For more details we refer the reader to the literature cited in this chapter. Instead, in

Table 9.4. *Influence functions $D_s^X(r)$ for boundary depth perturbations. U, V and W are normalized as in (9.9).*

Toroidal oscillations

$$2\omega D_s^{\mu} = 2\omega C_s^{\mu} - 2[\nu^2 - \frac{1}{2}s(s+1)]\frac{\dot{W}}{\nu^2}\left(\dot{W} - \frac{W}{r}\right)$$

$$2\omega D_s^{\rho} = 2\omega C_s^{\rho}$$

Spheroidal oscillations

$$2\omega D_s^{\kappa} = 2\omega C_s^{\kappa} - (\dot{U} + F)\left[2\dot{U} - s(s+1)\frac{V}{\nu r}\right]$$

$$2\omega D_s^{\mu} = 2\omega C_s^{\mu} - \frac{2}{3}\left(2\dot{U} - F\right)\left[2\dot{U} - s(s+1)\frac{V}{\nu r}\right]$$
$$-2\left[\nu^2 - \frac{1}{2}s(s+1)\right]\frac{\dot{V}}{\nu^2 r}\left(r\dot{V} - V + \nu U\right)$$

$$2\omega D_s^{\rho} = 2\omega C_s^{\rho}$$

the next section we shall present a more recent technique based on autoregressive filtering to recover $\boldsymbol{H}$ from a set of seismograms.

The selection rules also tell us about limitations in sensitivity of split spectra. For example, heterogeneity that is characterized by an odd spherical harmonic does not split the spectral line. Nor does small heterogeneity with $s > 2\ell$ split the line for angular order $\ell$. Since the spectral lines can only be observed for relatively low angular order, this illustrates the fact that the splitting functions have a resolution limited to very long wavelengths only.

### *Exercise*

**Exercise 9.4** For $s = 0$, it is easy to check that the expressions for $r^2 C_s^X$ reduce to those for $K_X$ in Table 9.2. Why would you expect that?

## 9.4 Observations of mode splits

Masters et al. [201] describe a technique to estimate the splitting matrix $\boldsymbol{H}$ based on the temporal evolution of the wavefield. We filter a large number of seismograms in a narrow band to isolate multiplet $k$. Using (9.26) we write the seismogram for

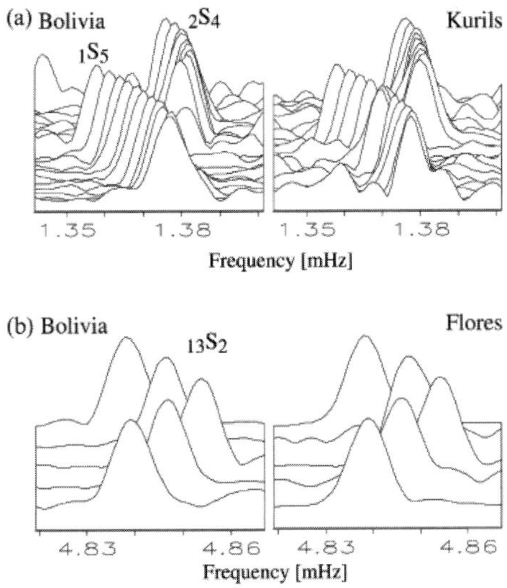


Fig. 9.5. Top: Receiver strips for spheroidal modes $n = 1, \ell = 5$ and $n = 2, \ell = 4$ from earthquakes in Bolivia and Kuril islands. Bottom: receiver strips for the spheroidal mode with $n = 13, \ell = 2$. In both cases there are few modes in the multiplet, and individual frequencies $\delta\omega_j$ become visible through the receiver stripping operation. From Masters et al. [201], reproduced with permission from Blackwell Publishing.

a component in the direction given by unit vector $\hat{\boldsymbol{n}}$ in station $j$ as:

$$\begin{aligned} s_j(t) = \hat{\boldsymbol{n}} \cdot \boldsymbol{u}(\boldsymbol{r}_j, t) &= \sum_{m'm} (\delta_{m'm} - \mathrm{i}t H_{m'm} + ...) a_m \hat{\boldsymbol{n}} \cdot \boldsymbol{u}_{m'}(\boldsymbol{r}_j) \mathrm{e}^{-\mathrm{i}\omega_k t} \\ &= \sum_{m'} a_{m'}(t) B_{jm'} \mathrm{e}^{-\mathrm{i}\omega_k t}, \end{aligned} \tag{9.30}$$

where

$$\begin{aligned} a_{m'}(t) &= \sum_{m} (\delta_{m'm} - \mathrm{i}t H_{m'm} + ...) a_m \\ B_{jm'} &= \hat{\boldsymbol{n}} \cdot \boldsymbol{u}_{m'}(\boldsymbol{r}_j) . \end{aligned}$$

We arrange all excitation coefficients $a_{m'}$ in a vector $\boldsymbol{a}$ of length $2\ell + 1$ that includes the time factor:

$$\boldsymbol{a}(t) = \mathrm{e}^{-\mathrm{i}t\boldsymbol{H}} \boldsymbol{a}(0) . \tag{9.31}$$

We now rank all seismograms in a time-dependent vector $\boldsymbol{s}(t) = [s_1(t), s_2(t), ...]$:

$$\boldsymbol{s}(t) = \boldsymbol{B} \cdot \boldsymbol{a}(t) \mathrm{e}^{-\mathrm{i}\omega_k t} . \tag{9.32}$$

Since $\omega_k$ and $\boldsymbol{B}$ are known, and $\boldsymbol{s}(t)$ are data, (9.32) represents an inverse problem for $\boldsymbol{a}(t)$ at every instant $t$. If we have more than $2\ell+1$ seismograms, we have an overdetermined system for the unknown coefficients $\boldsymbol{a}(t)$ at time $t$ and we can use the method of least squares to solve for the best fitting $\boldsymbol{a}$. Anticipating a more rigorous treatment of the method of least squares in Section 14.1, for now it suffices to say that we reduce (9.32) to a square matrix equation by multiplying both sides with the transpose matrix $\boldsymbol{B}^T$. We then invert the $(2\ell+1)\times(2\ell+1)$ matrix $\boldsymbol{B}^T\boldsymbol{B}$ to obtain an estimate $\tilde{\boldsymbol{a}}$:

$$\tilde{\boldsymbol{a}}(t) = (\boldsymbol{B}^T\boldsymbol{B})^{-1}\boldsymbol{B}^T\boldsymbol{s}(t)\mathrm{e}^{\mathrm{i}\omega_k t}.$$

In this way we obtain individual time series for each of the components of the multiplet, rather than the mix of components recorded in each receiver. Inverting for the effects of $\boldsymbol{B}$ means that we are using the amplitude differences in receivers to disentangle individual modes. The vector $\tilde{\boldsymbol{a}}$ (or $\tilde{\boldsymbol{a}}(t)\mathrm{e}^{\mathrm{i}\omega_k t}$ in the original formulation of the method) is therefore known as a 'receiver strip'. It contains as many time series as there are individual modes in the multiplet, generally $2\ell+1$ (Figure 9.5). If the data are noise-free and satisfy (9.32) exactly then $\tilde{\boldsymbol{a}}(t)=\boldsymbol{a}(t)$, so we expect $\tilde{\boldsymbol{a}}$ to behave like $\boldsymbol{a}$. Therefore, because of (9.31):

$$\tilde{\boldsymbol{a}}(t+\delta t) = \boldsymbol{P}(\delta t)\tilde{\boldsymbol{a}}(t)\,,$$

with

$$\boldsymbol{P}(\delta t) = \mathrm{e}^{\mathrm{i}\delta t\boldsymbol{H}}\,.$$

In practice, we work in the frequency domain, and use the fact that spectra over different time windows evolve in a simple manner. We divide each of the time series into time windows, and denote the spectrum of $\tilde{\boldsymbol{a}}$ over window $n$ by $\tilde{\boldsymbol{a}}_n(\omega)$. Then $\boldsymbol{P}(\delta t)$ is determined by the spectrum of the next window

$$\tilde{\boldsymbol{a}}_{n+1}(\omega) = \boldsymbol{P}(\delta t)\tilde{\boldsymbol{a}}_n(\omega)\,.$$

Because we estimate $\boldsymbol{P}$ – and with that $\boldsymbol{H}$ – from phase differences between different time windows, the determination of splitting coefficients is independent of the modal excitation coefficients, and knowledge of the source is not needed, in contrast to the earlier methods that rely on fitting the stripped multiplets. If we transpose the system, we obtain:

$$\boldsymbol{A}_n\boldsymbol{P}^T = \boldsymbol{A}_{n+1}\,, \tag{9.33}$$

where the matrix $\boldsymbol{A}$ has a column for each of the $2\ell+1$ modes in the multiplet and a row for each frequency:

$$\boldsymbol{A} = \begin{pmatrix} \tilde{a}_1(\omega_1) & \tilde{a}_2(\omega_1) & \dots & \tilde{a}_{2\ell+1}(\omega_1) \\ \tilde{a}_1(\omega_2) & \tilde{a}_2(\omega_2) & \dots & \tilde{a}_{2\ell+1}(\omega_2) \\ \vdots & & & \\ \tilde{a}_1(\omega_M) & \tilde{a}_2(\omega_M) & \dots & \tilde{a}_{2\ell+1}(\omega_M) \end{pmatrix}.$$

Equation (9.33) can be solved for every column of $\boldsymbol{A}_{n+1}$ as right-hand side, each producing an estimate of the corresponding column of $\boldsymbol{P}^T$. The system can be solved using singular value decomposition (see Section 14.3). Once $\boldsymbol{P}$ is known, we may diagonalize it:

$$\boldsymbol{P} = \boldsymbol{Z}\boldsymbol{D}\boldsymbol{Z}^{-1},$$

where we identify, (see Exercises 9.5 and 9.6):

$$\boldsymbol{D} = \exp[-\mathrm{i}\delta t\boldsymbol{\Omega}], \tag{9.34}$$

so that we can retrieve $\boldsymbol{H} = \boldsymbol{Z}\boldsymbol{\Omega}\boldsymbol{Z}^{-1}$ from the elements of $\boldsymbol{D}$.

We have assumed so far that the anelasticity in the Earth is spherically symmetric. In that case, Rayleigh's Principle can be used (with a small imaginary component to $V_\mathrm{P}$ and $V_\mathrm{S}$) to compute a small imaginary component to $\omega_k$. This takes care of the broadening of the spectral peaks. An alternative is to allow for 3D perturbations in $Q$. In that case $\boldsymbol{H}$ is not Hermitian and we must explicitly compute the inverse of $\boldsymbol{Z}$; but apart from this complication the methods outlined in this chapter are still useful.

## *Exercises*

**Exercise 9.5** using $\sum_j Z_{m'j}\delta\omega_j Z^*_{mj} = H_{m'm}$ and (9.22), show that

$$\sum_j Z^*_{mj}\mathrm{e}^{-\mathrm{i}\delta\omega_j t} Z_{m'j} = (\mathrm{e}^{-\mathrm{i}t\boldsymbol{H}})_{m'm}$$

by expansion in a Taylor series.

**Exercise 9.6** Prove (9.34) using (9.21).

# 10

# Surface wave interpretation: ray theory

In Chapter 9 we developed the wavefield into standing waves involving the whole planet (or Sun). We saw that localized heterogeneities influence the wavefield through the interaction of modes with different azimuthal order $m$. As the angular order $\ell$ grows, it quickly becomes impractical from a computational viewpoint to calculate the $2\ell + 1$ splitting coefficients, and from an observational viewpoint it becomes impossible to estimate their values from seismic recordings.

Recall that the displacements of normal modes are given by the spherical harmonics $Y_\ell^m(\theta, \phi)$ or their derivatives. As the angular order $\ell$ grows, such harmonics start to behave like ordinary harmonics, i.e. cosine and sine functions, except near the source or antipode. An example is shown in Figure 10.1. Clearly, the zero crossings of $Y_{20}^0$, given by the small circles in the figure, are very regularly spaced, at least away from the source. This is confirmed by the asymptotic expression for the associated Legendre function, originally due to Sommerfeld [337]:

$$P_\ell^m(\theta) \to \ell^m \left(\frac{2}{\pi \ell \sin\theta}\right)^{\frac{1}{2}} \cos\left[\left(\ell + \frac{1}{2}\right)\theta - \frac{\pi}{4} - \frac{m\pi}{2}\right]. \tag{10.1}$$

Such asymptotic properties allow for a different approach, one that involves travelling waves. This is true even though the wave pattern in Figure 10.1 is a standing wave pattern. In fact, if the wave did not have a beginning, and if we could assume it was not damped away with time, the viewpoint of the standing wave would be the most appropriate. But now consider the first half hour or so after the earthquake. It would violate the causal laws of physics if we assumed that the standing wave was established right at the origin time of the earthquake, because that would mean that the wave travels faster than the intrinsic seismic velocity of the rock. How can that be?

A deeper analysis involving the complex $\ell$-plane (upon which we briefly touch in Appendix C) shows that the causal behaviour is due to the summation of many modes. The modes interfere destructively, resulting in zero displacement, until the

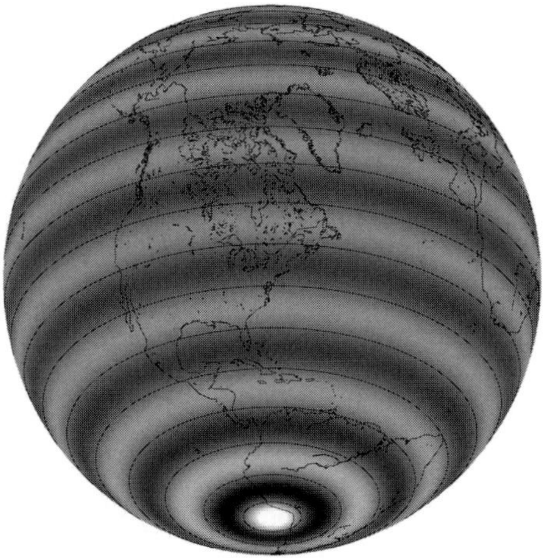

Fig. 10.1. Amplitude of the vertical displacement of mode ${}_0S_{20}^0$ from a hypothetical earthquake off the coast of Peru. Note the regularity of the nodal line spacing.

laws of causality allow the Earth to move. The further away one is from the source, the later that point in time is. Thus, the sum of waves exhibits a wavefront that travels. Reversely, each component in the sum also contains a travelling behaviour: a closer look at (10.1) shows that the cosine function introduces factors $e^{-i(\omega t-\ell\theta)}$ and $e^{-i(\omega t+\ell\theta)}$ when recombined with the time behaviour $e^{-i\omega t}$. But this means that the standing wave can be seen as a pair of travelling waves in positive and negative $\theta$-direction, travelling with a phase velocity $d\theta/dt = \omega/(\ell + \frac{1}{2})$. In fact, these are the waves we observe after an earthquake. The two waves meet after passing the antipode; they interfere and are able to set up a standing wave pattern. If we were to isolate a very narrow frequency band that only contains mode ${}_0S_{20}$, we should in fact be able to observe a standing wave pattern such as that in Figure 10.1 if we wait long enough for the standing wave to establish itself. Physicists will recognize the parallel with Bohr's analysis of the hydrogen atom in which the electron orbits a proton but can only occupy orbits that lead to constructive interference, leading to a discrete set of spectral lines.

In practice, we always consider a frequency band of finite width, and ${}_0S_{20}$ interferes with its neighbours ${}_0S_{19}$ and ${}_0S_{21}$. This interference will again introduce the impression of a travelling wave, but now governed by the group velocity, which – as we shall see later – is proportional to $d\omega/d\ell$. In this view, we see surface waves as the degeneration of a discrete normal mode spectrum into overlapping spectral peaks that are widened by attenuation and splitting. Except for some very high $Q$ modes, the degeneration is complete in the Earth (in the sense that it becomes very hard to isolate single modes) as $l \gtrsim 50$ or $\omega \gtrsim 0.1$ rad/s. Isolated solar modes have been observed for much higher $\ell$.

Because the eigenvectors such as ${}_nU_\ell^m(r)$ have most of their energy near the surface as $\ell$ increases, such high$-\ell$ modes usually go by the name of 'surface waves'. Seismic surface waves are the most prominent arrivals on a seismogram, but they easily fall between the cracks of a theoretical development: they do not offer the simplicity of a discrete spectrum, such as normal modes do, nor that of a simple wavefront, such as is the case for P- or S-waves. From the point of view of ray theory, in very heuristic terms, surface waves can also be seen as multiply reflected S-waves that arrive roughly at the same time and interfere (such as SSSS and SSSSS). As they traverse regions with different elastic properties they speed up and slow down much as we have seen for body waves. The theory of propagation of surface waves in laterally heterogeneous media is extensive, starting in the mid 1970s with a pioneering paper by Woodhouse [397] and a number of papers in the Eastern European literature that have received less notice in the West but are summarized in Keilis-Borok [157]. Much of the early theory deals with surface waves in a halfspace, but in this chapter we shall faithfully maintain the perspective of a spherical Earth.

## 10.1 The theory of surface waves

As we discussed in the introduction to this chapter, and show in more detail in Appendix C, the wavefield for large $\ell$ or $\omega$ can be expressed as a sum of surface wave modes with a continuous spectrum rather than a discrete one. For example, the spectrum of the vertical component of displacement at distance $\Delta$ and azimuth $\zeta_s$ from the source can be written as:

$$u_r(\Delta, \zeta_s, \omega) = \sum_n A_n(\omega, \Delta, \zeta_s) \exp[ik_n(\omega)\Delta] \,. \tag{10.2}$$

In the time domai:

$$u_r(\Delta, \zeta_s, t) = \sum_n \int A_n(\omega, \Delta, \zeta_s) \exp[i(k_n(\omega)\Delta - \omega t)] d\omega \,. \tag{10.3}$$

The integrand in this integral is oscillating heavily, and positive and negative contributions will cancel if $A_n$ is sufficiently smooth as a function of $\omega$. Only where the exponent is stationary as a function of frequency do we expect a significant contribution to the time signal. This occurs when:

$$\frac{d}{d\omega}[k_n(\omega)\Delta - \omega t] = 0 \,,$$

or when

$$\frac{\Delta}{t} = \left(\frac{dk_n}{d\omega}\right)^{-1} = \left(\frac{d\omega_n}{dk}\right) = \mathcal{U}_n \,. \tag{10.4}$$

Clearly, $\mathcal{U}_n$ has the dimension of velocity. Because it tells us at what time a wavegroup with frequency $\omega$ arrives, it is named the group velocity.

In contrast to the situation for body waves, where we usually express distances in degrees, the epicentral distance $\Delta$ for surface waves is given in units of m or km:

$$\Delta = a\theta,$$

with $a$ the radius of the Earth or the Sun. Comparing (10.2) with (10.1) we recognize

$$k = \frac{\ell + \frac{1}{2}}{a} = \frac{\nu}{a}.$$

Note that the discrete ${}_n\omega_\ell$ is replaced by a continuous function of $k$: $\omega_n(k)$. The distance travelled by a surface wave is normally expressed in km rather than radians, so $k$ and $\Delta$ take the role of $\nu$ (or $\ell$) and $\theta$. We will keep using $\theta$, though, when this gives an economy of notation, e.g. in the geometrical spreading that depends on $\sin\theta$.

The surface wave amplitude $A$ is the product of three terms: the excitation $\mathcal{S}$ at the source, the projection on the seismograph component at the site of the receiver, and a propagation term. Various derivations for the amplitude of surface waves in a spherical Earth exist in the literature, notably Snieder and Nolet [334], Dahlen and Tromp [78] and Zhou et al. [419]. The expressions given here are the complex conjugate of those given in these references because of a different Fourier transform convention.

The excitation term $\mathcal{S}$ for one particular multiplet is defined by the six elements of the moment tensor and the values of the mode amplitudes $U(r)$, $V(r)$ and $W(r)$ (see 9.2) at the source, and is given by:

$$\begin{aligned}
\omega\mathcal{S}_n(\omega) = {} & \mathrm{i}[M_{rr}\dot{U}_\mathrm{s} + (M_{\theta\theta} + M_{\phi\phi})r_\mathrm{s}^{-1}(U_\mathrm{s} - \frac{1}{2}\nu_n V_\mathrm{s})] \\
& + (-1)^{\wp}(\dot{V}_\mathrm{s} - r_\mathrm{s}^{-1}V_\mathrm{s} + \nu_n r_\mathrm{s}^{-1}U_\mathrm{s})(M_{r\phi}\sin\zeta_\mathrm{s} - M_{r\theta}\cos\zeta_\mathrm{s}) \\
& - \mathrm{i}\nu_n r_\mathrm{s}^{-1}V_\mathrm{s}[-M_{\theta\phi}\sin 2\zeta_\mathrm{s} + \frac{1}{2}(M_{\theta\theta} - M_{\phi\phi})\cos 2\zeta_\mathrm{s}] \\
& + (-1)^{\wp}(\dot{W}_\mathrm{s} - r_\mathrm{s}^{-1}W_\mathrm{s})(M_{r\theta}\sin\zeta_\mathrm{s} + M_{r\phi}\cos\zeta_\mathrm{s}) \\
& + \mathrm{i}\nu_n r_\mathrm{s}^{-1}W_\mathrm{s}[\frac{1}{2}(M_{\theta\theta} - M_{\phi\phi})\sin 2\zeta_\mathrm{s} + M_{\theta\phi}\cos 2\zeta_\mathrm{s}],
\end{aligned} \tag{10.5}$$

where $r_\mathrm{s}$ is the Earth's radius at the source depth, $M_{rr}$ etc. are the elements of the – frequency-dependent – moment tensor, $\dot{m}$ is the moment rate spectrum, $\zeta_\mathrm{s}$ is the azimuth as viewed from the source which is defined clockwise from North (see Eq. 3.10). The power $\wp$ denotes the number of transits of the surface wave through the antipode and the source. For minor arc traverses (epicentral distance less than 180°), $\wp = 0$. Surface waves that pass the antipode first before arriving at the station are said to travel the 'major arc' and have $\wp = 1$, or any odd $\wp$ if they circle the Earth more than once. The terms $U_\mathrm{s}$, $V_\mathrm{s}$, $W_\mathrm{s}$ are the eigenfunction amplitudes for mode $n$ at frequency $\omega$ evaluated at $r_\mathrm{s}$ and the dot indicates their derivative with respect to $r$. The eigenfunctions are normalized such that

$$\frac{\omega \mathcal{U}_n}{k_n} \int_0^a \rho(r)[U_n^2(r) + V_n^2(r) + W_n^2(r)] \left(\frac{r}{a}\right)^2 \mathrm{d}r = 1\,, \tag{10.6}$$

where $\mathcal{U}_n$ is in m/s and $k_n$ in rad/m. Equation (10.6) differs from the normalization used in the previous chapter but adheres to the current surface wave literature.

Note that a double couple source has only two nodal lines with azimuths in the direction of the force couples, and since the mode amplitude behaves as $\mathrm{e}^{im\phi}$, azimuthal orders with $m > 2$ are not excited. This explains that no terms with factors larger than $2\zeta_s$ occur in the excitation. The term $\mathcal{S}$ is complex and will add a constant initial phase $\phi_0$ to the propagation phase $k\Delta = \nu\theta$.

As the wave propagates, it suffers from anelastic damping and the amplitude is influenced by geometrical spreading. In addition, a polar phase shift of $-\pi/2$ is acquired by the wave upon every passage through the source or its antipode where the geometrical spreading goes to zero. Finally, the receiver polarization imposes a projection onto the direction of the component of the seismograph. Arranging the vertical, radial and transverse component in a vector amplitude $\boldsymbol{A}$, the full amplitude term can be expressed as:†

$$\boldsymbol{A}_n(\omega) = \mathcal{S}_n(\omega) \times \frac{\mathrm{e}^{\mathrm{i}(\pi/4 - \wp\pi/2) - k_n\Delta/2Q_n}}{\sqrt{8\pi\nu_n|\sin\theta|}} \times \begin{pmatrix} U_\mathrm{r} \\ iV_\mathrm{r} \\ -iW_\mathrm{r} \end{pmatrix}, \tag{10.7}$$

where $U_\mathrm{r}$ etc. are the eigenvectors for mode $n$ evaluated at the receiver r (normally the surface). The quality factor $Q_n$ measures the decay of the surface wave with time, as in (5.6). Since surface waves travel, it is more appropriate to characterize their decay with distance. The theory in Appendix C leads to a simple relationship between the two quality factors for surface waves, involving the phase velocity $C_n$ (which we define in the next section) and the group velocity $\mathcal{U}_n$:

$$Q_X = \frac{C_n}{\mathcal{U}_n} Q_T\,.$$

## 10.2 Love and Rayleigh waves

Spheroidal oscillations give rise to surface waves with motion predominantly in the vertical and radial directions, which are known as Rayleigh waves. Their solar

† In keeping with the global viewpoint of this book, we shall retain the spherical coordinate system, which is important for all but the most local surface waves. But it is easy to convert the theory of this chapter to a Cartesian coordinate system if this is more practical – e.g. for application in near-surface geophysics – by making the following substitutions:

$$\Delta = a\theta \rightarrow x, \qquad \sin\theta \rightarrow x/a$$

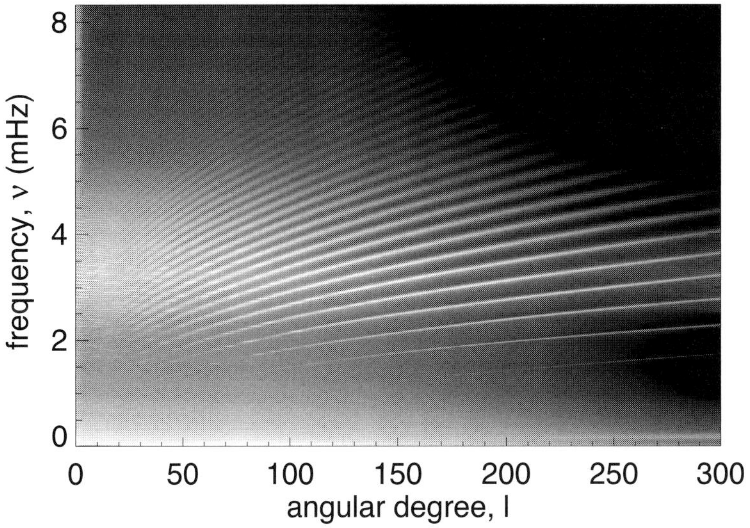


Fig. 10.2. Dispersion diagram of solar p-wave oscillations. Figure courtesy Alexander Kosovichev, SOHO/MDI research group (Stanford).

equivalents are purely acoustic and are called p-modes. Toroidal oscillations, which do not exist in the Sun, are at high frequency more adequately viewed as surface waves with transverse displacement known as Love waves. In studies that use more than the minor arc arrivals, a numbering system is used to denote the polar passages: R1 for the first arriving Rayleigh wave, R2 for the Rayleigh wave with one polar passage and so on. For Love waves we use G1, G2 etc.

Rayleigh and Love waves inherit their polarization, as well as their amplitude behaviour with depth, from the normal modes we studied in Chapter 9. Since $\ell$ is large, Rayleigh waves have a particle motion that is almost fully in the great circle plane. Love waves exhibit transverse motion, perpendicular to the great circle plane.

The significance of the wavenumber $k_n(\omega)$ is most easily grasped when combined with the time dependence $e^{-i\omega t}$. Surfaces of equal phase travel with a 'phase' velocity $C_n$ such that $k_n\Delta - \omega t$ is constant, or:

$$\frac{d\Delta}{dt} \equiv C_n(\omega) = \frac{\omega}{k_n(\omega)} .$$

Direct observation of the dispersion curves $k_n(\omega)$ for terrestrial surface waves is hampered by the lack of sensors since, ideally, one would like to Fourier transform from the distance $(\Delta)$ to the wavenumber $(k)$ domain. But in helioseismology, where we have Doppler observations at every pixel in the image, the transformation is possible, at least over the side of the Sun visible to the instrument. Figure 10.2

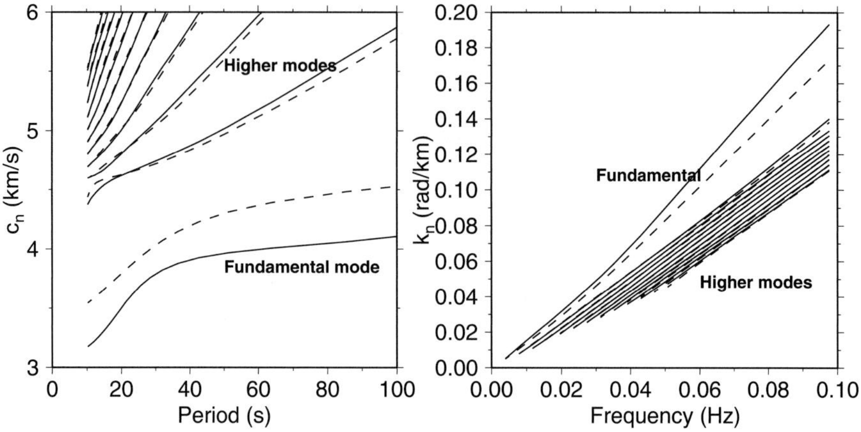


Fig. 10.3. Two representations of the same surface wave dispersion. The left figure shows the phase velocity $C_n = \omega / k_n(\omega)$ as a function of the period $2\pi/\omega$; the right figure plots $k_n$ as a function of frequency $2\pi\omega$. Rayleigh waves are shown with solid lines, Love waves by dashed lines. The model used for these calculations is the isotropic, continental model MC35 from van der Lee and Nolet [375]. Note that the higher mode dispersion for Rayleigh waves quickly approaches that for Love waves, a sign that both are dominated by interference of multiple S-waves.

shows a dispersion diagram obtained with the Michelson Doppler Imager on board the ESA/NASA SOHO (Solar & Heliospheric Observatory) mission.

For the Earth we have to resort to numerical predictions to produce similar graphs of surface wave dispersion. The equations are essentially the same as those for discrete modes, though the angular order $\ell$ takes on non-integer values,† and the gravity perturbation can be neglected except for very low frequency. Figure 10.3 shows the typical dispersion of surface waves calculated for a continental Earth model. Note that the distinction between Love and Rayleigh waves quickly disappears as the overtone number $n$ increases. Since Love waves are asymptotic toroidal modes, and only sensitive to the shear modulus (see Table 9.2), the higher modes of Rayleigh waves are obviously also dominated by S-wave energy.

Though we only established the connection between normal modes and surface waves, with Equation (10.2) we are now also in a position to analyse the relationship between S-waves and surface waves in a little more detail (see Figure10.4). Calculations of travel times for multiply reflected S-waves show that such waves arrive nearly at the same time. For the low frequencies in the S pulse, this means that the arrivals effectively overlap and interfere. One can indeed show that there is such a

† Programs `love.f` and `rayleigh.f` in the software repository calculate surface wave dispersion, neglecting the gravity perturbation.

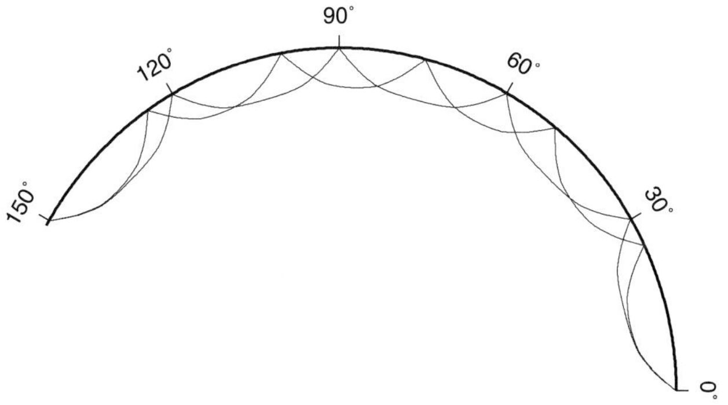


Fig. 10.4. Heuristic explanation of the ray–mode duality. The multiply reflected SSSSS- and SSSSSS-waves arrive almost at the same time and with almost the same angle to the vertical at a station with epicentral distance $\Delta = 150°$. Only for discrete frequencies will they be in phase. This imposes a dispersion relationship between the slowness $p$ and the frequency $\omega$.

close relationship between body waves and discrete or surface wave modes: Brune [34] shows that interference leads to conditions on the frequency of the S-wave, as a function of slowness and distance, that give a discrete spectrum and Nolet and Kennett [241] derive the S-wave travel time equations from the mode sum itself. Obviously, if one or more body waves constitute the same part of the wavefield as a selection of surface wave modes, their horizontal wavenumbers (or wavelengths) must be the same. We recognize $k_n(\omega)$ as the wavenumber of the surface wave. We could have found this wavenumber by differentiation of the phase $k_n\Delta$ in (10.2) with respect to $\Delta$. Similarly, the horizontal wavenumber of the body wave is found by differentiating its phase $\omega T$, where $T$ is the travel time, and using the definition of the ray parameter $p = \mathrm{d}T/\mathrm{d}\Delta$ (2.29):

$$k_\mathrm{S} = \frac{\mathrm{d}(\omega T)}{\mathrm{d}\Delta} = \omega \frac{\mathrm{d}T}{\mathrm{d}\Delta} = \omega p\,.$$

The condition that the wavenumbers must agree, $k_\mathrm{S} = k_n(\omega)$, leads to:

$$p = \frac{k_n(\omega)}{\omega} = \frac{1}{C_n(\omega)} \tag{10.8}$$

where the slowness $p$ is in s/km, and must be multiplied by the radius $a$ of the Earth or Sun to yield the more conventional spherical units of s/rad. The depth extent of the surface wave must also be the same as the depth extent of its constituency of body waves. From Snel's law (2.33), the body waves turn where the angle $i$ equals

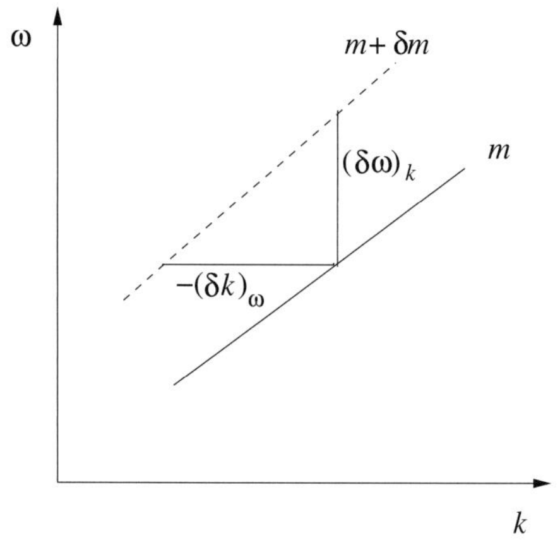


Fig. 10.5. A perturbation $\delta\omega$ at constant $k$, can be translated into the perturbation $\delta k$ at constant $\omega$. Solid line: dispersion of the background model $m$, dashed line: dispersion of the perturbed model $m + \delta m$.

$\pi/2$. This gives us the radius $r_b$ of the turning point through an implicit condition:

$$p = \frac{r_b}{V_S(r_b)},$$

where $p$ is now in s/rad. Equating the two slownesses in s/rad we find:

$$C_n(\omega) = \frac{a}{r_b} V_S(r_b), \tag{10.9}$$

which is an implicit equation for the depth extent of a surface wave mode with phase velocity $C_n$. A more thorough analysis of the differential equations (9.5–9.7) shows that below $r_b$ the wave amplitude decreases exponentially with depth.

As is evident from Figure 10.3, the fundamental modes behave somewhat differently from the higher modes. In fact, the simple body wave analogy breaks down for the lowest overtones, especially the fundamental mode ($n = 0$). An in-depth analysis shows that the fundamental Love and Rayleigh modes are strongly influenced by their evanescent part below the turning point. For shallow earthquakes, the fundamental mode ($n = 0$) is strongly excited; it dominates the seismogram and arrives later than the higher modes, such that it can be separated in the time domain.

A linear relationship between small wavenumber perturbations $\delta k$ and small perturbations in the model parameters exists that is closely related to Rayleigh's Principle. Though Rayleigh's Principle (9.15) was derived to compute the change in eigenfrequency of a discrete mode at constant angular order, thus at constant wavenumber $k$, it is easy to mould it into a perturbation theory for $k$ at constant $\omega$ (Figure 10.5) using the group velocity $\mathcal{U}_n$ for mode $n$:

$\mathcal{U}_n = \mathrm{d}\omega_n(k)/\mathrm{d}k$, or

$$(\delta k_n)_\omega = -(\delta\omega)_k/\mathcal{U}_n\,.$$

The relevant expressions for the partial derivatives of the wavenumber can thus be easily derived from the expressions in Table 9.2; for model parameter $m$ (where $m$ is $V_S$, $V_P$ or $\rho$):

$$\left(\frac{\partial k_n}{\partial m}\right)_\omega = -\frac{1}{\mathcal{U}_n}\left(\frac{\partial \omega_n}{\partial m}\right)_k\,, \tag{10.10}$$

whereas for the phase velocity partial derivatives we use:

$$\left(\frac{\partial C_n}{\partial m}\right)_\omega = \frac{\partial}{\partial m}\left(\frac{\omega}{k_n}\right)_\omega = -\frac{C_n}{k_n}\left(\frac{\partial k_n}{\partial m}\right)_k\,. \tag{10.11}$$

The lowest frequency surface wave modes are able to penetrate as deep as the transition zone. As is clear from the discussion in Section 9.3, one should take the influence of rotation of the Earth into account when interpreting such deeply reaching surface waves with periods larger than 250 s.

## 10.3 Measuring fundamental mode dispersion

Surface waves are characterized by amplitudes that are large near the surface, and that damp exponentially with depth below $r_b$ where the S-wave become evanescent. Since the phase velocity increases with period, long-period surface waves reach deeper than their short-period equivalents. At the same frequency, higher modes reach deeper than the fundamental mode. Below a period of about 20 s, a fundamental mode is mostly confined to the Earth's crust if it travels a continental path. Because of the large heterogeneity of the crust it is therefore subject to strong scattering.

Early methods to exploit the dispersive properties of Love and Rayleigh waves almost all relied upon the ability to isolate the fundamental mode in the time domain, since a single mode signal is far easier to analyse than a sum like (10.2). The spectrum of the fundamental mode at a station/component combination indicated with index $j$ is written as (see 10.2):

$$u_j(\omega) = A_{0j}(\omega)\mathrm{e}^{\mathrm{i}k_0(\omega)\Delta_j} = |A_{0j}(\omega)|\mathrm{e}^{\mathrm{i}\psi_j(\omega)}\,,$$

where $\psi_j(\omega)$ includes the initial phase $\phi_0^{(j)}$ for the wave to station $j$:

$$A_{0j}(\omega) = |A_{0j}(\omega)|\mathrm{e}^{\mathrm{i}\phi_0^{(j)}(\omega)}$$

$$\psi_j(\omega) = k_0(\omega)\Delta_j + \phi_0^{(j)}(\omega)\,.$$

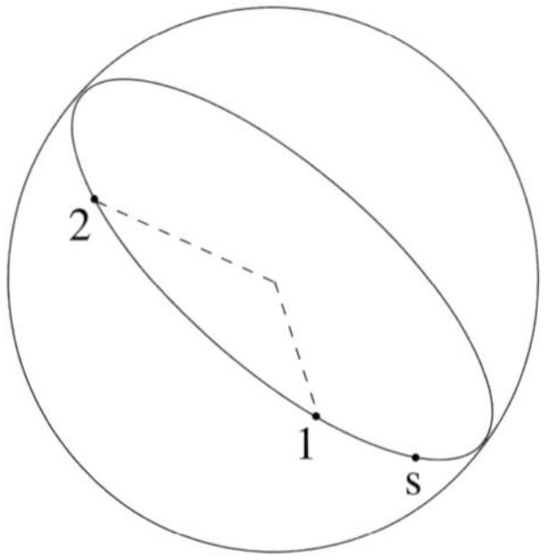


Fig. 10.6. If two stations are located on the same great circle path as the source s, this allows one to find the phase velocity for the segment between them.

From (10.5) we see that the initial phase $\phi_0^{(j)}$ depends on the azimuth of the path from the source. But if we have two stations on the same propagation path (i.e. on the same great circle that passes through the source, see Figure 10.6), the $\phi_0^{(j)}$ are the same for the two spectra, and we obtain the wavenumber by subtracting the phase $\psi_2$ from that of $\psi_1$ to eliminate the initial phase:

$$\psi_1(\omega) - \psi_2(\omega) = k_0(\omega)(\Delta_1 - \Delta_2)\,,$$

a procedure commonly known as the *two-station* method. The requirement that the two stations should be on the great circle is a severe limitation from an experimental point of view, but has the advantage that the initial phase need not be known. If we know the initial phase $\phi_0^{(1)}$ from accurate knowledge of the source orientation and depth, we can of course determine the phase shift $k_0(\omega)\Delta_1$, acquired over the path between s and station 1, from the spectrum in the first station only. For very long wavelength surface waves, which do not attenuate away within their first orbit, multiple passages of the wave in the same station can be handled just as in the two-station method.

Phase velocity measurements by themselves are not local, but represent the phase increase between source and station (or stations 1 and 2 in a two-station method). To translate this into a localized value one assumes that ray theory is valid, i.e. that the total phase is the cumulative phase along the great circle between $\Delta = 0$ and the station distance $\Delta_j$:

$$\psi_{nj}(\omega) = \int_0^{\Delta_j} k_n(\omega, \theta, \phi)\mathrm{d}\Delta + \phi_0^{(j)}(\omega)\,, \qquad (10.12)$$

where $k_n(\omega, \theta, \phi)$ is understood as the wavenumber for a surface wave in a 1D Earth with the properties of the 3D Earth at location $(\theta, \phi)$.

One commonly followed procedure is to construct (2D) maps of the local dispersion at a series of frequencies. When constructed for $C_0(\omega) \equiv \omega/k_0(\omega)$ these are known as 'phase velocity' maps. The maps themselves only allow for a qualitative interpretation: lower frequencies reach deeper than high frequencies. A precise depth dependence can only be obtained by inverting the local phase velocities for many frequencies for a local (1D) model. While this approach has the advantage of separating the large 3D inverse problem into many smaller ones in one depth dimension, the evaluation of the ability of the final model to satisfy the original measurements is rather cumbersome. As we shall see in the next chapter, the construction of phase velocity maps rests also on the assumption that ray theory is valid and is fundamentally incompatible with a finite-frequency approach to surface wave tomography. Since far more transparent methods now exist, one should in general not use phase velocity maps for inversion.

To measure group velocity – only feasible for fundamental modes because higher modes tend to arrive in the same time window – the seismogram can be sent through a series of narrow band-pass filters. The time of the maximum of the envelope determines the group arrival time $\Delta/\mathcal{U}_0(\omega)$ for the dominant frequency in each band. Again, ray theory must be assumed to link the observed group arrival times $T_j(\omega)$ to the local group velocity $\mathcal{U}_0(\omega, \theta, \phi)$:

$$T_j(\omega) = \int_0^{\Delta_j} \frac{\mathrm{d}\Delta}{\mathcal{U}_0(\omega, \theta, \phi)} .$$

This can be used to construct 2D group velocity maps for a number of frequencies. However, such maps are even more questionable than phase velocity maps. The reason is that the concept of 'group velocity' as an interference phenomenon between neighbouring frequencies only makes sense in laterally homogeneous structures. In a heterogeneous Earth, interference between waves following multiple raypaths to a station may dominate over the frequency interference, and move the maximum of an envelope far away from the time predicted by $\Delta/\mathcal{U}$. Intensive averaging among many paths helps to alleviate this problem.

Although analytical expressions for group velocity partial derivatives have been derived by Gilbert [114], a more practical procedure is to use finite differences on the partial derivatives for the phase velocity $C_n$. Rodi et al. [283] find:

$$\frac{\partial \mathcal{U}_0}{\partial m} = \frac{\mathcal{U}_0}{C_0}\left(2 - \frac{\mathcal{U}_0}{C_0}\right)\frac{\partial C_0}{\partial m} + \frac{\mathcal{U}_0^2}{C_0^2}\frac{1}{\delta}\left(\frac{\partial C_\delta}{\partial m} - \frac{\partial C_{-\delta}}{\partial m}\right),$$

where $C_{\pm\delta} = C_0(\omega \mathrm{e}^{\pm\delta})$.

## Exercises

**Exercise 10.1** If $C_n(\omega) = c_n$ is constant, what is the group velocity for this mode?

**Exercise 10.2** If $C_n(\omega) = C_0 + a/\omega$ for some constant $a$, what is the group velocity? How does this compare to the phase velocity? In Figure 10.3, what would $a$ be for the first higher mode near a period of 50s? What do you predict for the group velocity of this mode?

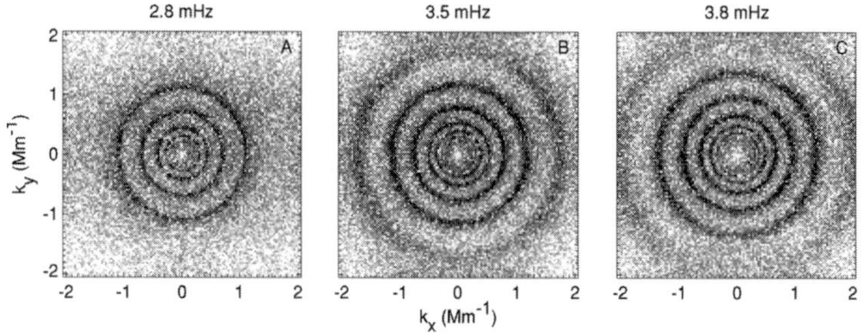


Fig. 10.7. The power density spectrum of the Doppler velocities in a region of the solar surface for three different frequencies. The highest wavenumbers $\boldsymbol{k}$ are found for the $f$ mode, a surface gravity mode. The circles at progressively smaller radii represent the $p_1$, $p_2$, ... modes. Figure courtesy Brad Hindman [133], reproduced with permission from University of Chicago Press.

**Exercise 10.3** Compare an SSSS-wave with an ScS-wave arriving at the same distance from the source. Since we can always describe the Earth's wavefield in terms of a sum of modes, both these body waves must be contributing to at least one mode, satisfying $p = \mathrm{d}\Delta/\mathrm{d}T = 1/C_n$. Which of the two body waves has the highest equivalent phase velocity? Assuming that they must also come in at the appropriate group velocity of their mode 'family', which of the two body waves has the highest group velocity? (Hint: the group velocity is determined by the actual arrival time of the wave energy).

**Exercise 10.4** Surface waves are most useful for the investigation of the upper mantle. At the 410 km discontinuity, where $r = 5961$ km, $V_S$ jumps from 4.87 to 5.08 km/s. At what phase velocity $C_n$ do the surface wave modes become evanescent below 410 km depth? For the modes shown in Figure 10.3, what is the period range of these modes? Is this in a high- or low-noise window in the seismic spectrum of the Earth?

## 10.4 Measuring higher mode dispersion

Higher mode dispersion is routinely observed in helioseismology by the method of 'ring-diagram analysis'. The Doppler wavefield is observed over a rectangular patch on the solar surface that is tracked to compensate for the Sun's rotation. This yields data $u(x, y, t)$ as a function of time and two spatial coordinates. Fourier transforming both time and space yields a power density spectrum $P(k_x, k_y, \omega) = |u(k_x, k_y, \omega)|^2$, in which the modes show up as maxima along rings (Figure 10.7). If the Sun was homogeneous and static, the rings would be perfectly circular and centred at $k_x = k_y = 0$. Flow displaces the circles from the origin, and inhomogeneities deform the circular shape. By repeating the measurement over many patches, the necessary spatial information is obtained to construct tomographic maps of flow (Figure 2.11).

On Earth we lack the ability to observe seismograms on a dense, equidistantly spaced grid, and a similar Fourier-based technique is not possible. Direct measurement of the phase of the higher modes such as we do for the fundamental mode is normally also out of reach, mostly because group velocities of higher modes overlap and the signals cannot be separated in the time domain, except for waves travelling specific continental paths where the first higher mode arrives isolated from the others in a narrow frequency band (Levshin et al. [184]). Instead, $k_n$ at a particular frequency must be determined by inverting (10.2) for observed values of spectra $u_j(\omega)$. Since the observation of one seismogram at one frequency is insufficient to resolve more than one wavenumber at that frequency, Nolet and Panza [243] and Cara [41] use arrays of stations to attempt to perform an approximate spatial Fourier transform by stacking spectra at distance $\Delta_j$ with phase delays $ik\Delta_j$. This method works if many stations are available over a large distance, and if the phase velocity does not change drastically over the array.

To eliminate this requirement of homogeneity under the array, several methods have been proposed to estimate higher mode phase or group velocity from one station only. Cross-correlation of the seismogram with a synthetic for one mode only will enhance the power of that mode with respect to all others (Lerner-Lam and Jordan [181]) and this can be used to adjust the dispersion of that mode by fitting a synthetic cross-correlogram $\gamma_{\rm synth}$ to the observed correlogram $\gamma_{\rm obs}$ (see also Section 7.2). To this end one defines a measure $I$ of the data fit as a function of $k_n$ – discretized at a number of frequencies, for example:

$$I = \int [\gamma_{\rm obs}(t) - \gamma_{\rm synth}(t)]^2 {\rm d}t \,,$$

(or similarly in the frequency domain). The most sophisticated of these methods is by van Heijst and Woodhouse [378], who 'strip' the seismogram starting with the highest amplitude mode, by adjusting $k_n$ for that mode until a good fit to the cross-correlogram is obtained, and subtracting the mode out of the seismogram before cross-correlating for the next mode. To reduce non-uniqueness in the determination of $k_n$ one may select several events in a small region with different modal excitations $A_n(\omega)$. When doing this, and inverting for relative dispersion perturbations that are forced to be correlated among neighbouring frequencies, one may obtain a well-posed inverse problem to find the dispersion functions $k_n(\omega)$ that are valid for the average structure between a station and the source region. Even so, the system is highly nonlinear, many local minima in the data fit functional $I$ as a function of the $k_n$ are present, and consequently an extensive exploration of all possible solutions is needed (Beucler et al. [20]).

An alternative approach correlates the unknown $k_n(\omega)$ by making them a-priori functionals of a 1D model that represents the source–receiver path. In this case, the unknowns to invert for are not the $k_n$ at a number of frequencies for a series of $n$, but the average velocity perturbations at a small number of depth levels along the wavepath. This method is described in detail in the following section.

## 10.5 Waveform fitting

Instead of trying to disentangle wavenumbers of higher modes one may attempt to fit the seismogram or its spectrum directly by designing a 1D model for the structure between the source and the receiver (Nolet et al. [244]). The starting point is again (10.2) with the ray-theoretical assumption (10.12) in one station, indexed $p$:

$$u_p(\omega) = \sum_{n=0}^{\infty} A_{np}(\omega) \exp\left[ i \int_0^{\Delta_p} k_n(\omega, \theta, \phi) d\Delta \right] . \qquad (10.13)$$

We start with a 'best guess' model for each path that is one-dimensional, velocities depending only on the depth coordinate. The heterogeneity along the propagation path then causes a local perturbation to the wavenumber for the starting model:

$$k_n(\omega, \theta, \phi) = k_n(\omega) + \delta k_n(\omega, \theta, \phi) .$$

The model may consist of the shear velocity as a function of depth or be multi-dimensional, so we shall write model perturbations in vector form: $\delta \boldsymbol{m}(\boldsymbol{r})$, and

$$\int_0^{\Delta_p} k_n(\omega, \theta, \phi) d\Delta = k_n^0(\omega)\Delta_p + \int_0^a \int_0^{\Delta_p} \boldsymbol{G}_n(\omega, r) \cdot \delta \boldsymbol{m}(\boldsymbol{r}) d\Delta dr ,$$

with a double integral over radius (or depth) and along the great-circle path $p$ between source and receiver. The kernel $\boldsymbol{G}_n$ contains the partial derivatives. For example, if the model contains parameters for both shear velocity $V_S$ and density $\rho$, $\boldsymbol{G}_n$ is the vector $(\partial k_n / \partial V_S, \partial k_n / \partial \rho)$ and depends on both $\omega$ and $r$.

We now make the nonlinear waveform inversion very efficient by exploiting ray theory: we make use of the fact that the kernel $\boldsymbol{G}_n$ does not depend on $\theta$ and $\phi$ and can therefore be taken out of the path integral:

$$\delta k_n(\omega) \equiv \int_0^{\Delta_p} k_n(\omega, \theta, \phi) d\Delta - k_n^0(\omega)\Delta_p = \int_0^a \boldsymbol{G}_n(\omega, r) \cdot \left[ \int_0^{\Delta_p} \delta \boldsymbol{m}(\boldsymbol{r}) d\Delta \right] dr .$$

The path integral yields averages of the model over the path as a function of $r$. It is thus simply a 1D model representative for that path and we can parametrize it in terms of a handful of layers. In Chapter 12 we shall see that parameters can be formulated in terms of a basis of functions $\boldsymbol{h}_i(r)$. Here it may be helpful to think of a basis in terms of layers, i.e. $\boldsymbol{h}_i(r) = 1$ for $r_{i-1} < r \leq r_i$ and 0 elsewhere. Expanding the model, averaged over the path, in such a basis:

$$\frac{1}{\Delta_p}\int_0^{\Delta_p} \delta\boldsymbol{m}(\boldsymbol{r})\mathrm{d}\Delta = \sum_{i=1}^{M} \gamma_i \boldsymbol{h}_i(r)\,, \tag{10.14}$$

gives us only $M$ model parameters (typically 10 or 20) to invert for by changing $\delta k_n$ until we have found a satisfactory fit to the observed waveform:

$$\delta k_n(\omega) = \sum_{i=1}^{M} \gamma_i \int_0^a \boldsymbol{G}_n(\omega, r)\cdot \boldsymbol{h}_i(r)\mathrm{d}r\,.$$

We may fit either the time series $u_p(t)$ or its Fourier transform $u_p(\omega)$ but one shall eventually wish to check data fits in the time domain because it is easier to interpret waveform misfits in terms of the physical causes than to do so for misfits of spectra. The parameters $\gamma_i$ are determined by minimizing a weighted least squares norm of the fit of a synthetic seismogram $u_p(t)$ to an observed seismogram $d_p(t)$ along path $p$:

$$F(\boldsymbol{\gamma}) = \int [\boldsymbol{R}d_p(t) - \boldsymbol{R}u_p(t, \boldsymbol{\gamma})]^2\mathrm{d}t + \boldsymbol{\gamma}\cdot \boldsymbol{C}_\gamma^{-1}\boldsymbol{\gamma}\,, \tag{10.15}$$

where $\boldsymbol{R}$ is a filtering and windowing operator that isolates and weights the relevant parts of the seismogram, and $\boldsymbol{C}_\gamma$ is the prior covariance matrix of the model (used for regularization of the inverse problem, to be discussed in detail in Chapter 14). Often $\boldsymbol{R}$ includes a time-dependent weighting to keep high energy signals – such as the fundamental mode – from overwhelming the equally informative higher modes or regional S-waves.

The most efficient way to perform the nonlinear optimization is to use the nonlinear conjugate gradient method. In order to find the optimum $\boldsymbol{\gamma}_{\mathrm{opt}}$ that minimizes $F(\boldsymbol{\gamma})$, one performs a search along a gradient direction in parameter space. The first direction is the direction of steepest descent:

$$\boldsymbol{v}_0 = -\nabla_\gamma F(0) \equiv -\boldsymbol{g}_0\,,$$

and subsequent directions are found using a simple recursive scheme (Polak and Ribiere [264]):

$$\boldsymbol{v}_k = -\boldsymbol{g}_k + \frac{(\boldsymbol{g}_k - \boldsymbol{g}_{k-1})\cdot \boldsymbol{g}_k}{\boldsymbol{g}_{k-1}\cdot \boldsymbol{g}_{k-1}}\,.$$

In practice, $\boldsymbol{g}_k$ is easiest to compute using finite differences of $F$ around the central point $\boldsymbol{\gamma}_k$; $F$ is then evaluated along the line $\boldsymbol{v}_k$ until a minimum is passed, when a simple quadratic interpolation refines the location of the optimum $\boldsymbol{\gamma}_{k+1}$. Generally, the decrease of $F$ becomes small after two or three iterations in model spaces

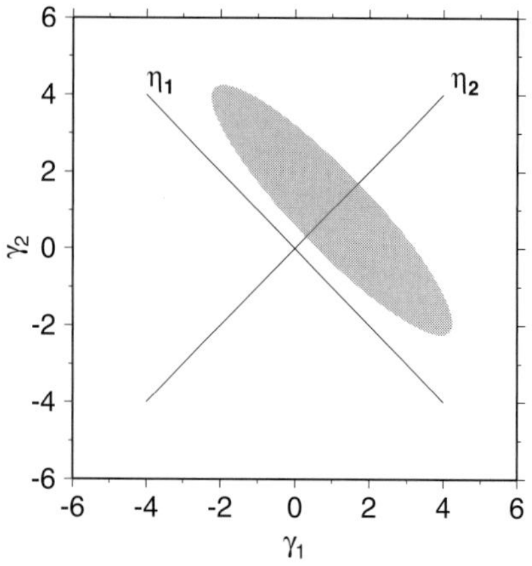


Fig. 10.8. A hypothetical example of a misfit function. The grey area indicates those parameter combinations $(\gamma_1, \gamma_2)$ where the misfit function $F$ has an acceptably low value. In this case $\gamma_1 + \gamma_2$ is much better constrained than $\gamma_1 - \gamma_2$. After transformation to a rotated coordinate system $(\eta_1, \eta_2)$ the uncertainties in the parameters $\eta_1 = \gamma_1 - \gamma_2$ and $\eta_2 = \gamma_1 + \gamma_2$ can be correctly determined.

of small $M$. Convergence to local minima can be avoided by including a strong low-pass filter in $\boldsymbol{R}$ which is relaxed in later iterations.

Strong nonlinearities exist, and the fitting can be greatly speeded up by selecting a starting model with a structure close to the (guessed) real Earth. The ability to select a different starting model for every path is a crucial element in the method of partitioned waveform inversion (PWI) developed by Nolet [236].

## 10.6 Partitioned waveform inversion (PWI)

The term 'partitioned' in PWI comes from the fact that, once the parameters $\gamma_i$ are determined using nonlinear optimization, the right-hand side of (10.14) is known and provides a horizontal average of the model over the path at each depth $r$. We have effectively partitioned the inverse problem in two stages: first a nonlinear inversion of each seismogram separately, then a joint linear inversion of the resulting averages to obtain a tomographic image. However, one cannot simply feed these averages at a series of depths as constraints ('data') into the joint linear inverse problem. This is easy to understand if we realize that the dispersion of a surface wave is an integral property of the model, so that averages over layers (say, $\gamma_1 + \gamma_2$) are usually determined much more accurately than their differences ($\gamma_1 - \gamma_2$). The situation is illustrated in Figure 10.8, where the value of $F$ is sketched as a function of these two layer parameters. The uncertainty in the parameters $\gamma_1$ and $\gamma_2$ is strongly correlated: at any particular level of $\gamma_1$ the allowable variation in $\gamma_2$ is small, but since $\gamma_1$ can vary widely in value, so can $\gamma_2$. By rotating the parameters to a coordinate system that aligns with the major axes of the ellipse, errors in $\eta_1$ and $\eta_2$ are uncorrelated, allowing us to identify and isolate the most powerful linear constraints with their error estimates for subsequent tomographic inversion.

To accomplish such coordinate transformation we need the ('Hessian') matrix of the second derivatives of $F$ with respect to $\boldsymbol{\gamma}$ in $\boldsymbol{\gamma}_{\text{opt}}$:

$$H_{ij} = \frac{\partial^2 F}{\partial \gamma_i \partial \gamma_j} .$$

We develop the misfit function $F$ in a second-order Taylor expansion:

$$F(\boldsymbol{\gamma}) \approx F(\boldsymbol{\gamma}_{\text{opt}}) + \frac{1}{2}(\boldsymbol{\gamma} - \boldsymbol{\gamma}_{\text{opt}}) \cdot \boldsymbol{H}(\boldsymbol{\gamma} - \boldsymbol{\gamma}_{\text{opt}}) ,$$

and transform:

$$\boldsymbol{\gamma} = \boldsymbol{S}\boldsymbol{\eta}$$
$$\boldsymbol{\eta} = \boldsymbol{S}^T\boldsymbol{\gamma} ,$$

where $\boldsymbol{S}$ is the orthogonal transformation matrix that diagonalizes the symmetric matrix $\boldsymbol{H}$:

$$\boldsymbol{H} = \boldsymbol{S}\boldsymbol{\Lambda}\boldsymbol{S}^T .$$

If we allow a misfit of magnitude $\epsilon$, the confidence region of acceptable solutions with a deviation $\Delta\boldsymbol{\gamma} = \boldsymbol{\gamma} - \boldsymbol{\gamma}_{\text{opt}}$ from the optimal value (the grey area in Figure 10.8) is defined by:

$$\frac{1}{2}\Delta\boldsymbol{\gamma} \cdot \boldsymbol{H}\Delta\boldsymbol{\gamma} = \frac{1}{2}\Delta\boldsymbol{\eta} \cdot \boldsymbol{S}^T\boldsymbol{H}\boldsymbol{S}\Delta\boldsymbol{\eta} = \frac{1}{2}\Delta\boldsymbol{\eta} \cdot \boldsymbol{\Lambda}\Delta\boldsymbol{\eta} = \frac{1}{2}\sum_{i=1}^{M} \lambda_i \Delta\eta_i^2 < \epsilon .$$

The threshold $\epsilon$ is usually determined 'by eye', comparing various misfits and judging the influence of errors in instrument calibration, scattered energy neglected by ray theory, effects of anisotropy and possibly non-modelled phases such as PcP. Once a suitable value of $\epsilon$ has been determined, we obtain a formal error estimate for the parameters $\eta_i$.

Since the new parameters $\eta_i$ are independent (there are no cross-terms between $\eta_i$ and $\eta_j$ in this constraint), individual parameters must satisfy:

$$|\Delta\eta_i| < \sqrt{\frac{2\epsilon}{\lambda_i}} . \tag{10.16}$$

The Earth must now satisfy, instead of (10.14):

$$\frac{1}{\Delta_p}\int_p \delta\boldsymbol{m}(\boldsymbol{r})\mathrm{d}\Delta = \sum_{i=1}^{M}\gamma_i \boldsymbol{h}_i(r) = \sum_{i=1}^{M}\sum_{j=1}^{M} S_{ij}\eta_j \boldsymbol{h}_i(r) = \sum_{j=1}^{M}\eta_j \boldsymbol{g}_j(r), \quad (10.17)$$

where we use new basis functions defined as

$$\boldsymbol{g}_j(r) = \sum_{i=1}^{M} S_{ij}\boldsymbol{h}_i(r).$$

The radius $r$ in (10.17) can be specified arbitrarily. It seem as if we have an infinite number of constraints! However, these are not independent. To reduce them to a set of independent linear constraints we introduce the dual basis $\tilde{\boldsymbol{g}}$ that satisfies:[†]

$$\int_0^a \tilde{\boldsymbol{g}}_i(r)\cdot\boldsymbol{g}_j(r)\mathrm{d}r = \delta_{ij}.$$

The dual basis can easily be found by expanding $\tilde{\boldsymbol{g}}_i(r) = \sum_j a_{ij}\boldsymbol{g}_j(r)$ and imposing the orthogonality conditions to determine the coefficients $a_{ij}$. The final linear constraints on the Earth structure are obtained by multiplying (10.17) with $\tilde{\boldsymbol{g}}_i(r)$ and integrating over $r$:

$$\frac{1}{\Delta_p}\int_p\int_0^a \delta\boldsymbol{m}(\boldsymbol{r})\cdot\tilde{\boldsymbol{g}}_k(r)\mathrm{d}r\mathrm{d}\Delta = \eta_k \pm \Delta\eta_k. \quad (10.18)$$

The final term $\pm\Delta\eta_k$ expresses the uncertainty in the right hand side. The perturbation $\delta\boldsymbol{m}(\boldsymbol{r})$ is with respect to a background (or 'starting') model, which is adapted for every path to minimize nonlinearity. The existence of different background models complicates the interpretation of constraints such as (10.18) when we combine them in a large linear inversion for 3D structure. But for every path we can easily make the constraint independent of the starting model $\boldsymbol{m}_0$ by adding the integral over the starting model to the constraint (10.18):

$$\frac{1}{\Delta_p}\int_p\int_0^a \boldsymbol{m}(\boldsymbol{r})\cdot\tilde{\boldsymbol{g}}_k(r)\mathrm{d}r\mathrm{d}\Delta = \eta_k \pm \Delta\eta_k + \frac{1}{\Delta_p}\int_p\int_0^a \boldsymbol{m}_0(\boldsymbol{r})\cdot\tilde{\boldsymbol{g}}_k(r)\mathrm{d}r\mathrm{d}\Delta. \quad (10.19)$$

This allows us to use different starting models for different paths. This feature is especially important because, at higher frequencies, the nonlinearity of the problem becomes too large for a linear constraint like (10.18) to hold for one general starting model in a global inversion or an inversion in an area where crustal thickness varies considerably (in which case one should still avoid fitting paths that have segments with crustal thickness that vary too wildly). The feature to overcome

[†] If the $\boldsymbol{h}_k$ are an orthogonal set of functions (e.g. layers) the $\boldsymbol{g}_k$ are already orthogonal, because $\boldsymbol{S}$ is an orthogonal transformation. In that case $\tilde{\boldsymbol{g}}_i(r) = \boldsymbol{g}_i(r)$.

nonlinearity and the ability to provide formal estimates for uncorrelated errors is what distinguishes PWI from non-partitioned inversion methods.

Because the nonlinear waveform fitting can be done one seismogram at a time the method is also very efficient and requires little memory. Each such waveform fit results in $M$ constraints (10.18) with known errors – some of which may be too large to be of much use as a constraint, in which case we can reduce the size of the subsequent linear inversion of constraints by discarding them. Van der Lee and Nolet [375] extend the method to include variations in depth of discontinuities, minimizing the effects of nonlinearity in Moho-depth inversions by computing the partial derivatives with respect to Moho depth using finite differences over rather large intervals.

Lebedev et al. [176] develop a very efficient automatic inversion program for the waveform fitting step. Lebedev and van der Hilst [177] present the result of a benchmark test using synthetic seismograms and show that the method works very well for the large anomalies in this test (size in excess of 2000 km). Yet PWI shares a disadvantage with all other methods described in this chapter in that it relies on the validity of ray theory for surface waves, which becomes more questionable as the size of heterogeneities is smaller. For the fundamental mode, the validity generally becomes problematic for longer paths at frequencies above 30 mHz, when scattered energy and bending of raypaths leads to multiple arrivals of the same wavefront, and interference of the mode with itself makes the simple representation (10.2) invalid, as shown by Kennett and Nolet [163]. For higher modes, or at low frequencies for the fundamental mode, PWI provides a powerful method to extract information on $V_S$ from surface waves or multiple S reflections.

### *Exercise*

**Exercise 10.5** Formulate the equations that determine the dual basis $\tilde{g}_k(r)$ .

## 10.7 Appendix C: Asymptotic theory

In this appendix we explain the theoretical foundation of asymptotic mode theory and the surface wave ray approximation in some more detail. We first give a formal derivation of (10.2). In the second part of this appendix we explore several alternative formulations that the reader may encounter in the literature.

### *Travelling waves*

Recall the expression for the normal mode wavefield:

$$\begin{aligned}
{}_n\boldsymbol{u}_\ell^m &= {}_nU_\ell^m(r)Y_\ell^m(\theta,\phi)\,\hat{\boldsymbol{r}} \\
&+ \nu^{-1}[{}_nV_\ell^m(r)\partial_\theta Y_\ell^m(\theta,\phi) + (\sin\theta)^{-1}\,{}_nW_\ell^m(r)\partial_\phi Y_\ell^m(\theta,\phi)]\,\hat{\boldsymbol{\theta}} \\
&+ \nu^{-1}[{}_nV_\ell^m(r)(\sin\theta)^{-1}\partial_\phi Y_\ell^m(\theta,\phi) - {}_nW_\ell^m(r)\partial_\theta Y_\ell^m(\theta,\phi)]\,\hat{\boldsymbol{\phi}}, \qquad \text{(9.2 again)}.
\end{aligned}$$

For large $\ell$ and $m \ll \ell$, the asymptotic theory for Legendre functions gives with (10.1):

$$Y_\ell^m(\theta,\phi) \to \frac{1}{\pi \sin^{\frac{1}{2}}\theta} \cos\left[\nu\theta - \frac{\pi}{4} + \frac{m\pi}{2}\right] \mathrm{e}^{\mathrm{i}m\phi},$$

where $\nu \equiv \sqrt{\ell(\ell+1)} \approx \ell + \frac{1}{2}$; the factor $(-1)^m$ has been absorbed by changing the sign of $m\pi/2$, We immediately see that, again for $\ell \gg 1$:

$$\partial_\theta Y_\ell^m(\theta,\phi) \to -\frac{\nu}{\pi \sin^{\frac{1}{2}}\theta} \sin\left[\nu\theta - \frac{\pi}{4} + \frac{m\pi}{2}\right] \mathrm{e}^{\mathrm{i}m\phi}, \tag{10.20}$$

$$\partial_\phi Y_\ell^m(\theta,\phi) \to \frac{\mathrm{i}m}{\pi \sin^{\frac{1}{2}}\theta} \cos\left[\nu\theta - \frac{\pi}{4} + \frac{m\pi}{2}\right] \mathrm{e}^{\mathrm{i}m\phi}. \tag{10.21}$$

The multiplication with the large factor $\nu$ in the asymptotic expression for $\partial_\theta Y_\ell^m$ shows that the motion in the radial ($\theta$) direction is dominated by ${}_nV_\ell^m$, whereas that in the transverse ($\phi$) direction is governed by the displacement ${}_nW_\ell^m$. In the following, we shall neglect the small horizontal motions associated with the terms not multiplied by $\nu$.

To simplify the analysis, we focus on the vertical displacement field in a spherically symmetric Earth or Sun. In principle this field can be found by summing all normal modes. In the time domain:

$$u_r(\theta,\phi,t) = \sum_{m,\ell,n} Y_\ell^m(\theta,\phi)_n A_\ell^m \exp[\mathrm{i}_n\omega_\ell t - {}_n\alpha_\ell t],$$

where ${}_nA_\ell$ represents the excitation at the source and includes the amplitude of the eigenvector ${}_nU_\ell^m(a)$ at the surface $r = a$, and ${}_n\alpha_\ell$ is related to the quality factor ${}_nQ_\ell$ of the mode:

$${}_n\alpha_\ell = \frac{{}_n\omega_\ell}{2_n Q_\ell}.$$

As we see from (10.5), which has only terms going up to twice the azimuth angle $\zeta_s$, ${}_nA_\ell^m = 0$ for $m > 2$ if the source can be approximated by a point source at the 'North Pole' $\theta = 0$, such that $\phi$ can be identified with the azimuth $\zeta_s$. Thus, we can satisfy the condition $m \ll \ell$ by rotating the spherical coordinate system such that the $(\theta,\phi)$ origin is at the epicentre.

Nolet [232] transforms the sum over $\ell$ to an integral in the complex plane using the Watson transform, in which the sum over $\ell$ is constructed to be the sum over residues of $\tan \pi\nu$. The function $\tan \pi\nu$ has simple poles in $\nu = (\ell + \frac{1}{2})\pi$, with

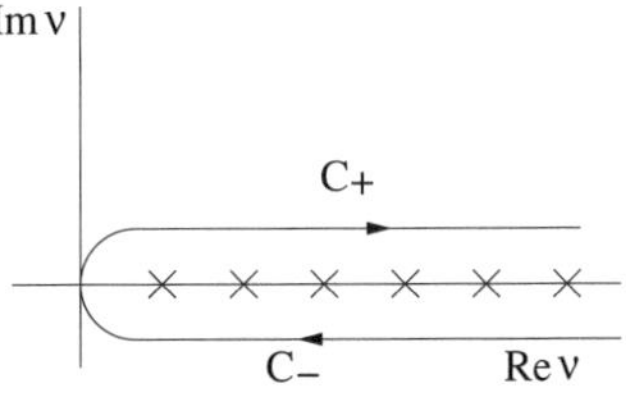


Fig. 10.9. Contour of integration for the Watson transform in the $\nu$-plane. The crosses denote poles of $\tan \pi \nu$ where $\ell = \nu - \frac{1}{2}$ is integer.

residues $-1/\pi$. Therefore the sum over $\ell$ can be replaced by the following integral:

$$u_r(\theta, \phi, t) = \sum_{m,n} \frac{\mathrm{i}}{2} \int_C {}_nA^m_{\nu-\frac{1}{2}} \exp(\mathrm{i}m\phi) P^m_{\nu-\frac{1}{2}}(\cos\theta) \times \exp[\mathrm{i}_n\omega_{\nu-\frac{1}{2}}t - {}_n\alpha_{\nu-\frac{1}{2}}t] \tan \pi\nu \, \mathrm{d}\nu .$$

The contour of integration is as in Figure 10.9. The transform can be verified by using the theorem of residues. Gilbert [113] and Snieder and Nolet [334] circumvent the Watson transform by a direct application of Poisson's sum formula, which is not fundamentally different from what we do here.

We now use the fact that the operator $\mathcal{L}$ for the normal modes depends on $\ell(\ell+1)$ rather than $\ell$ as seen from (9.5–9.7). The numerical value of $\ell(\ell+1)$ does not change if we replace $\ell$ by $-\ell - 1$. We also note that $-\nu = (-\ell - 1) + \frac{1}{2}$. Thus, all terms in the mode sum are an even function of $\nu$, except $\tan \pi\nu$, which is an odd function of $\nu$, so that the integrand itself is an odd function of $\nu$. This enables us to replace the path $C-$ by one that runs from $\nu = -\infty$ to 0, slightly above the real axis.

We next remove the poles in the tangent by expanding it in a Taylor series:

$$\tan \pi\nu = -\mathrm{i}\frac{\mathrm{e}^{\mathrm{i}\pi\nu} - \mathrm{e}^{-\mathrm{i}\pi\nu}}{\mathrm{e}^{\mathrm{i}\pi\nu} + \mathrm{e}^{-\mathrm{i}\pi\nu}} \approx \mathrm{i}[1 - 2\exp(2\mathrm{i}\pi\nu) + ...] ,$$

and use the asymptotic expression for the associated Legendre function for $\nu \gg 1$ and $m \ll \nu$ to find:

$$u_r(\theta, \phi, t) = -\frac{1}{2} \sum_{n,m} \int_{-\infty}^{\infty} \frac{1}{\sin^{\frac{1}{2}}\theta} \, {}_nA^m_{\nu-\frac{1}{2}} \exp(\mathrm{i}m\phi), \times$$

$$\exp(\mathrm{i}_n\omega_{\nu-\frac{1}{2}}t - {}_n\alpha_{\nu-\frac{1}{2}}t) \cos\left[\nu\theta - \left(m + \frac{1}{2}\right)\frac{\pi}{2}\right] [1 - 2\exp(2\mathrm{i}\pi\nu) + ...] \, \mathrm{d}\nu .$$

This integral contains terms of the form:

$$\exp\left[\mathrm{i}\left(\pm_n\omega_{\nu-\frac{1}{2}}t \pm \nu\theta + 2\pi N\nu\right)\right] .$$

In general, the real and imaginary terms oscillate strongly, with positive contributions cancelling negative contributions. Such strongly oscillating functions will only add up effectively to the integral where the exponent is stationary with respect to $\nu$, even if briefly. From this we see that $N$ corresponds to the number of orbits that the surface wave has completed around the Earth, since if the term is stationary for one combination of $\theta$ and $\nu$, it will again be stationary at $\theta + 2N\pi$. With some algebra, restricting ourselves to the direct arrival ($N = 0$) and using the fact that odd terms in $\nu$ yield zero upon integration over both positive and negative $\nu$, we find:

$$u_r(\theta, \phi, t) = \sum_n \int_{-\infty}^{\infty} A'_n(\nu) \cos({}_n\omega_{\nu-\frac{1}{2}} t - \nu\theta) \exp(-{}_n\alpha_{\nu-\frac{1}{2}} t) \, d\nu \,, \qquad (10.22)$$

where we define

$$A'_n(\nu) = -\sum_m \frac{1}{2\pi \sin^{\frac{1}{2}} \theta} \, {}_nA^m_{\nu-\frac{1}{2}} \exp(im\phi) \cos\left[\left(m - \frac{1}{2}\right)\frac{\pi}{2}\right].$$

Taking the Fourier transform of (10.22), and neglecting the peak at negative frequency which is of the order $(\omega + {}_n\omega_{\nu-\frac{1}{2}})^{-1}$ when $\omega > 0$:

$$u_r(\theta, \phi, \omega) = \sum_n \int_{-\infty}^{\infty} A'_n(\nu) \frac{\exp(i\nu\theta)}{i(\omega - {}_n\omega_{\nu-\frac{1}{2}}) - {}_n\alpha_{\nu-\frac{1}{2}}} \, d\nu \,.$$

We introduce a continuous wavenumber $k$, related to $\nu$ by:

$$k = \frac{\nu}{a} = \frac{l + \frac{1}{2}}{a},$$

and the epicentral distance $\Delta = a\theta$ in units of m or km, where $a$ is the radius of the Earth or Sun. From now on we write $\omega_n(k)$ instead of ${}_n\omega_{\nu-\frac{1}{2}}$ to recognize explicitly that $\omega$ is a continuous variable of the wavenumber – or vice versa: $k_n(\omega)$. The integrand has a factor:

$$\frac{1}{i(\omega - {}_n\omega_{\nu-\frac{1}{2}}) - {}_n\alpha_{\nu-\frac{1}{2}}}.$$

For fixed $\omega = \omega_0$, this term has a maximum at $k = k_0$ where $\omega_n(k_0) = \omega_0$. If damping is small, the term is sharply peaked around $k_0$ and we can assume both $A'_n$ and $\alpha_n$ to be constant and be brought outside of the frequency integral, so that:

$$u_r(\Delta, \phi, \omega_0) = \sum_n A'_n(\omega, \Delta, \phi) D(\Delta, \omega_0) \exp(ik_0\Delta) \,, \qquad (10.23)$$

with

$$D(\Delta, \omega_0) = \int_{-\infty}^{\infty} \frac{\exp[\mathrm{i}(k - k_0)\Delta]}{\mathrm{i}[\omega_0 - \omega_n(k)] - \alpha_n(k_0)} \, \mathrm{d}k \, . \tag{10.24}$$

Equation (10.24) is essentially a damping term: for large distance $\Delta$ the numerator ($\exp \mathrm{i}k\Delta$) oscillates many times within the peak region and the signal fades out.

We obtain the final expression for a surface wave by expanding $\omega_n(k)$ in a Taylor series:

$$\omega_n(k) = \omega_0 + (k - k_0)\mathcal{U}_n(k_0) + \ldots$$

where we define the *group velocity*:

$$\mathcal{U}_n = \frac{\mathrm{d}\omega_n(k)}{\mathrm{d}k}. \qquad \text{(10.4 again)}$$

Physically, the group velocity represents the time of arrival of the energy packet with frequency $\omega$. Since a single frequency stretches out over infinite time, the packet must theoretically have a small bandwidth ($\mathrm{d}\omega$) to display a maximum of its envelope. The interference between neighbouring frequencies within this band dictates the velocity with which this maximum travels, which is generally less than the phase velocity.

If we truncate the Taylor series after the first-order term, the integrand in (10.24) has a pole in $k = k_0 + \mathrm{i}\alpha_n(k_0)/\mathcal{U}_n(k_0)$ and $D$ can be evaluated using a contour over the positive imaginary $k$-plane:

$$D(\omega_0, \Delta) = \frac{-2\pi}{\mathcal{U}_n(k_0)} \exp\left[-\frac{\alpha_n(k_0)\Delta}{\mathcal{U}_n(k_0)}\right].$$

Substituting $\alpha_n = \omega/2Q_n$ and writing both $Q_n$ and $\mathcal{U}_n$ as functions of $\omega$ rather than $k$ leads to a Fourier spectrum given by:

$$u_r(\Delta, \phi, \omega) = \sum_n A_n(\omega, \Delta, \phi) \exp[\mathrm{i}k_n(\omega)\Delta] \qquad \text{(10.2 again).,}$$

where

$$A_n = -\frac{2\pi A_n'}{\mathcal{U}_n} \exp\left[-\frac{\omega\Delta}{2Q_n(\omega)\mathcal{U}_n(\omega)}\right].$$

A comparison of the damping exponent $-\alpha_n\Delta/\mathcal{U}_n = -\omega\Delta/2\mathcal{U}_n Q_n$ with the damping of the normal mode as $-\omega t/2Q_T$ shows that the $Q_n$ introduced here for the surface wave is a 'distance' $Q_X$. Setting $t = \Delta/C_n$ in the phase factor of the normal mode, we find $Q_X = (C_n/\mathcal{U}_n)Q_T$.

Though the derivation has been given here for the vertical component $u_r$, Equations (10.20) and (10.21) show that the same asymptotic can be applied to the radial and transverse components, the only difference being in the expressions for the amplitude factor $A_n$. But as a formal expression, (10.2) can be used for any surface wave component. Note that $\partial_\theta Y_\ell^m$ is proportional to $\nu$, which cancels the factor $\nu^{-1}$ in the expression for the $\theta$- and $\phi$-components (9.2) for the displacement $\boldsymbol{u}$.

### *The normal mode connection*

To help readers find their way among the publications dealing with the asymptotic properties of normal modes, and the various dialects of tomographic inversion associated with them, we explore some connections between surface waves and the normal mode splitting theory of Chapter 9, with a slight change in notation to adhere to conventions more commonly encountered in the literature.

Because seismic waves attenuate with time, we can never obtain the long time windows that are needed to observe the splitting of a multiplet by Fourier-transforming a single seismogram except for the lowest $\ell$ such as in Figure 9.4. For higher $\ell$, the only evidence for splitting comes from the fact that the mode peak location varies slightly with geographical location, because the amplitudes of the singlets differ among stations. Such peak shifts were first observed by Silver and Jordan [315]; Masters et al. [199] discovered that the geographical dependence of peak shifts reveals a strong degree-2 component that is in fact anti-correlated with what can be predicted from the Earth's ellipticity.

Dahlen and Tromp [78] show that the shift $\Delta\bar{\omega}_k$ of the mode peak for multiplet $k$ can be obtained by averaging individual singlet shifts $\delta\omega_j$ with their amplitudes $A_j$ as weights. With the amplitudes from (9.24) and the receiver component direction given by the unit vector $\hat{\boldsymbol{n}}$:

$$\Delta\bar{\omega}_k = \frac{\sum_j A_j \delta\omega_j}{\sum_j A_j} = \frac{\sum_j (\sum_m Z^*_{mj} a_m)\delta\omega_j [\sum_{m'} Z_{m'j} \boldsymbol{u}_{m'}(\boldsymbol{r}_\mathrm{r}) \cdot \hat{\boldsymbol{n}}]}{\sum_j (\sum_m Z^*_{mj} a_m)[\sum_{m'} Z_{m'j} \boldsymbol{u}_{m'}(\boldsymbol{r}_\mathrm{r}) \cdot \hat{\boldsymbol{n}}]} .$$

Because $\boldsymbol{Z}$ is unitary, the sum over singlets $j$ in the denominator results in $\delta_{mm'}$. Using $\boldsymbol{H} = \boldsymbol{Z}\boldsymbol{\Omega}\boldsymbol{Z}^{-1}$ we get:

$$\Delta\bar{\omega}_k = \frac{\sum_{mm'} a_m H_{m'm} \boldsymbol{u}_{m'}(\boldsymbol{r}_\mathrm{r}) \cdot \hat{\boldsymbol{n}}}{\sum_m a_m \boldsymbol{u}_m(\boldsymbol{r}_\mathrm{r}) \cdot \hat{\boldsymbol{n}}} . \tag{10.25}$$

With a change of notation, this equation is often found in the following form:

$$\Delta\bar{\omega}_k = \frac{\sum_{mm'} S_k^m(\theta_\mathrm{s}, \phi_\mathrm{s}) H_{m'm} R_k^{m'}(\theta_\mathrm{r}, \phi_\mathrm{r})}{\sum_m S_k^m(\theta_\mathrm{s}, \phi_\mathrm{s}) R_k^m(\theta_\mathrm{r}, \phi_\mathrm{r})} , \tag{10.26}$$

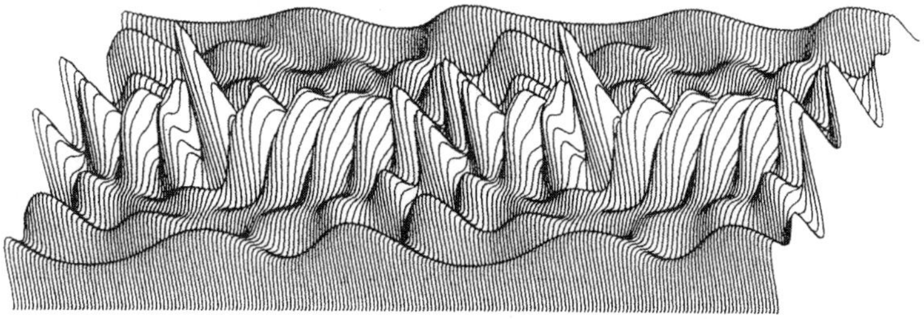

Fig. 10.10. Example of the shape of the sensitivity function $K(\theta, \phi)$ for the vertical component and a strike-slip earthquake source at a distance of 108° for the spheroidal mode ${}_0S_{10}$. The horizontal plane is a Mercator projection of the Earth, centred around the great-circle path as the long axis. The source is located at the rightmost end, the two largest peaks indicate the location of the receiver and its antipode. From Woodhouse and Girnius [401], reproduced with permission from Blackwell Publishing.

If we write out $H_{m'm}$ and $\gamma_s^{mm't}$ using (9.27) and (9.28) without the rotation and ellipticity terms, we transform (10.26) into another expression:

$$\Delta\bar{\omega}_k = \frac{\sum_{mm'} S_k^m \sum_{st} c_{st} \int_0^{2\pi} \int_0^{\pi} Y_\ell^m Y_s^t Y_\ell^{m'*} \sin\theta \mathrm{d}\theta \mathrm{d}\phi R_k^{m'}}{\sum_m S_k^m R_k^m}$$
$$= \int_0^{2\pi} \int_0^{\pi} K^{(k)}(\theta, \phi)\eta^{(k)}(\theta, \phi) \sin\theta \mathrm{d}\theta \mathrm{d}\phi ,$$

where

$$\eta^{(k)}(\theta, \phi) = \sum_{s,t} c_{st} Y_s^t(\theta, \phi) ,$$

is known as the splitting function for multiplet $k$. When combined with (9.29) we see that $\eta$ represents the heterogeneity of the Earth after integrating over radius or depth, the 'horizontal heterogeneity'. The kernel $K^{(k)}(\theta, \phi)$ collects the remaining terms and is essentially showing the sensitivity of the frequency shift to the horizontal heterogeneity ($K$ depends on the multiplet index $k$ because the summation is over $2\ell + 1$ angular orders $s$ for which the $\gamma_s^{mm't}$ are nonzero). Woodhouse and Girnius [401] were the first to compute $K$, and the images of $K$ are the first sign of a 'finite-frequency' sensitivity away from the great-circle path (Figure 10.10).

In Exercise 9.4 we have already noted that, in the case of perturbations in density or elastic parameters with angular order $s = 0$, the expressions (9.29) for the coefficients $c_{st}$ that define $H_{m'm}$ reduce to the partial derivatives for $\delta\omega$ in a spherically symmetric Earth. We carry this argument a little further and assume that for large $\ell$, the terms with $\nu^2$ dominate over those with $s(s + 1)$ in Table 9.3; this has an interesting effect on the expression that defines $H_{m'm}$. For example, the expression for $H_{m'm}$ due to a perturbation $\delta\mu$ can be approximated by (compare 9.27 and 10.29):

$$\sum_{st}\gamma_s^{mm't}\int C_s^{\mu}(r)\delta\mu_{st}(r)r^2\mathrm{d}r \approx \int C_0^{\mu}(r)\left(\sum_{st}\delta\mu_{st}\int Y_\ell^m Y_s^t Y_\ell^{m'*}\right)\mathrm{d}Sr^2\mathrm{d}r$$

$$= \int Y_\ell^m(\sum_{st}\delta\omega_{st}^{(k)}Y_s^t)Y_\ell^{m'*}\mathrm{d}S = \int Y_\ell^m\delta\omega_k\, Y_\ell^{m'*}\mathrm{d}S,$$

where we define

$$\delta\omega_{st}^{(k)} = \int C_0^{\mu}(r)\delta\mu_{st}(r)r^2\mathrm{d}r\,,$$

and where we interpret the perturbation

$$\delta\omega_k(\theta,\phi) \equiv \sum_{st}\delta\omega_{st}^{(k)}Y_s^t(\theta,\phi)\,,$$

as though the local perturbation $\delta\mu(r)$ beneath $(\theta,\phi)$ is valid for the whole Earth. The same can be done for perturbations in $\rho$ and $\lambda$; in general we can write:

$$H_{m'm} = \int Y_\ell^m(\theta_\mathrm{x},\phi_\mathrm{x})\delta\omega_k(\theta_\mathrm{x},\phi_\mathrm{x})Y_\ell^{m'*}(\theta_\mathrm{x},\phi_\mathrm{x})\mathrm{d}S\,. \qquad (10.27)$$

Mochizuki [209] and Tanimoto [347] combine the asymptotic expression (10.1) for $Y_\ell^m$ with this 'local' approximation for $H_{m'm}$ to derive a path-averaged approximation for a travelling surface wave from a discrete sum of standing modes. We shall show this for a slightly simplified case, that of an explosion source at $\theta = 0$ for which all $a_m$ are zero except $a_0$, and the vertical component at a receiver such that $\boldsymbol{u}_{m'}(\boldsymbol{r}_\mathrm{r})\cdot\hat{\boldsymbol{n}} \propto Y_\ell^{m'}(\theta_\mathrm{r},\phi_\mathrm{r})_n U_\ell(r_\mathrm{r})$. Since $a_0$ and ${}_nU_\ell(r_\mathrm{r})$ divide out in (10.25) we get:

$$\Delta\bar{\omega}_k = \frac{\sum_{m'} H_{m'0}Y_\ell^{m'}(\theta_\mathrm{r},\phi_\mathrm{r})}{Y_\ell^0(\theta_\mathrm{r},\phi_\mathrm{r})}$$

$$= \frac{\int Y_\ell^0(\theta_\mathrm{x},\phi_\mathrm{x})\delta\omega_k(\theta_\mathrm{x},\phi_\mathrm{x})\sum_{m'} Y_\ell^{m'*}(\theta_\mathrm{x},\phi_\mathrm{x})Y_\ell^{m'}(\theta_\mathrm{r},\phi_\mathrm{r})\mathrm{d}S}{Y_\ell^0(\theta_\mathrm{r},\phi_\mathrm{r})}\,.$$

The addition theorem for spherical harmonics takes care of the summation over $m'$:

$$\sum_{m'} Y_\ell^{m'}(\theta_\mathrm{r},\phi_\mathrm{r})Y_\ell^{m'*}(\theta_\mathrm{x},\phi_\mathrm{x}) = \left(\frac{2\ell+1}{4\pi}\right)P_\ell(\cos\theta_\mathrm{rx})\,,$$

where $\theta_{\mathrm{rx}}$ is the epicentral distance between scatterer and receiver. Without loss of generality we can choose the coordinate system such that $\phi_{\mathrm{r}} = 0$. Substituting this:

$$\Delta\bar{\omega}_k = \left(\frac{2\ell+1}{4\pi}\right) \frac{\int Y_\ell^0(\theta_{\mathrm{x}}, \phi_{\mathrm{x}})\delta\omega_k(\theta_{\mathrm{x}}, \phi_{\mathrm{x}}) P_\ell(\cos\theta_{\mathrm{rx}})\mathrm{d}S}{Y_\ell^0(\theta_{\mathrm{r}}, \phi_{\mathrm{r}})} . \tag{10.28}$$

We now substitute the asymptotic expressions for large $\ell$:

$$Y_\ell^0(\theta, \phi) \approx \frac{1}{\pi \sin^{\frac{1}{2}}\theta} \cos(\nu\theta - \pi/4) .$$

The asymptotic expression for the Legendre polynomial $P_\ell(\cos\theta_{\mathrm{xr}})$ is a special case ($m = 0$) of (10.1):

$$P_\ell(\cos\theta) \approx \left(\frac{2}{\pi\nu\sin\theta}\right)^{\frac{1}{2}} \cos(\nu\theta - \pi/4) .$$

The cosine terms contain exponentials $\mathrm{e}^{\pm\mathrm{i}\nu\theta}$. Combined with a time dependence $\mathrm{e}^{-\mathrm{i}\omega t}$ for $\theta > 0$ and $t > 0$, only the positive exponentials $\mathrm{e}^{+\mathrm{i}\nu\theta}$ will have a phase that is stationary with respect to changes in $\omega$ when we evaluate the Fourier integral over frequency. The negative exponentials contribute to the major arc arrivals R2, G2, etc. which we do not consider here. Substituting the asymptotic expressions into (10.25), and retaining only the stationary terms, we obtain:

$$\Delta\bar{\omega}_k = \frac{1}{2\pi} \int \left[\frac{\nu\sin\theta_{\mathrm{rs}}}{2\pi\sin\theta_{\mathrm{xr}}\sin\theta_{\mathrm{xs}}}\right]^{\frac{1}{2}} \mathrm{e}^{\mathrm{i}\nu(\theta_{\mathrm{xr}}+\theta_{\mathrm{xs}}-\theta_{\mathrm{rs}})+\mathrm{i}\pi/4} \delta\omega_k(\theta_{\mathrm{x}}, \phi_{\mathrm{x}})\mathrm{d}S ,$$

where x denotes the location $(\theta_{\mathrm{x}}, \phi_{\mathrm{x}})$ on the spherical surface $\mathrm{d}S$; $\theta_{\mathrm{xr}}$ and $\theta_{\mathrm{xs}}$ are the epicentral distance between x and receiver and source, respectively. We may develop the detour $\delta\Delta = \theta_{\mathrm{xr}} + \theta_{\mathrm{xs}} - \theta_{\mathrm{rs}}$ in powers of $q = \pi/2 - \phi$, the distance between point x and the closest point on the ray path. Since $\delta\Delta$ has a minimum on the ray path, the lowest-order term must be quadratic. Tanimoto [347] finds:

$$\delta\Delta = \theta_{\mathrm{xr}} + \theta_{\mathrm{xs}} - \theta_{\mathrm{rs}} \approx \frac{q^2\sin\theta_{\mathrm{rs}}}{2\sin\theta_{\mathrm{xs}}\sin\theta_{\mathrm{xr}}} ,$$

so that

$$\Delta\bar{\omega}_k = \frac{1}{(2\pi)^{\frac{3}{2}}} \int_0^{2\pi} \int \frac{1}{\mathcal{W}} \exp\left[\frac{\mathrm{i}q^2}{2\mathcal{W}^2} - \frac{\mathrm{i}\pi}{4}\right] \delta\omega_k(\theta_{\mathrm{x}}, q)\cos q \,\mathrm{d}q\mathrm{d}\theta ,$$

where $\theta$ integrates along the raypath, and where

$$\mathcal{W} = \left[\frac{\sin\theta_{\mathrm{xr}}\sin\theta_{\mathrm{xs}}}{\nu\sin\theta_{\mathrm{rs}}}\right]^{\frac{1}{2}} .$$

The factor $e^{iq^2/2\mathcal{W}}$ oscillates strongly for large $q$ such that positive and negative contributions to the integral cancel, except near the raypath itself where the exponent is stationary. This behaviour depends on the value of $\mathcal{W}$ which determines the 'width' of the surface wave. Yoshizawa and Kennett [411] use this expression for $\Delta\bar{\omega}_k$ to study the lateral range of influence of a surface wave. We further recognize that $\delta\Delta$ is only stationary near that part of the great circle that is located between the source and receiver, and we can limit the range of integration of $\theta$ to $0 \le \theta \le \theta_{\mathrm{rs}}$. The approximation needed to get a ray-theoretical expression was first made by Jordan [151], who assumed that $\delta\omega_k(\theta_{\mathrm{x}}, q)$ is constant near the raypath and equal to its value at the raypath itself. Also, $\cos q \approx 1$ near the raypath where we expect the major contribution. Neglecting therefore $\cos q$, extending the integration over $q$ to infinity and using the integral identity:

$$\int_{-\infty}^{\infty} e^{iaq^2} dq = \sqrt{\frac{\pi}{a}} e^{i\pi/4},$$

we obtain:

$$\Delta\bar{\omega}_k = \frac{1}{2\pi} \int_0^{\theta_{\mathrm{rs}}} \delta\omega_k(\theta_{\mathrm{x}}, 0) d\theta \,. \tag{10.29}$$

If we retain both positive and negative exponentials $e^{\pm i\nu\theta}$ we must integrate over the full great circle $0 \le \theta < 2\pi$. The frequency shift then includes the contributions of the full wavefield and is an approximation for the frequency shift of the discrete normal mode; the shift is mostly influenced by heterogeneity along the great circle but we lose information about the distribution of anomalies along the great circle. In this form, $\Delta\bar{\omega}_k$ was named the 'location parameter' by Jordan. Heterogeneities corresponding to odd $\ell$ harmonics average out over the full great circle and result in a zero location parameter, in accordance with the selection rules we encountered in Section 9.3. The limitation of surface wave observations to minor arc measurements thus has a beneficial effect in that it introduces an added geographical sensitivity by ignoring the later arrivals of the surface wave that have been influenced by other segments of the great circle.

To keep the math as simple as possible we assume an explosive source. It is left to the reader to show that non-explosive sources with $a_1$ and $a_2 \neq 0$ lead to the same result: though the amplitude factors do not cancel out initially, once we restrict the integration to the great circle, only scatterers close to the great circle are considered and the wave that hits the scatterer has approximately the same initial amplitude as the direct wave.

Equation (10.29) is essentially the same as (10.12), even though here it is formulated for the frequency shift, rather than for a phase shift due to a shift in frequency or wavenumber. It formed the basis for several inversion schemes using long-period waveforms, starting with Woodhouse and Dziewonski [400], in which the seismogram is modelled as the real part of:

$$u(t) = \sum_k A_k(\omega) e^{-i\tilde{\omega}_k t} .$$

This approach is often named the 'path average approximation' or PAVA. Apart from the summation over discrete modes, it relies on the same approximation as the surface wave methods discussed in this chapter, in particular as PWI. One shortcoming of this approximation is that the averaging over the ray path seems to destroy the spatial sensitivity in the ray plane itself, that is a property of body waves like P and S: to the wave it makes no difference where along the path the major contribution to the shift $\Delta\bar{\omega}_k$ comes from. Yet, if one sums enough modes, the P and S pulses show up clearly in the seismogram. One would thus expect that anomalies far away from the actual body wave ray path would not have much effect on those parts of the seismogram.

Li and Tanimoto [186] show that one can restore the spatial sensitivity, at least in the ray plane, by summing (10.26) over more than one multiplet: if one includes coupling between more than one mode branch into the splitting matrix $\boldsymbol{H}$, one includes waves with slightly different phase velocity, and the effect of coupling becomes dependent on the distance that the wave travels. Li and Romanowicz [185] designed a new inversion method that takes this coupling into account and that improves the validity at longer times by redefining $\Delta\bar{\omega}_k$ as the difference between the frequency perturbations defined by (10.26) and the fully ray-theoretical (10.29). The theory behind it is referred to as 'non-linear asymptotic coupling theory' or NACT. Marquering and Snieder [197] developed a version of a coupling theory, working with travelling waves rather than discrete modes which leads to a somewhat higher efficiency for short, band-limited waveforms. By offering spatial sensitivity within a ray plane, these methods are a compromise between ray theory and a fully fledged finite-frequency theory, with some advantages in the efficiency of waveform inversions.

# 11

# Surface waves: finite-frequency theory

Because of the stronger heterogeneity near the surface of the Earth, surface waves are even more prone to the effects of scattering than the teleseismic P- and S-waves. Only at rather low frequencies is it safe to assume that scattering can be ignored, but this is of course also the frequency band where the approximations of ray theory become questionable. Detailed studies on the validity of asymptotic approximations by Park [253], Kennett and Nolet [163] and Clévédé et al. [62] show that the approximations of 'ray theory' for surface waves can be problematic even for rather smooth Earth models. Lateral heterogeneity poses even stronger problems for the inversion of group velocity, which is theoretically an interference phenomenon between neighbouring frequencies, as expressed by the differentiation with respect to frequency in (10.4). When the Earth is laterally heterogeneous, waves travelling different paths may equally well interfere and significantly perturb the time of arrival of the maximum energy at a particular frequency, robbing the group arrival time of its conventional interpretation.

Phase velocity measurements also display significant oscillations due to interference of multipathed arrivals. When using ray theory (Chapter 10) there is no other solution but to average over many observations (Pedersen, [258]). First-order scattering should at least be able to model some of the multipathed energy, and this has led to efforts to extend finite-frequency theory to surface wave observations.

The mode perturbation theory that we outlined in Chapter 9 is a first-order scattering theory, and has been used by Woodhouse and others to model low-frequency seismograms. In Appendix C we encountered attempts to carry the mode perturbation theory to higher frequencies. Li and Tanimoto [186] and Li and Romanowicz [185] introduced mode coupling across mode branches. This allows for the inclusion of scattering within the surface wave 'ray plane' (the plane that contains the great circle), but not yet for off-path scattering. Li's approach had mode splitting theory as its starting point. Capdeville [39] gives efficient numerical approaches to compute synthetic seismograms and their sensitivity kernels in three dimensions using first-order scattering.

Twenty years ago, Snieder [327, 328] and Yomogida and Aki [410] had already applied 3D Born theory directly to surface waves, and attempted to invert the scattered coda of fundamental modes to locate strong scatterers. These pioneering imaging attempts were still severely hampered by a lack of sufficiently dense data, but the density of broadband seismic stations has significantly increased to make imaging of scatterers a realistic possibility. Revenaugh [274] uses a somewhat simpler method to locate scatterers in the crust using the dense station network in California.

In this chapter we shall describe a first-order scattering method that deals with the more commonly available phase velocity and amplitude data and which is applicable at very high frequencies. It follows the same philosophy as used in Chapter 7 for travel times: we use ray theory to model the perturbations $\delta u$ to a wavefield $u$, though in this case the 'rays' are surface waves travelling along the great circle between source, scatterer and receiver. Much of this chapter is based on work by Zhou and Dahlen.

## 11.1 Phase and amplitude perturbations

As usual, we assume a laterally homogeneous background model in which the dispersion of the surface waves does not depend on location. In that case the phase $\phi_n(\omega) = k_n(\omega)\Delta + \phi_{n0}(\omega)$ increases linearly with the distance $\Delta$. Heterogeneities will introduce scattered wave energy, which will perturb the phase and amplitude and make these dependent on the location of the source and receiver as well as on the radiation pattern from the source. The situation is best analysed in the frequency domain, adding a perturbation $\delta u(\omega)$ to the unperturbed spectrum $u(\omega)$ that influences both amplitude $A$ and phase $\phi$:

$$u(\omega) + \delta u(\omega) = \sum_n (A_n + \delta A_n)\mathrm{e}^{\mathrm{i}(\phi+\delta\phi)} . \tag{11.1}$$

Without loss of generality – since we are only interested in the differences due to $\delta u$ – we assume that the phase of the unperturbed seismogram is zero. Figure 11.1 shows that, to first order in $\delta u \ll u$, the phase angle perturbation is given by:

$$\delta\phi = \mathrm{Im}\left(\frac{\delta u}{u}\right) , \tag{11.2}$$

and for the amplitude perturbation:

$$\frac{\delta A}{A} = \delta \ln A = \mathrm{Re}\left(\frac{\delta u}{u}\right) . \tag{11.3}$$

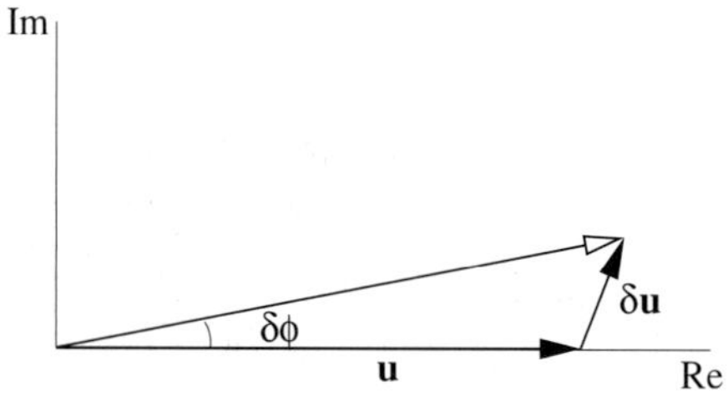


Fig. 11.1. If we plot $u$ and $\delta u$ as vectors in the complex plane to sum to $u + \delta u = (A + \delta A)e^{i\delta\phi}$ we can show that for small $\delta u$ both $\delta\phi$ and the amplitude change $\delta A$ are approximately linearly related to $\delta u$.

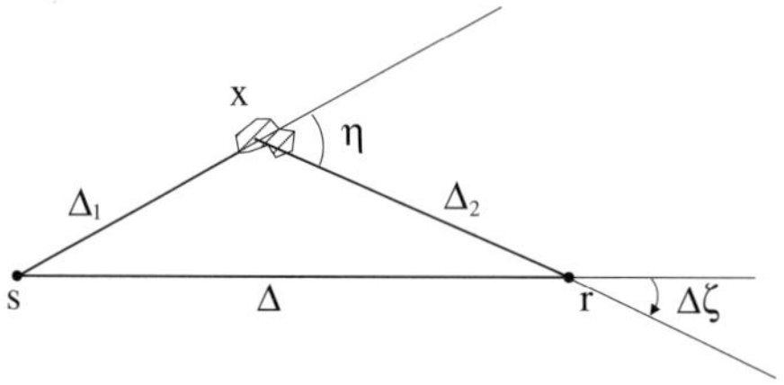


Fig. 11.2. Surface wave scattering geometry.

Though at first sight it may seem that the approximation is limited to phase perturbations $\delta\phi \ll \pi/2$, the optimistic note voiced at the end of Section 7.5 is equally well applicable to (11.2): the saturation near $\pi/2$ implicit in this equation has an effect that is of the same sign as the second-order term neglected in the Born approximation.

The derivation of the integral kernels for perturbation of a surface wave is very similar to the derivation of the kernels for P- or S-waves. We already have an expression for the displacement $\boldsymbol{u}(\omega)$ for mode $n$:

$$\boldsymbol{u}_n(\omega) = \boldsymbol{A}_n(\omega)\exp(ik_n\Delta)\,, \tag{11.4}$$

where $\boldsymbol{A}_n$ is given in (10.7). The scattered wave consists of a wave travelling between the source s and the scatterer x followed by a wave travelling from x to the receiver r (Figure 11.2). Scattering coefficients for spherical models were derived by Snieder and Nolet [334] and Zhou et al. [419] – the latter containing second-order terms – and are summarized in Table 11.1. Similar coefficients for use with Cartesian coordinates, e.g. in near-surface seismics, are give by Snieder [326], or may simply be derived from Table 11.1 by making the substitutions $\nu \to ka$, $a/r \to 1$ and neglecting the second-order terms of order $r^{-1}$.

Table 11.1. *Surface wave interaction coefficients. Subscripts 1 and 2 denote incoming and outgoing waves, respectively; $U$, $V$, $W$ are normalized using (10.6). $\eta$ is the scattering angle (Figure 11.2), $\nu = ak$ with $k$ in rad/m if $U$, $V$, $W$ and $r$ are in m.*

| | |
|---|---|
| | $\Omega_P$ |
| Rayleigh → Rayleigh | $-2(\lambda+2\mu)(\dot{U}_1+2r^{-1}U_1-\nu_1 r^{-1}V_1)(\dot{U}_2+2r^{-1}U_2-\nu_2 r^{-1}V_2)$ |
| Rayleigh → Love | 0 |
| Love → Rayleigh | 0 |
| Love → Love | 0 |
| | $\Omega_S$ |
| Rayleigh → Rayleigh | $4\mu(\dot{U}_1+2r^{-1}U_1-\nu_1 r^{-1}V_1)(\dot{U}_2+2r^{-1}U_2-\nu_2 r^{-1}V_2)$ |
| | $-\,2\mu[2\dot{U}_1\dot{U}_2+r^{-2}(2U_1-\nu_1 V_1)(2U_2-\nu_2 V_2)]$ |
| | $-\,2\mu(\dot{V}_1-r^{-1}V_1+\nu_1 r^{-1}U_1)(\dot{V}_2-r^{-1}V_2+\nu_2 r^{-1}U_2)\cos\eta$ |
| | $-\,2\mu\nu_1\nu_2 r^{-2}V_1V_2\cos 2\eta$ |
| Rayleigh → Love | $-\,2\mu(\dot{V}_1-r^{-1}V_1+\nu_1 r^{-1}U_1)(\dot{W}_2-r^{-1}W_2)\sin\eta$ |
| | $-\,2\mu\nu_1\nu_2 r^{-2}V_1W_2\sin 2\eta$ |
| Love → Rayleigh | $2\mu(\dot{W}_1-r^{-1}W_1)(\dot{V}_2-r^{-1}V_2+\nu_2 r^{-1}U_2)\sin\eta$ |
| | $+\,2\mu\nu_1\nu_2 r^{-2}W_1V_2\sin 2\eta$ |
| Love → Love | $-\,2\mu(\dot{W}_1-r^{-1}W_1)(\dot{W}_2-r^{-1}W_2)\cos\eta$ |
| | $-\,2\mu\nu_1\nu_2 r^{-2}W_1W_2\cos 2\eta$ |
| | $\Omega_\rho$ |
| Rayleigh → Rayleigh | $\rho\omega^2(U_1U_2+V_1V_2\cos\eta)$ |
| | $-\,\mu[2\dot{U}_1\dot{U}_2+r^{-2}(2U_1-\nu_1 V_1)(2U_2-\nu_2 V_2)]$ |
| | $-\,\lambda(\dot{U}_1+2r^{-1}U_1-\nu_1 r^{-1}V_1)(\dot{U}_2+2r^{-1}U_2-\nu_2 r^{-1}V_2)$ |
| | $-\,\mu(\dot{V}_1-r^{-1}V_1+\nu_1 r^{-1}U_1)(\dot{V}_2-r^{-1}V_2+\nu_2 r^{-1}U_2)\cos\eta$ |
| | $-\,\mu\nu_1\nu_2 r^{-2}V_1V_2\cos 2\eta$ |
| Rayleigh → Love | $\rho\omega^2 V_1W_2\sin\eta$ |
| | $-\,\mu(\dot{V}_1-r^{-1}V_1+\nu_1 r^{-1}U_1)(\dot{W}_2-r^{-1}W_2)\sin\eta$ |
| | $-\,\mu\nu_1\nu_2 r^{-2}V_1W_2\sin 2\eta$ |
| Love → Rayleigh | $-\,\rho\omega^2 W_1V_2\sin\eta$ |
| | $+\,\mu(\dot{W}_1-r^{-1}W_1)(\dot{V}_2-r^{-1}V_2+\nu_2 r^{-1}U_2)\sin\eta$ |
| | $+\,\mu\nu_1\nu_2 r^{-2}W_1V_2\sin 2\eta$ |
| Love → Love | $\rho\omega^2 W_1W_2\cos\eta$ |
| | $-\,\mu(\dot{W}_1-r^{-1}W_1)(\dot{W}_2-r^{-1}W_2)\cos\eta$ |
| | $-\,\mu\nu_1\nu_2 r^{-2}W_1W_2\cos 2\eta$ |

A point scatterer of unit volume will generally have perturbations in both $V_P$, $V_S$ and $\rho$. We define the 'interaction coefficient' $\Omega$ as:

$$\Omega = \Omega_P \frac{\delta V_P}{V_P} + \Omega_S \frac{\delta V_S}{V_S} + \Omega_\rho \frac{\delta \rho}{\rho} .$$

The scattered wave does not arrive from the same azimuth as the direct wave, which follows a great circle (the shortest path between source and receiver over the surface of the Earth), but from a 'back azimuth' that is different by $\Delta\zeta$, the sign convention of which is shown in Figure 11.2. Since spectral analysis requires the presence of at least a few wavelengths in the time window to yield accurate estimates, time windows are rather large and we have less of an opportunity to exclude late scatterers from large angles than we may have in the body wave case. This leads to a mix of transverse and radial components in the final expressions for $\delta u$. At this point we introduce a simplification in notation and omit the higher mode index $n$ from the expressions for $u$ and $\delta u$. For the scattered wave, we replace this index by a '1' for the incoming wave and a '2' for the scattered wave (the mode numbers $n$ for 1 and 2 need not be the same, nor need they be equal to that of the direct wave). The scattering angle $\eta$ is measured counterclockwise between the unperturbed trajectory and the direction of the scattered wave (Figure 11.2). Incorporating the interaction coefficient $\Omega$ into the expression for the surface wave amplitude, we find for the scattered wave from a volume of size $\mathrm{d}V$:

$$\delta \boldsymbol{u}(\omega) = \mathcal{S}_1(\omega) \left( \frac{\mathrm{e}^{\mathrm{i}(k_1\Delta_1 - \wp_1\pi/2 + \pi/4)}}{\sqrt{8\pi\nu_1|\sin\theta_1|}} \right) \Omega \mathrm{d}V \times \left( \frac{\mathrm{e}^{\mathrm{i}(k_2\Delta_2 - \wp_2\pi/2 + \pi/4)}}{\sqrt{8\pi\nu_2|\sin\theta_2|}} \right) \begin{pmatrix} U_2 \\ +\mathrm{i}V_2\cos\Delta\zeta - \mathrm{i}W_2\sin\Delta\zeta \\ -\mathrm{i}W_2\cos\Delta\zeta - \mathrm{i}V_2\sin\Delta\zeta \end{pmatrix}_{\mathrm{r}} . \quad (11.5)$$

Equation (11.5) can be understood, from left to right, as a term describing the source excitation in the direction of the wave, a propagation term from source to scatterer, the strength of the scattering $\Omega\mathrm{d}V$, the propagation from scatterer to receiver, and finally a polarization term to be evaluated at the receiver r. We have simplified the notation by using $\nu$ and $\theta$ rather than $k/a$ and $\Delta/a$ in the denominator that describes the geometrical spreading on the sphere; $\Delta\zeta$ is the difference between the arrival azimuths $\zeta_2 - \zeta$ between the scattered and direct wave (azimuth again measured clockwise from North). The index $\wp$ is equal to the number of polar crossings. It takes care of the phase shift that arises directly from the $\pi/4$ phase shift acquired at the source that appears in the asymptotic expression (10.1). That it is twice the $\pi/4$ phase shift can be understood if one realizes that an antipodal passage implies that the wave first converges onto the antipode before diverging again. This polar phase shift was first discovered by Brune [35]. Wielandt [393] notes that for a more precise expression of the polar phase shift one should replace:

$$\frac{\pi}{4} \to \frac{\pi}{4} - \frac{(2m-1)(2m+1)\cot\theta}{8\nu}.$$

The last term – which is largest if the azimuthal order $m$ equals 2 – can be used to screen the data for distances too close to a polar crossing. Using such data in inversions would require applying the correction directly into the expressions for $u(\omega)$ or $\delta u(\omega)$, which is more involved. For details and explicit expressions see Romanowicz and Roult [293]. We also note that data near the antipode have the additional complication that their Fréchet kernels extend over all azimuths.

When calculating the perturbation in the phase caused by a point heterogeneity, we have to select one of the three components from (11.4) and (11.5). Inserting the expressions for that component into (11.2) and (11.3), we find:

$$\delta\phi = \int \left( K_{\mathrm{P}}^{\phi} \frac{\delta V_{\mathrm{P}}}{V_{\mathrm{P}}} + K_{\mathrm{S}}^{\phi} \frac{\delta V_{\mathrm{S}}}{V_{\mathrm{S}}} + K_{\rho}^{\phi} \frac{\delta\rho}{\rho} \right) \mathrm{d}^3 \boldsymbol{r}_{\mathrm{x}}\,, \tag{11.6}$$

and

$$\delta \ln A = \int \left( K_{\mathrm{P}}^{A} \frac{\delta V_{\mathrm{P}}}{V_{\mathrm{P}}} + K_{\mathrm{S}}^{A} \frac{\delta V_{\mathrm{S}}}{V_{\mathrm{S}}} + K_{\rho}^{A} \frac{\delta\rho}{\rho} \right) \mathrm{d}^3 \boldsymbol{r}_{\mathrm{x}}\,, \tag{11.7}$$

with the Fréchet kernels

$$K_X^{\phi} = \mathrm{Im}\left( \sum_{\mathrm{mode1}} \sum_{\mathrm{mode2}} N \Omega_X \mathcal{P} \right)$$

$$K_X^{A} = \mathrm{Re}\left( \sum_{\mathrm{mode1}} \sum_{\mathrm{mode2}} N \Omega_X \mathcal{P} \right)$$

$$N = \frac{\mathcal{S}_1}{\mathcal{S}} \times \begin{cases} U_2/U, & \text{vertical} \\ (V_2 \cos\Delta\zeta - W_2 \sin\Delta\zeta)/V, & \text{radial} \\ (W_2 \cos\Delta\zeta + V_2 \sin\Delta\zeta)/W, & \text{transverse} \end{cases} \tag{11.8}$$

$$\mathcal{P} = \frac{\exp(\mathrm{i}[k_1\Delta_1 + k_2\Delta_2 - k\Delta - (\wp_1 + \wp_2 - \wp)\pi/2 + \pi/4]}{|8\pi\nu_1\nu_2 \sin\theta_1 \sin\theta_2/(\nu\sin\theta)|^{\frac{1}{2}}}, \tag{11.9}$$

and $\Omega_X$ from Table 11.1. The eigenvector amplitudes in (11.8) are evaluated at the receiver depth (usually the surface), and $U$ etc. denote the unperturbed mode (if there is no mode conversion $U_2 = U$). The summation should be done over incoming and scattered modes that contribute significantly within the time window of interest. For large-scale global inversions of fundamental modes, interactions with higher modes and between Love and Rayleigh waves can often be neglected since only long wavelengths are considered. The factor $N$ is similar to the $N$ for body waves defined by (7.25), but in the case of surface waves the approximation $N \approx 1$ is generally not allowed: the longer time span between the arrival of the direct and scattered wave allows for large scattering angles $\eta$ and thus significantly different excitation factors $\mathcal{S}_1$ and $\mathcal{S}$ for the scattered and direct wave, respectively; at the receiver end $\Delta\zeta$ may be significantly different from 0. Conversion between Love and Rayleigh modes should not be neglected if the station is near a node of the source radiation pattern and $\mathcal{S}$ is small.

Boundary layer perturbations such as the depth to the Moho or upper mantle discontinuities are treated by Zhou et al. [420]; since their main use is for the application of crustal corrections we list the relevant equations in Chapter 13.

The effects of attenuation can be introduced by the addition of an imaginary component to the phase. Thus we use (11.6) to model the effects of energy loss. A straightforward application of the relationship between a change in intrinsic attenuation $\delta Q_X^{\mathrm{S}}$ and in the imaginary component $\delta \mathrm{Im}\, V_{\mathrm{S}}$, and identifying $\exp(-k\Delta/2Q_n) = \exp(-\mathrm{Im}\,\phi)$ shows that the Fréchet kernel for surface wave $Q$ is proportional to the phase kernel:

$$
\begin{aligned}
\delta Q_n^{-1} &= \frac{2}{k\Delta} \int \left( K_{\mathrm{P}}^{\phi} \frac{\delta \mathrm{Im} V_{\mathrm{P}}}{\mathrm{Re} V_{\mathrm{P}}} + K_{\mathrm{S}}^{\phi} \frac{\delta \mathrm{Im} V_{\mathrm{S}}}{\mathrm{Re} V_{\mathrm{S}}} \right) \mathrm{d}^3 \boldsymbol{r}_{\mathrm{x}} \\
&= -\frac{1}{k\Delta} \int \left[ K_{\mathrm{P}}^{\phi} \delta \left(Q_X^{\mathrm{P}}\right)^{-1} + K_{\mathrm{S}}^{\phi} \delta \left(Q_X^{\mathrm{S}}\right)^{-1} \right] \mathrm{d}^3 \boldsymbol{r}_{\mathrm{x}} \,. \qquad (11.10)
\end{aligned}
$$

### Exercise

**Exercise 11.1** If $u$ has a nonzero phase $\phi_0$, show that (11.2) and (11.3) are formulated in such a way that they remain correct (Hint: multiply both $u$ and $\delta u$ by $\mathrm{e}^{\mathrm{i}\phi_0}$).

## 11.2 Practical considerations

The way we measure the phase velocities does potentially influence the kernel, because the windowing in the time domain influences the spectral averaging for the unperturbed signal $u$ and both the spectrum and the time duration of the scattered signal $\delta u$. With (2.76):

$$
\int_{-\infty}^{\infty} u(t) w(t) \mathrm{e}^{\mathrm{i}\omega t} \mathrm{d}t = U(\omega) * W(\omega) = \int_{-\infty}^{\infty} U(\sigma) W(\omega - \sigma) \mathrm{d}\sigma \,,
$$

which shows that $U(\omega)$ is only perfectly estimated if $W(\omega) = \delta(\omega)$, i.e. if the time window has infinite duration. All other windows influence the estimate by averaging over some finite spectral bandwidth and it is important to pay attention to the properties of $w(t)$ in the frequency domain.

The desire to have a narrowband $W(\omega)$ often clashes with the requirement that the window should isolate only the seismic phase of interest, and thus have a short time duration. We shall discuss this in detail for the phase of most practical interest, the fundamental mode of either Love or Rayleigh waves, which can be isolated by time windowing. This time window is chosen such that it excludes all but the fundamental mode itself and its scattered energy (if a wave like ScS enters the time window with appreciable energy in the surface wave frequency band, the window should be rejected as too noisy).

Since we try to estimate a difference with respect to the signal predicted for the background model, we must apply the window to both the observed and the predicted signal. Gomberg et al. [120] use a cross-spectral technique to estimate the phase velocity and amplitude between two stations, that can be adapted to estimate phase and amplitude differences with respect to a synthetic. We introduce the transfer function $T(\omega)$ between observed signal $u + \delta u$ and the predicted signal $u$ for the background model:

$$u_{\mathrm{obs}}(\omega) = u(\omega) + \delta u(\omega) = T(\omega)u(\omega)\,.$$

Comparison with (11.1) shows that

$$T(\omega) = \frac{A + \delta A}{A} \mathrm{e}^{\mathrm{i}\delta\phi}\,,$$

so that we can estimate both $\delta A/A$ and $\delta\phi$ by estimating $T$. We must find the transfer function $T(\omega)$ that minimizes:

$$\int [u(\omega) + \delta u(\omega) - T(\omega)u(\omega)]^*[u(\omega) + \delta u(\omega) - T(\omega)u(\omega)]\mathrm{d}\omega\,.$$

Differentiating this with respect to $T$ we find the cross-spectral estimate:

$$T(\omega) = \frac{u(\omega)^*[u(\omega) + \delta u(\omega)]}{u(\omega)^*u(\omega)} = \frac{u(\omega)^*u_{\mathrm{obs}}(\omega)}{u(\omega)^*u(\omega)}\,.$$

To compute the spectrum we apply the fast fourier transform (FFT) after applying a time window. But what kind of window should we choose? The most intuitive window is a boxcar window that simply selects the data in a window $T_1 \le t < T_2$. The peak of its spectrum comes closest to a delta-function but its sidelobes spread over a large band of frequencies and as a result it is actually an inferior window. A cosine taper:

$$w(t) = \frac{1}{2}\left(1 - \cos\frac{2\pi(t - T_1)}{T_2 - T_1}\right)\,,$$

appreciably reduces the sidelobes at the expense of a widening of the spectral peak. Harris [129] reviews a large number of other possible windows and discusses the tradeoff between peak width and sidelobe power in $W(\omega)$. If a very high precision for the spectral estimate is desired (and warranted), there exists an even more powerful approach than searching for the best single window: using multiple windows. Laske and Masters [171] apply the multitaper technique of Thomson [358], which averages spectra over $K$ different (orthogonal) windows. This also allows for an estimate of the error by investigating how much the estimate changes if one of the windows is left out ('cross validation'):

$$\sigma_{\tilde{T}}^2 = \frac{1}{K(K-1)} \sum_{i=1}^{K} (E_i - \tilde{T})^2,$$

where

$$\tilde{T} = \frac{1}{N} \sum_{i=1}^{K} E_i\,,$$

$$E_i = KT - (K-1)\hat{T}_i\,,$$

and $\hat{T}_i$ the estimates of $T$ with the $i$-th window omitted. However, this error does not include any systematic errors, e.g. because of the presence of a weak but unmodelled phase, so care must be taken to acccpt $\sigma_T$ without further modification.

The derivation of kernels for tapered windows must take the convolution with $w(\omega)$ into account. Instead of (11.2), we use:

$$\delta\phi = \mathrm{Im}\left(\frac{\delta u * w}{u * w}\right), \tag{11.11}$$

and similarly for $\delta A/A$. For multitaper measurements we replace the single signal convolution $u * w$ with the sum $\sum_j u * w_j$. A simple alternative to (11.11) is given by Zhou et al. [419] who finds that an approximate but acceptable way of dealing with tapered windows is to taper the kernels themselves, with the same taper $w(t)$ sampled at the group arrival time of the scattered wave:

$$K_n(\boldsymbol{r}_{\mathrm{x}}, \omega) \to K_n(\boldsymbol{r}_{\mathrm{x}}, \omega)\, w\left(\frac{\Delta_1 + \Delta_2}{\mathcal{U}_n(\omega)}\right). \tag{11.12}$$

The length of the taper determines primarily the width of the kernel (Figure 11.3). This is different from the situation for body wave kernels, where the width of the kernel depends mostly on the width of the bandpass filter or of the spectrum of $\dot{m}(\omega)$.

As was already mentioned in the previous chapter, the measurement of higher mode phase anomalies is appreciably more complicated, certainly in cases where a simple ray-theoretical expression like (10.2) breaks down. Nor is partitioned

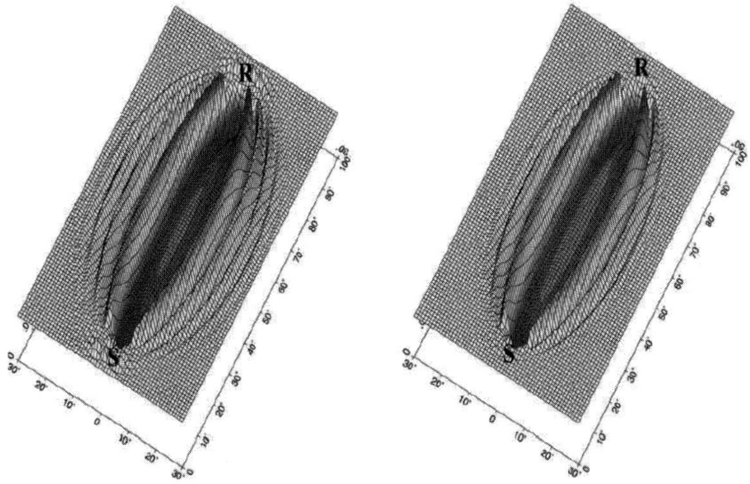


Fig. 11.3. Map view of the kernel $K_S^\phi$ at a depth of 108 km for the Rayleigh fundamental mode at 10 mHz, ignoring interaction with other modes. The receiver R is at an epicentral distance $\theta = 90°$ from the source S. The seismogram was windowed with a cosine taper with a length of 1200 s (left) and 600 s (right) respectively. Source: Zhou et al. [419], reproduced with permission from Blackwell Publishing.

waveform inversion easily applicable since this also relies on ray theory to limit the number of variable parameters to be used in the nonlinear optimization.

### *Exercises*

**Exercise 11.2** Consider a boxcar window equal to 1 for $-T/2 \leq t < T/2$ and 0 elsewhere. Find its Fourier transform and sketch the power as a function of $\omega$.

**Exercise 11.3** Similarly for the cosine taper between $-T/2 \leq t < T/2$. Compare the power spectra for the two windows.

## 11.3 Phase velocity maps: an incompatibility

As we saw in Chapter 10, the ray-theoretical approach allows for the assignment of a local phase velocity $c(\boldsymbol{r})$, such that the total phase of the mode is a path integral of $\mathrm{d}s/c(\boldsymbol{r})$ along the great circle connecting source and receiver. Physically, we can interpret the 'local' phase velocity by considering the depth-dependent properties beneath location $\boldsymbol{r}$ and assuming these to be valid everywhere so we can compute the dispersion using a spherically symmetric model with those properties.

Both Zhou et al. [419] and Yoshizawa and Kennett [412] derive partial derivatives for 'phase velocity maps', using the Born approximation. These phase velocity maps are maps of perturbations $\delta c(\omega)$ to the background model phase velocity $c(\omega)$. They are constructed from all phase observations at one frequency $\omega$, and serve as an intermediate stage in a 3D tomography experiment: the depth dependence must be found by combining different frequencies, each with its own depth sensitivity. For this, one would again appeal to Rayleigh's Principle and set up an inverse problem using the 1D partial derivatives in Table 9.2.

A closer look at (11.5), however, quickly shows that phase velocity maps are not a substitute for a full 3D interpretation, as convincingly shown by Zhou et al. [420]. The main problem is that a finite-frequency treatment does not lead to a unique, local phase velocity unless fairly draconic approximations are made. This is understandable when one realizes that the sum of a wave $u(\omega)$ and a perturbation $\delta u(\omega)$ coming from a scatterer off the great-circle path has a phase that depends on the direction of the scatterer, i.e. on the past history of the wave. It requires the rare condition that only scatterers near the great circle are important, and that mode conversions from these scatterers can be neglected, to obtain a 'local' phase velocity.

In no case should one use the single frequency phase velocity maps to compare the resolving power of finite-frequency with ray-theoretical interpretations: such a comparison not only ignores the conceptual problem that the phase velocity cannot be properly localized, it also can never represent the effect of variability in horizontal sensitivity of the finite-frequency kernels, since the width of the kernel and the location of kernel sidelobes change appreciably for different frequencies.

# 12

# Model parametrization

So far we have treated the model parameters as continuous functions in three-dimensional space, e.g. $\rho(\boldsymbol{r})$ for the density at location $\boldsymbol{r}$. Sooner or later, however, we must represent the model by a finite set of numbers in order to perform the direct and inverse calculations. One could, of course, simply *discretize* the model by sampling it at a sufficiently dense set of pixels (sometimes called 'voxels' in 3D). This has the advantage that one does not restrict the smoothness of the model, but the price to be paid is a significant loss of computational efficiency, and this is something we can ill afford. The proper approach is to *parametrize* the model – taking care, however, that the imposed smoothness does not rule out viable classes of models. In addition, the model parametrization should allow for the data to be fit to the error level attributed to them. Note that these two conditions are not identical! In practice, one does well to overparametrize and allow for more parameters than can be resolved. This reduces the risk that the limitations of the parameter space appreciably influence the inversion. Overparametrization poses some problems to the inverse problem, but these can be overcome. We shall deal with that in Chapter 14. If one is forced to underparametrize, effects of bias can be suppressed by using an 'anti-leakage' operator such as proposed by Trampert and Snieder [365].

The formal expression of parametrization is through a set of basis functions $h_j(\boldsymbol{r})$, $j = 1, ..., N$. If the model consists of more than one parameter (e.g. $V_S$ and $Q_X^S$), we rank the model in a vector, e.g. $(V_S, Q_X^S)$, and use vector basis functions $\boldsymbol{h}_j(\boldsymbol{r})$. Since the extension to multi-parameter models is straightforward, we treat only a single parameter model in this chapter.

Once a choice of basis functions is made, the model is defined by a finite set of numbers, the $M$ model parameters $m_j$:

$$m(\boldsymbol{r}) = \sum_{j=1}^{M} m_j h_j(\boldsymbol{r}),$$

which allows us to formulate the inverse problem for the $N$ data $d_i$ in matrix form:

$$\begin{aligned} d_i &= \int_V K_i(\boldsymbol{r})m(\boldsymbol{r})\mathrm{d}^3\boldsymbol{r} \\ &= \sum_{j=1}^{M} m_j \int K_i(\boldsymbol{r})h_j(\boldsymbol{r})\mathrm{d}^3\boldsymbol{r} \\ &= \sum_{j=1}^{M} A_{ij}m_j \,, \end{aligned}$$

where the integral is over the total volume $V$ of Earth or Sun. In matrix notation:

$$\boldsymbol{A}\boldsymbol{m} = \boldsymbol{d} \,, \tag{12.1}$$

with the matrix elements given by

$$A_{ij} = \int_V K_i(\boldsymbol{r})h_j(\boldsymbol{r})\mathrm{d}^3\boldsymbol{r} \,. \tag{12.2}$$

## 12.1 Global parametrization

When the basis functions $h_j$ are nonzero over all or most of space, the parametrization is called 'global'. A frequently used global parametrization is in terms of spherical harmonics:

$$h_{k\ell m}(\boldsymbol{r}) = f_k(r)Y_\ell^m(\theta, \phi) \,, \tag{12.3}$$

with $Y_\ell^m$ a fully normalized spherical harmonic introduced in Section 9.1:

$$Y_\ell^m(\theta, \phi) = (-1)^m \left[ \left( \frac{2\ell + 1}{4\pi} \right) \frac{(\ell - m)!}{(\ell + m)!} \right]^{\frac{1}{2}} P_\ell^m(\cos\theta)\mathrm{e}^{\mathrm{i}m\phi} \,.$$

Here $P_\ell^m(\cos\theta)$ is the associated Legendre function and the $f_k(r)$ form a set of functions over the depth region of interest. The $Y_\ell^m$ form an orthogonal set of functions over the surface spanned by co-latitude $0 \le \theta \le \pi$ and longitude $0 \le \phi < 2\pi$. Since the model values are real, the $\pm m$ terms must combine with coefficients that are complex conjugates of each other, so that the sum of the $\mathrm{e}^{\mathrm{i}m\phi}$ and $\mathrm{e}^{-\mathrm{i}m\phi}$ terms is real. It is often more practical to do this from the beginning and use real spherical harmonics:

$$h_{k\ell m}(\boldsymbol{r}) = f_k(r) \begin{cases} \sqrt{2}X_\ell^{|m|}(\theta)\cos m\phi & -\ell \le m < 0 \\ X_\ell^0 & m = 0 \\ \sqrt{2}X_\ell^m \sin m\phi & 0 < m \le \ell \end{cases} \,, \tag{12.4}$$

where the $X_\ell^m$ are the (real) colatitudinal harmonics:

$$X_\ell^m = (-1)^m \left[\frac{(2\ell+1)(l-m)!}{4\pi(l+m)!}\right]^{\frac{1}{2}} P_\ell^m(\cos\theta),$$

with the orthogonality property:

$$\int_0^\pi X_\ell^m X_{\ell'}^m \sin\theta \, \mathrm{d}\theta = \frac{1}{2\pi}\delta_{\ell\ell'}.$$

The radial basis functions $f_k$ are often chosen to be orthogonal, although there is no compelling reason to do so, except that it allows for an easy decomposition (see Exercise 12.2). Masters et al. [198] use natural cubic splines for the radial parametrization.

The spherical harmonic parametrization was first used by Dziewonski [90] and others in the pioneering days of seismic tomography. It has the advantage that it allows for an easy low-pass filtering of the data and comparison with similarly filtered maps of the geoid, the gravity field, or the heat flux, all of which are available as spherical harmonic expansions. For each $\ell = 0, 1, \ldots$, there are $2\ell + 1$ zonal harmonics with $m$ ranging from $-\ell$ to $+\ell$. As is immediately evident from the term $e^{im\phi}$, the smallest wavelength resolvable is therefore $2\pi/\ell_{\max}$ radians, or $40\,030/\ell_{\max}$ km at the Earth's surface. Because every $m$ effectively provides two basis functions, with a dependence $\sin m\phi$ and $\cos m\phi$ respectively, the smallest structure that can be resolved is equal to *half* the wavelength: about $20\,000/\ell_{\max}$ km for the Earth and $2.2 \times 10^6/\ell_{\max}$ km for the Sun, which has a radius of about 700 000 km.

The disadvantage of spherical harmonics is that many basis functions are needed to resolve features of geodynamic interest in the Earth. To obtain a horizontal resolution of 100 km, $\ell_{\max} = 200$ and the total number of spherical harmonics is about $2\ell_{\max}^2$ or 80 000. This has to be multiplied by the number of depth basis functions $f_k(r)$, so one easily ends up with more than $10^6$ basis functions that are global: to compute the model value at one particular location all spherical harmonics must be evaluated at that location. As the resolution of seismic tomography increased, attention shifted to local parametrizations instead. The choice need not be absolute: Kuo et al. [169] use a hybrid approach, in which they invert first for a low-order spherical harmonic parametrization, then use a local parametrization for the remaining data residuals.

### *Exercises*

**Exercise 12.1** Show that the basis defined by (12.4) is orthogonal on the space defined by $0 \le r \le a$, $0 \le \theta \le \pi$ and $0 \le \phi < 2\pi$ if the radial functions $f_k(r)$ are chosen to be orthogonal.

**Exercise 12.2** If we expand a model $m(\boldsymbol{r})$ into an orthogonal basis $h_k(\boldsymbol{r})$:

$$m(\boldsymbol{r}) = \sum_k a_k h_k(\boldsymbol{r}),$$

show that we can find the coefficients $a_k$ from:

$$a_k = \int_V m(\boldsymbol{r}) h_k(\boldsymbol{r}) \mathrm{d}^3\boldsymbol{r}.$$

## 12.2 Local parametrization

The simplest example of a local parametrization is to divide the earth up into cells, e.g.:

$$h_j(\boldsymbol{r}) = \Delta V_j^{-\frac{1}{2}} \quad \text{if } \boldsymbol{r} \text{ in cell } i \tag{12.5}$$

$$= 0 \quad \text{elsewhere}, \tag{12.6}$$

where $\Delta V_j$ is the volume of cell $j$. Homogeneous cell parametrizations were applied in the very first local studies, but quickly found their way into more global inversions. If the cells do not overlap, the basis functions scaled in this way are orthonormal:

$$\int_V h_i(\boldsymbol{r}) h_j(\boldsymbol{r}) \mathrm{d}^3\boldsymbol{r} = \delta_{ij}.$$

Often, the cells are equidistant in the latitude and longitude directions, at least over wide latitude bands. Cells can then be uniquely ordered in order of increasing coordinate and it is easy to find the cell that contains a specific location $\boldsymbol{r}$.

Instead of homogeneous cells, it is preferable to use a smooth interpolation rule between model nodes. A set of grid nodes at the corners of a rectangular box allows for easy multilinear interpolation. For example, Thurber [359] uses the Lagrangian interpolation rule:

$$m(x, y, z) = \sum_{i,j,k=1}^{2} b_{ijk} m(x_i, y_j, z_k), \tag{12.7}$$

with

$$b_{ijk} = \left(1 - \frac{|x - x_i|}{x_2 - x_1}\right)\left(1 - \frac{|y - y_j|}{y_2 - y_1}\right)\left(1 - \frac{|z - z_k|}{z_2 - z_1}\right).$$

The equidistant longitude/latitude grid, however, has a disadvantage that is also hampering spherical harmonic parametrizations: the resolvable detail obtainable with the basis functions decreases in length with depth, whereas the minimum size of features resolvable by the data generally increases with depth. This necessitates corrective action at the time of inversion, in the form of regularizations (Chapter 14), or a change in node distance with depth with associated complicated bookkeeping.

Wang and Dahlen [389] propose a horizontal parametrization of the spherical surface using B-splines. By combining this with a vertical parametrization using radial functions $f_k(r)$ as in (12.3) one obtains a more flexible, local parametrization in spherical coordinates. The starting point of their scheme is a spherical icosahedron: a set of 20 equilateral spherical triangles defined by 12 nodes on the spherical surface. Connecting the midpoints of a triangle, one obtains four smaller triangles. The unit vector $\hat{\boldsymbol{r}}$ to the midpoint between two points given by unit vectors $\hat{\boldsymbol{r}}_i$ and $\hat{\boldsymbol{r}}_j$ is given by $\hat{\boldsymbol{r}} = (\hat{\boldsymbol{r}}_i + \hat{\boldsymbol{r}}_j)/|\hat{\boldsymbol{r}}_i + \hat{\boldsymbol{r}}_j|$. Similar schemes can be designed using an $n$-fold subdivision of each triangle. Though these subtriangles are not equilateral, dimensional differences are small and the spherical tessellation thus obtained is very regular. If the average node spacing between neighbours is $\Delta_{\rm av}$, and $\Delta_i$ is the distance between location $(\theta, \phi)$ and node $i$ at $(\theta_i, \phi_i)$, a B-spline interpolant is obtained with:

$$m(\theta, \phi) = \sum_i b_i(\Delta_i) m(\theta_i, \phi_i) \tag{12.8}$$

with interpolation weights

$$b_i(\Delta_i) = \begin{cases} \frac{3}{4}p^3 - \frac{3}{2}p^2 + 1 & \text{if } \Delta_i \le \Delta_{\rm av} \\ -\frac{1}{4}q^3 + \frac{3}{4}q^2 - \frac{3}{4}q + \frac{1}{4} & \text{if } \Delta_{\rm av} < \Delta_i < 2\Delta_{\rm av} \\ 0 & \text{if } \Delta_i \ge 2\Delta_{\rm av} \end{cases},$$

$$p = \frac{\Delta_i}{\Delta_{\rm av}}, \qquad q = \frac{\Delta_i - \Delta_{\rm av}}{\Delta_{\rm av}}.$$

The subdivision can also be applied locally to create a grid that is adapted to geographical variations in resolving power. Nolet and Montelli [242] use repeated subdivision by halving of the sides in each triangle until a desired grid spacing is obtained, given a local estimate of the expected resolution. By combining spheres with different radii the volume of the Earth can be filled with a grid of points, in between which one interpolates to find model parameter values. Four such ‘interpolation supports’ span a tetrahedron, and linear interpolation can be applied to find the model values within each tetrahedron.

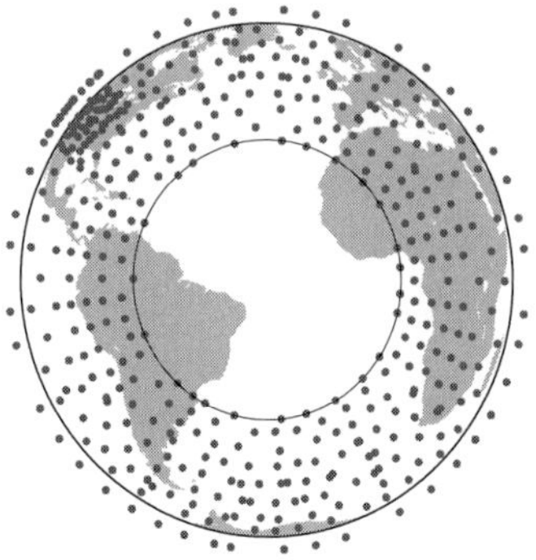

Fig. 12.1. Cross-section through the Earth showing a set of node locations within a slice of finite thickness in the mantle. The subregion with a denser set of nodes visible near the surface enables one to obtain a high resolution beneath a dense network deployment. The set of outer nodes represents the convex hull. These nodes are located outside the Earth to ensure that the tetrahedra that contain the hull nodes always enclose the Earth's surface.

There are many different ways in which one can combine a given set of interpolation supports into tetrahedra, even if one specifies that no tetrahedron may contain internally a vertex of another tetrahedron (the Delauney criterion). It is desirable to minimize the difference in edge lengths for each tetrahedron, so that the interpolation is done among supports that are at comparable distance for each space direction. Commercial and open source software exists to create meshes that are optimized in this way; these were introduced in seismic tomography by Sambridge et al. [299] and Sambridge and Gudmundsson [300]. The command `triangulate` in GMT† and the `qhull` program distributed by the Geometry Center of Minneapolis are the most readily available.‡ The outer shell of such a 'Delauney' mesh must be convex and these interpolation supports are actually located outside of the Earth; it is called the convex hull (Figure 12.1). Linear interpolation between cells of the tetrahedra is usually sufficient for tomography purposes, because the tomography problem is governed by integral equations, and insensitive to the precise interpolation rule. Given four vertices with function values $m_1, ..., m_4$, we interpolate:

$$f(x, y, z) = \sum_{k=1}^{4} b_k(x, y, z) m_k \, .$$

† Generic Mapping Tools, Wessel and Smith [392]. See `http://gmt.soest.hawaii.edu/`
‡ See Barber et al. [17] and `www.qhull.org`

We find the coefficients $b_k$ by imposing a simple function: $m(x, y, z) = x - x_1$, for which:

$$x - x_1 = b_2(x_2 - x_1) + b_3(x_3 - x_1) + b_4(x_4 - x_1)\,,$$

and similarly for $y - y_1$ and $z - z_1$. Working with differences such as $x - x_1$ instead of $x$ itself improves the numerical stability. A fourth equation is found by interpolating a constant function $f(x, y, z) = 1$ which yields $b_1 + b_2 + b_3 + b_4 = 1$. At location $(x, y, z)$ we thus find the coefficients by solving:

$$\begin{pmatrix} 0 & x_2 - x_1 & x_3 - x_1 & x_4 - x_1 \\ 0 & y_2 - y_1 & y_3 - y_1 & y_4 - y_1 \\ 0 & z_2 - z_1 & z_3 - z_1 & z_4 - z_1 \\ 1 & 1 & 1 & 1 \end{pmatrix} \begin{pmatrix} b_1 \\ b_2 \\ b_3 \\ b_4 \end{pmatrix} = \begin{pmatrix} x - x_1 \\ y - y_1 \\ z - z_1 \\ 1 \end{pmatrix}.$$

Or, **Tb=x**. Precomputing and storing the inverse of **T** minimizes the steps needed for the computation of the $b_k$, and is more efficient than using the analytical solution for the coefficients for $b_k$, e.g. for $b_4$:

$$b_4(\boldsymbol{r}) = \frac{(\boldsymbol{r} - \boldsymbol{r}_1) \cdot [(\boldsymbol{r}_2 - \boldsymbol{r}_1) \times (\boldsymbol{r}_3 - \boldsymbol{r}_1)]}{(\boldsymbol{r}_4 - \boldsymbol{r}_1) \cdot [(\boldsymbol{r}_2 - \boldsymbol{r}_1) \times (\boldsymbol{r}_3 - \boldsymbol{r}_1)]}\,. \tag{12.9}$$

Since the numbering of the vertices is arbitrary, permutation of the indices allows this expression to be used for each of the four nodes. Note that finite difference equations require much smoother interpolations and for such the B-splines (12.8) should be preferred.

To find the tetrahedron that encloses a particular location, a clever strategy was used by Sambridge and Gudmundsson [300] and Menke [207]. One uses the property that point $\boldsymbol{r}$ is located in a tetrahedron if every plane of the tetrahedron is at the opposite side of the one vertex that is not in that plane. Define the test function $s(\boldsymbol{r})$ which is 0 for a point $\boldsymbol{r}$ on the plane spanned by points $\boldsymbol{r}_1$, $\boldsymbol{r}_2$ and $\boldsymbol{r}_3$:

$$s(\boldsymbol{r}) = (\boldsymbol{r} - \boldsymbol{r}_1) \cdot [(\boldsymbol{r}_2 - \boldsymbol{r}_1) \times (\boldsymbol{r}_3 - \boldsymbol{r}_1)]\,,$$

then $\boldsymbol{r}_4$ and $\boldsymbol{r}$ are on the same side of the plane if $s(\boldsymbol{r}_4)s(\boldsymbol{r}) > 0$. To find the tetrahedron one starts with a first guess, and tests the location of $\boldsymbol{r}$ with respect to the four faces. As soon as the test is failed, however, one moves on to the neighbouring tetrahedron attached to this face, until the test is satisfied for every face. Since most calculations move in a rational way through the model space, the result for a previous location is an obvious guess for the next point.

Strongly related to Delauney tetrahedra are Voronoi polyhedra; a Voronoi polyhedron is defined as the volume of all locations in space closest to one particular grid point (Figure 12.2). Though such parametrization is still viable in two dimensions, where they offer some advantage in case the velocity is constant within the

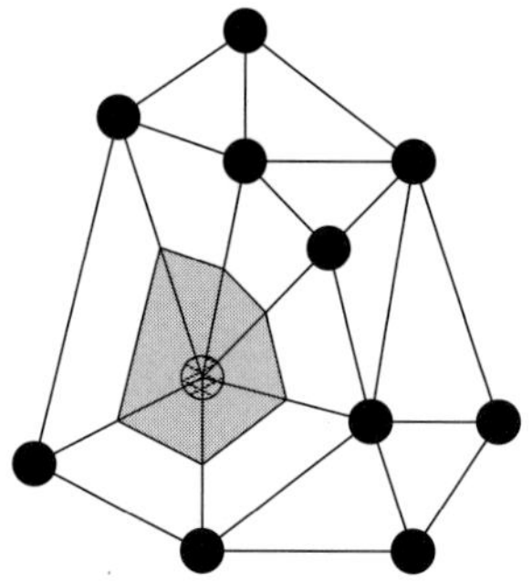

Fig. 12.2. A Voronoi polyhedron or cell for a node (grey) is bounded by the midpoint of each connection to that node, here shown for a triangulation in two dimensions.

volume (Böhm et al. [26]), the flexibility of linear interpolation within tetrahedra is much to be preferred for the 3D case.

Several efforts to parametrize the Earth using wavelet decomposition have been successful for two-dimensional models: Chiao and Kuo [55] use a global scheme of triangularly distributed nodes that can be subdivided into smaller scales. Loris et al. [188] show that the local expansion of the model into 2D wavelets leads to advantages in the regularization of the inverse problem. Extension of these first attempts to three dimensions, however, is still in its infancy, but Chevrot and Zhao [54] have parametrized sensitivity kernels in 3D using Haar wavelets, and this scheme could equally well be applied to the model itself.

Sometimes it may be advisable to refine the grid further than can be obtained with subdivision. The benefits of this must be weighed against the danger that Delauney tetrahedralization results in a number of ill-configured tetrahedra. This problem is well known in the literature of mesh optimization and computational geometry, but experience with such methods in seismic tomography is still rare. Montelli et al. [215] use a random packing method, which rejects nodes that are too close. Nolet and Montelli [242] adapt a grid with nodes $(x_i, y_i, z_i)$ to a given distribution of resolving lengths in the Earth by optimization, defining a penalty function $E$:

$$E = \sum_{i=1}^{N} \sum_{j \in \mathcal{N}_i} \frac{(L_{ij} - \ell_{ij})^2}{\ell_{ij}^2}, \tag{12.10}$$

where $L_{ij}$ is the actual distance between nodes $i$ and $j$, $\mathcal{N}_i$ is the set of nearest (or natural) neighbours of node $i$ – all nodes in all tetrahedra that have node $i$ as a vertex – and $\ell_{ij}$ the average resolving length between $i$ and $j$. One can see $E$ as the potential energy of a system of springs of rest length $\ell_{ij}$. The minimum can be

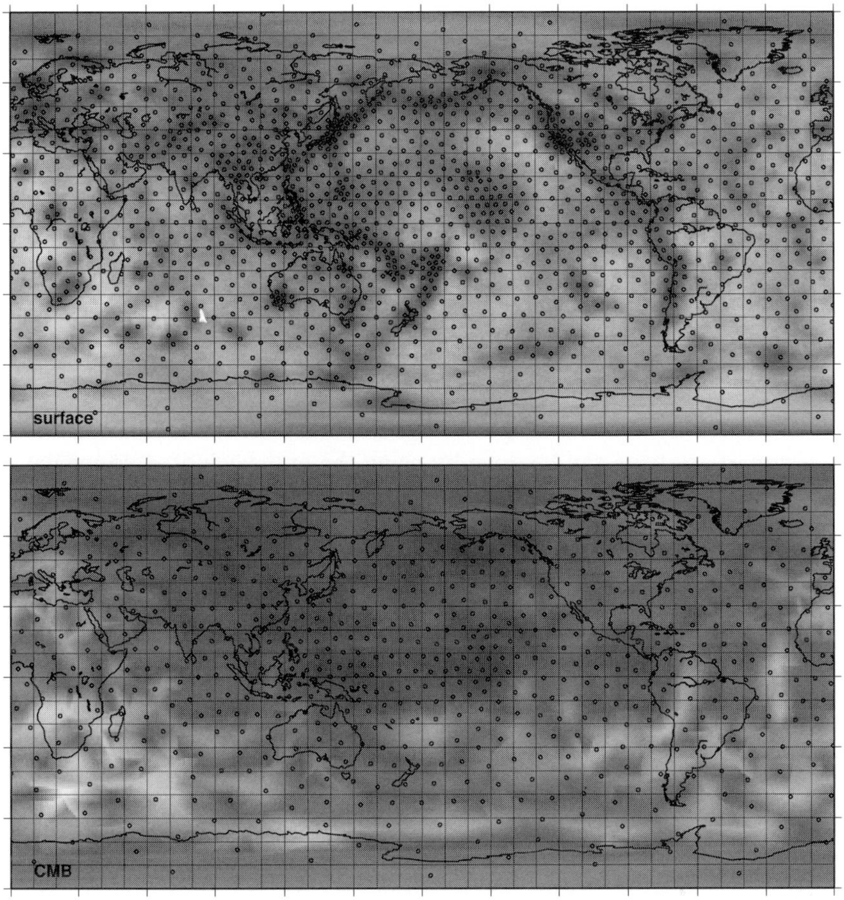


Fig. 12.3. Example of a set of optimum node locations adapted to the ray density obtainable with teleseismic S, SS and ScS travel times. The greyscale represents the expected resolution length on a scale from 0 km (black) to 1500 km (white). Top: near the Earth's surface, bottom: near the core–mantle boundary. From Nolet and Montelli [242], reproduced with permission from Blackwell Publishing.

found by a gradient search algorithm using:

$$\frac{\partial E}{\partial x_k} = \sum_{j \in N_k} 4 \left(1 - \frac{\ell_{kj}}{L_{kj}}\right) \frac{(x_k - x_j)}{\ell_{kj}^2}, \tag{12.11}$$

and similar equations for the derivatives with respect to node coordinates $y_k$ and $z_k$. During the optimization the convex hull is kept fixed and node migration outside the hull is inhibited. An example of an optimal distribution for a particular global tomography experiment is shown in Figure 12.3.

An alternative strategy to optimize the grid configuration for surface gridding is proposed by Debayle and Sambridge [84]. Starting from a dense grid, one constructs its Voronoi polyhedra. A quality criterion is defined, e.g. one requests that at least one surface wave path be present in each 36° azimuth bin. One then randomly removes a small number of nodes belonging to Voronoi cells that do not satisfy the quality criterion, and tests the new configuration in the same way. This is repeated until all Voronoi cells satisfy the quality criterion.

If the dimensions of structures to be imaged are close to the grid spacing, artefacts due to the gridding will inevitably show up in plots. This is evident if one parametrizes the model using homogeneous cells as in (12.6). Plotted pure and simple, such tomographic images look like chessboards, which is far from geophysical reality. Commonly accepted practice is therefore to interpolate linearly between the centres of the cells. The integral (12.2) is usually not very sensitive to the exact nature of the interpolation; another way of saying this is that the suppression of artefacts can be done by using the nullspace of $\boldsymbol{A}$ so the data fit is not, or minimally, affected. Even a tetrahedral grid may lead to plotting artefacts, which are easily removed by using mild image smoothing (e.g. using GMT's `grdfilter` command).

### *Exercises*

**Exercise 12.3** In the cell interpolation given by (12.7), verify that the interpolated value for $m$ in each of the corners of the box equals the assigned value $m(x_i, y_j, z_k)$.

**Exercise 12.4** To derive (12.11) note that every node pair occurs twice in the sum over pairs, such that

$$\frac{\partial E}{\partial x_k} = \sum_{j\in\mathcal{N}_k} 4\frac{(L_{kj}-\ell_{kj})}{\ell_{kj}^2}\frac{\partial L_{kj}}{\partial x_k}.$$

Complete the derivation of (12.11).

## 12.3 Numerical considerations

The computation of the matrix $\boldsymbol{A}$ in (12.1) takes a considerable amount of CPU time. The generic tomographic inverse problem deals with integral equations of the form:

$$d_i = \int_V K_i(\boldsymbol{r})m(\boldsymbol{r})\mathrm{d}^3\boldsymbol{r}\,. \tag{12.12}$$

If we adopt a parametrization in the form of tetrahedra, there are $N$ model nodes, for which we seek the model parameters $m_j$, $j = 1, ..., N$. After Delaunay tetrahedralization, each of these nodes is part of more than one tetrahedron, and therefore identified with several $m_k^t$, $k = 1, ..., 4$, the four vertices of tetrahedron $t$. Thus, we can split the space integral up into small volumes $\Delta V_p$ around location $\boldsymbol{r}_p$ located in tetrahedron $t_p$ and sum:

$$d_i = \sum_p K_i(\boldsymbol{r}_p) m(\boldsymbol{r}_p) \Delta V_p = \sum_p K_i(\boldsymbol{r}_p) \sum_{k=1}^{4} b_k^{t_p}(\boldsymbol{r}_p) m_k^{t_p} \Delta V_p \,.$$

To obtain sufficient accuracy, the size of the volumes $\Delta V_p$ should be an order of magnitude smaller than the size of the tetrahedron over which one interpolates. If we map each vertex back to its original parameter index $j$, we see that every volume element adds to four different elements of row $i$ in the matrix, i.e.

$$A_{ij} \leftarrow A_{ij} + K_i(\boldsymbol{r}_p) b_k^{t_p}(\boldsymbol{r}_p) \Delta V_p \qquad (k = 1, ..., 4) \,,$$

where $j$ is the model index belonging to vertex node $m_k^{t_p}$ and where the $A_{ij}$ are initially zero. If we interpret seismic travel times using ray theory, the volume element $\Delta V_p$ must be replaced by a ray segment length $\Delta \ell_p$, but otherwise the formalism is the same.

As discussed in Chapter 7, we may use ray theory near source and receiver to avoid the singularities that occur in finite-frequency kernels for travel times and amplitudes of body waves. Isolating a homogeneous volume $V_0$ around the singularity, its contribution to the travel time delay will be a sum over $p = 1, ..., N_p$ ray elements of length $\Delta \ell_p$:

$$\delta T_{\text{singularity}} \approx -\int_{\mathrm{d}\ell \in V_0} \frac{1}{c_0} \left( \frac{\delta c}{c_0} \right) \mathrm{d}\ell = -c_0^{-1} \sum_{p=1}^{N_p} \sum_{k=1}^{4} b_k^{t_p}(\boldsymbol{r}_p) m_k^{t_p} \Delta \ell_p \,.$$

For amplitude kernels, the assumption of homogeneity implies that the amplitude is not influenced by the structure within $V_0$, so that its contribution to the matrix is zero.

## 12.4 Spectral analysis and model correlations

We have already mentioned that many properties of the Earth – such as its gravity field or its magnetic field – are potentials for which the differential equations are solved in terms of spherical harmonics, and such fields are often specified simply by their spherical harmonic coefficients. At low order $\ell$ it is therefore useful to expand tomographic models on a spherical harmonics basis – even if the model itself was obtained using a different basis – to allow for a direct comparison. Similarly, if two tomographic models differ by eye, it may be useful to investigate if the difference is confined to a specific wavelength domain or if differences exist across the board. One obvious way to compare models is by the spectral power, i.e. the strength of the heterogeneity as a function of the wavelength.

For models expanded in spherical harmonics we can investigate the power present in selected wavelength (angular order) bands. At a fixed depth, or radius $r'$:

$$m(r',\theta,\phi) = \sum_k f_k(r') \sum_\ell [a_{k\ell 0} X_\ell^0(\theta) + \sqrt{2} \sum_{m=1}^{\ell} X_\ell^m(\theta)(a_{k\ell m}\cos m\phi + b_{k\ell m}\sin m\phi)],$$

where we split the coefficients $h_{k\ell m}$ in (12.4) into coefficients $a$ and $b$ so that the sum is only over positive $m$. We can write the expansion even more economically introducing

$$a_{\ell m} = \sum_k f_k(r') a_{k\ell m}, \qquad b_{\ell m} = \sum_k f_k(r') b_{k\ell m},$$

so that

$$m(r',\theta,\phi) = \sum_\ell [a_{\ell 0} X_\ell^0(\theta) + \sqrt{2} \sum_{m=1}^{\ell} X_\ell^m(\theta)(a_{\ell m}\cos m\phi + b_{\ell m}\sin m\phi)]. \tag{12.13}$$

Because of the orthogonality of the spherical harmonics, the coefficients are simply:

$$a_{\ell 0} = \int_0^{2\pi}\int_0^{\pi} m(\boldsymbol{r}') X_\ell^0(\theta)\sin\theta \, \mathrm{d}\theta \mathrm{d}\phi$$

and for $m > 0$:

$$a_{\ell m} = \int_0^{2\pi}\int_0^{\pi} m(\boldsymbol{r}')\sqrt{2} X_\ell^m(\theta)\cos m\phi \sin\theta \, \mathrm{d}\theta \mathrm{d}\phi$$

$$b_{\ell m} = \int_0^{2\pi}\int_0^{\pi} m(\boldsymbol{r}')\sqrt{2} X_\ell^m(\theta)\sin m\phi \sin\theta \, \mathrm{d}\theta \mathrm{d}\phi .$$

The total 'power' of the model at $r = r'$ is given by:

$$\int_0^{2\pi}\int_0^{\pi} m(\boldsymbol{r}')^2 \sin\theta \, \mathrm{d}\theta \mathrm{d}\phi = \sum_\ell [a_{\ell 0}^2 + \sum_m (a_{\ell m}^2 + b_{\ell m}^2)],$$

and the average power $E_\ell$ per unit area and per degree over all $2\ell + 1$ model components at angular order $\ell$ (representing a wavelength equal to $2\pi r/\ell$) can be found by summing the power of all components for that angular order and dividing by $2\ell + 1$:

$$E_\ell = \frac{1}{2\ell + 1}\left[a_{\ell 0}^2 + \sum_{m=1}^{\ell}(a_{\ell m}^2 + b_{\ell m}^2)\right].$$

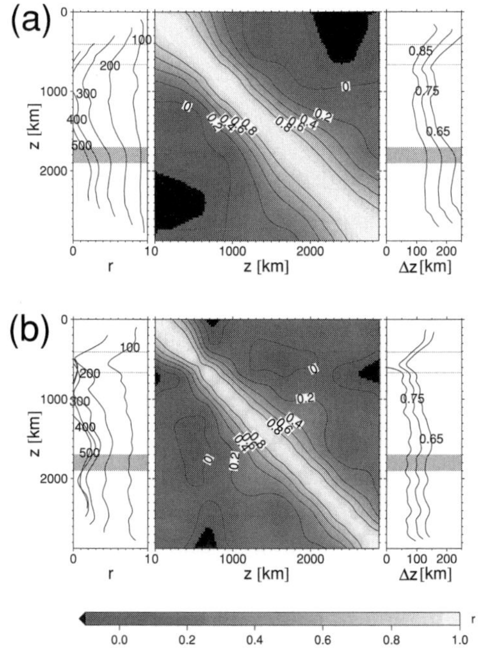


Fig. 12.4. Example of the radial correlation function for two global P-wave models. (a) The finite-frequency model from Montelli et al. [214], (b) the ray-theoretical model from Soldati and Boschi [335]. The centre plots show $R(r, r')$ as a function of depth $z = 6371 - r$. On the left is plotted the radial correlation for fixed values of $\frac{1}{2}|r - r'|$ of 100–500 km. On the right the value of $\Delta z = \frac{1}{2}|r - r'|$ for fixed values of the radial correlation $R = 0.65$, 0.75 and 0.85. From Boschi et al. [29], reproduced with permission from the AGU.

The 3D correlation coefficient between two models $m$ and $m'$ is defined as:

$$C_{3D} = \frac{\int_V [m(\boldsymbol{r}) - \bar{m}][m'(\boldsymbol{r}) - \bar{m}']\mathrm{d}^3\boldsymbol{r}}{\left(\int_V [m(\boldsymbol{r}) - \bar{m}]^2\mathrm{d}^3\boldsymbol{r} \int_V [m'(\boldsymbol{r}) - \bar{m}']^2\mathrm{d}^3\boldsymbol{r}\right)^{\frac{1}{2}}}, \tag{12.14}$$

where $\bar{m}$ is the average model value over the volume of the Earth (which can be simply removed by setting $a_{00} = 0$). This coefficient is rarely used, but the 2D variant, $C_{2D}$, in which we integrate over a surface at $r = r'$, directly compares two models at that depth level and is often given as a function of $r'$ to identify regions where models agree or disagree. Such an analysis can be further refined by expanding $m$ and $m'$ into spherical harmonics with coefficients $a, b$ and $a', b'$ respectively. The orthogonality of the spherical harmonics implies that:

$$C_{2D} = \sum_{\ell} C_{\ell}, \tag{12.15}$$

with

$$C_\ell = \frac{a_{\ell 0}a'_{\ell m} + \sum_{m=1}^{\ell}(a_{\ell m}a'_{\ell m} + b_{\ell m}b'_{\ell m})}{\left(a_{\ell 0}^2 + \sum_{m=1}^{\ell}(a_{\ell m}^2 + b_{\ell m}^2)\right)^{\frac{1}{2}}\left(a_{\ell 0}'^2 + \sum_{m=1}^{\ell}(a_{\ell m}'^2 + b_{\ell m}'^2)\right)^{\frac{1}{2}}}.$$

To compare the correlation at different depth levels for the same model, Jordan et al. [152] introduced the radial correlation function $R(r, r')$:

$$R(r, r') = \frac{4\pi \int_0^{2\pi}\int_0^{\pi} m(r, \theta, \phi)m(r', \theta, \phi)\sin\theta \, \mathrm{d}\theta\mathrm{d}\phi}{\left(\int_0^{2\pi}\int_0^{\pi} m(r, \theta, \phi)^2 \sin\theta \, \mathrm{d}\theta\mathrm{d}\phi \int_0^{2\pi}\int_0^{\pi} m(r', \theta, \phi)^2 \sin\theta \, \mathrm{d}\theta\mathrm{d}\phi\right)^{\frac{1}{2}}},$$

which is symmetric: $R(r, r') = R(r', r)$ and which can be used to study the vertical continuity of tomographic features. An example is shown in Figure 12.4. Though both models were constructed from a data set that provides resolution in the lower mantle, the radial correlation functions differ. This reflects the fundamental shortcoming of seismic data to provide a unique image of the Earth, which necessitates the making of choices (known as 'regularization' – see Chapter 14). Both models were regularized by selecting the 'smoothest' model. For the lower mantle, the difference in the width of the diagonal band of high values of $R$ in Figure 12.4 may be due to the fact that the model in (a) was inverted with no preferred direction for the smoothing operator, whereas the inversion for the model in (b) favoured horizontal over vertical smoothing, which leads to diminished vertical correlations.

Though correlations averaged over the entire surface of the sphere are easily expressed using (12.15), correlations over a selected province of limited area require us to taper the model at the edges of the selected region. This complicates the statistical interpretation of such estimates, as Simons et al. [319] show in a planar approximation for Australia. Simons et al. [316] provide a complete formalism for tapering of regions on a spherical surface.

## *Exercise*

**Exercise 12.5** The excess topography $t_e$ due to the Earth's ellipticity is given by:

$$t_e(\theta) = \epsilon\left(\frac{1}{3} - \cos^2\theta\right) = -\epsilon\frac{2}{3}P_2^0(\cos\theta).$$

For a basis given by (12.13), using only one radial function $f_1(r) \equiv 1$ to eliminate the $r$-dependency, find all nonzero coefficients $a_{\ell m}$ and $b_{\ell m}$ if we expand $t_e$ into this basis.

# 13

# Common corrections

From a philosophical point of view, it would be desirable to image the Earth exactly as it is. From a practical point of view however, it is inescapable that we simplify the Earth model. For example, it is impractical to incorporate changes in crustal properties with length scales of a few km into a global tomographic model where the resolving length is a few hundred km. Yet the crustal properties near each source and station affect the travel time in a way that is not negligible. The crustal structure may often be known a priori. In this case, we prefer to apply 'crustal corrections'.

It is very much a matter of taste whether one adds the correction to the predicted model values, or subtracts them from the observed data. In both cases the correction will end up in the 'delay' $\delta t$, i.e. the difference between observed and predicted travel times:

$$\begin{aligned} \delta t &= T_{\text{obs}} - (T_{\text{pred}} + \delta t_{\text{cor}}) \\ &= (T_{\text{obs}} - \delta t_{\text{cor}}) - T_{\text{pred}} \,. \end{aligned}$$

It is good experimental practice to leave the observations untouched (except perhaps for corrections that are directly related to the instrument recording the datum we wish to interpret), and apply corrections only to the theoretical predictions. In this way one avoids proliferation of different versions of the same data set, with a new data set for every different correction model applied to them.

## 13.1 Ellipticity corrections

Almost all travel time predictions are calculated using spherical models. But the Earth rotates and its shape is deformed by the centrifugal force, to the effect that the Poles are about 22 km closer to the centre than a location at the Equator. To first order, we can approximate the shape of the Earth by an ellipsoid of revolution or a *Maclaurin ellipsoid*. The best-fitting ellipsoid to the Earth has a flattening

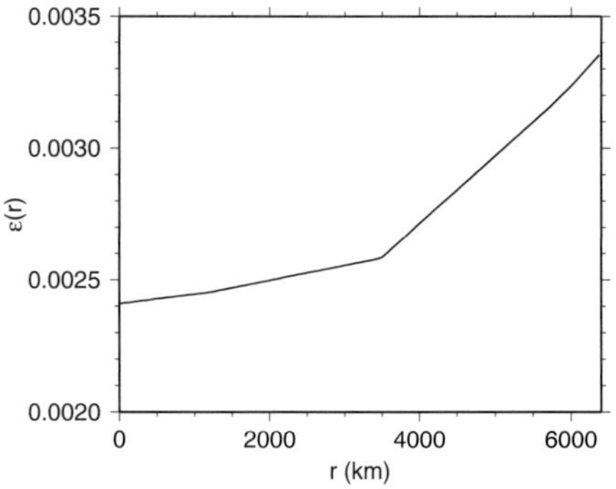


Fig. 13.1. Increase of ellipticity inside the Earth as a function of radius, for model PREM in hydrostatic equilibrium (Huang, personal communication, 2007).

$f = 1/298.3$ or $0.3352 \times 10^{-2}$. The ellipticity of the Sun is difficult to measure and may even vary over the solar cycle (Kuhn et al., [168]) but is an order of magnitude smaller ($0.5 \times 10^{-3}$) and it is usually neglected in time–distance helioseismology, since it has only a minor influence on the shallow images obtained.

In the geophysical and geodetic literature the shape of the ellipsoid is described in spherical coordinates $r$ (distance from the centre of the Earth) and $\theta$ (co-latitude, or angle with the North pole axis: $z = r\cos\theta$). The deviation $\delta r$ from the spherical Earth is then given by

$$\frac{\delta r}{r} \approx \epsilon(r)\left(\frac{1}{3} - \cos^2\theta\right). \tag{13.1}$$

The depth dependence of the ellipticity $\epsilon(r)$ can be computed from Clairaut's equation if we assume that the Earth has the shape of a rotating fluid under hydrostatic equilibrium:

$$\frac{\mathrm{d}^2\epsilon}{\mathrm{d}r^2} + \frac{8\pi G\rho}{g}\left(\frac{\mathrm{d}\epsilon}{\mathrm{d}r} + \frac{\epsilon}{r}\right) - \frac{6\epsilon}{r^2} = 0,$$

where $G$ is the constant of gravity, $g$ the acceleration of gravity, and $\rho$ the density – all well known for standard Earth models such as PREM (Dziewonski and Anderson [91]). At the surface $\epsilon(a) = f$ (see Figure 13.1), whereas $\mathrm{d}\epsilon/\mathrm{d}r = 0$ at $r = 0$. A modern solution, which is not hydrostatic everywhere, but that gives a better fit to the free core nutation period is given by Huang et al. [136].

The ellipsoidal deflection $\delta r$ has two nodal lines ('zeroes') at latitudes of about $35^\circ$ in the northern and southern hemisphere. Equation (13.1) implies that the surfaces of constant velocity are ellipsoids, not the spherical surfaces that allow us to use the efficient theory of Section 2.6. Since we always perform the computations for a spherical Earth, we must correct these for the effects of ellipticity.

A first-order correction, originally due to Bullen, can be obtained when one calculates the extra travel time needed to cross the height of the ellipsoid above the sphere, but is not sufficiently precise for seismic tomography. A more accurate method by Dziewonski and Gilbert [92] takes the full effect of the deflection $\delta r$ along the ray into account. We parametrize the ray by the epicentral angle $0 \le \varphi \le \Delta$. At the same ray location given by $\varphi$, the elliptical Earth will have a slightly different velocity because the surfaces of equal velocity have been deformed into ellipsoids according to (13.1). Because of Fermat's Principle (Section 2.9) we can neglect the change in raypath location for a stationary time path. We must take care that we explicitly take into account the extra travel times at the ray endpoints as well as the change in velocity because of the displaced velocity discontinuities. According to (7.1) the difference in travel time along the same raypath in an elliptical Earth with the time $T_0$ in a spherical Earth is:

$$\delta t_{\rm el} = T - T_0 \approx -\int_{P_0} \frac{\delta c}{c^2} \mathrm{d}s = \frac{1}{p}\int_0^{\Delta} \frac{r^3}{c^3}\frac{\delta r}{r}\frac{\mathrm{d}c}{\mathrm{d}r}\mathrm{d}\varphi\,, \tag{13.2}$$

where we use $\mathrm{d}s = r\mathrm{d}\varphi/\sin i$ and $p = r\sin i/c$. Since the ray stays put when we impose an elliptical shape and bring the Earth up to the ray depth, the change in velocity involves a minus sign: $\delta c = -(\mathrm{d}c/\mathrm{d}r)\delta r$. Though $\delta r$ is a function of radius and the colatitude $\theta$, not explicitly of the ray distance $\varphi$, the known ray geometry provides the necessary link when the integral is evaluated. To find the colatitude of the ray we can use the transformation matrix $\boldsymbol{T}$ given by (3.20). Thus, as we trace the ray, we compute $z = \cos\theta$, the vertical component of the unit vector $\hat{\boldsymbol{r}}$ pointing to the ray location. The deflection $\delta r$ then follows from (13.1).

To the integral (13.2) should be added the delay acquired by the deflection of the source and receiver ends of the ray, and of velocity discontinuities in the Earth. We discuss the passage of a discontinuity first. The situation is sketched in Figure 13.2. Since the deflection is small, we can locally ignore the sphericity of the interface. To correct the travel time for the effects of ellipticity, we add the difference between the two rays at the wavefront shown by the dotted line in Figure 13.2:

$$\begin{aligned}\delta t_{\rm el} &= \frac{AB}{c_1} - \frac{AC}{c_2} = AB\left(\frac{1}{c_1} - \frac{\cos(i_1 - i_2)}{c_2}\right)\\ &= -\delta r\left(\frac{\cos i_2}{c_2} - \frac{\cos i_1}{c_1}\right),\end{aligned}$$

where we use Snel's law and $AB = \delta r/\cos i_1$. Index 1 is for the incident wave, index 2 for the transmitted wave. A similar reasoning for reflections yields (with

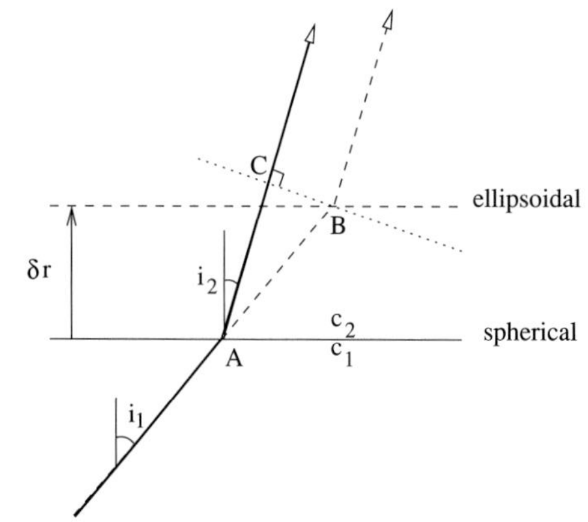


Fig. 13.2. Rays in the spherical Earth (solid line) and elliptical Earth (dashed line) acquire a time difference upon transmission through a deflected velocity discontinuity. When the wavefront for the elliptical Earth crosses the discontinuity at B, the time difference across the wavefront (dotted line) is complete.

index 2 now denoting the reflected wave)

$$\delta t_{\mathrm{el}} = \pm \delta r \left( \frac{\cos i_1}{c_1} + \frac{\cos i_2}{c_2} \right) , \tag{13.3}$$

with the positive sign for an underside reflection. The velocities $c_1$ and $c_2$ are the same for incident and reflected waves if there is no wave conversion.

Contributions from the ray endpoints follow in the same way, i.e. for a source at $r = r_s$: $\delta t_{\mathrm{el}} = -\delta r_s \cos i_s / c_{\mathrm{s}}$, and a similar term for the receiver at $r_r$: $\delta t_{\mathrm{el}} = +\delta r_r \cos i_r / c_{\mathrm{r}}$. Because the ray length is multiplied by the cosine of the ray angle, we see that the correction is exactly equal to the difference in *vertical* travel times of the wavefront! A more insightful derivation with the same result will be given in the section on topographic corrections.

Program `raydyntrace.f` (Tian et al., [361]), available from the software repository, allows for the computation of ellipticity corrections of rays of arbitrary complexity. A fast routine that interpolates tables for selected phases is `ellip.f` (Kennett and Gudmundsson [162]), also available in the public domain.† The ellipticity corrections are only very weakly dependent on the velocity model $c(r)$ that is used to compute them. An example is given in Figure 13.3. The ellipticity correction is important, as it can easily approach 1 s for a P-wave.

Kennett and Gudmundsson [162] show that the ellipticity correction for a diffracted phase is well approximated by the ellipticity correction calculated for the raypaths down to and up from the diffracting surface, since the lengthening of the raypath along that surface is itself of second order.

† `www.rses.anu.edu.au/seismology/ttsoft.html`

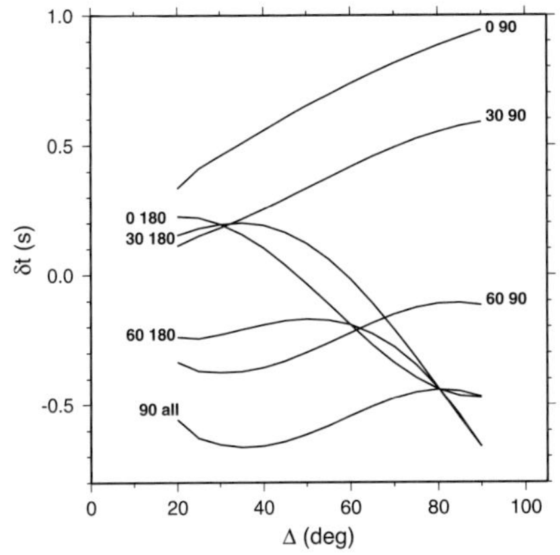


Fig. 13.3. Ellipticity corrections for P-waves in the Earth from a surface source as a function of epicentral distance Δ. The numbers next to each curve indicate the source latitude and source azimuth of the ray (90 = East, 180 = South), respectively. This figure was computed with Brian Kennett's `ellip.f`.

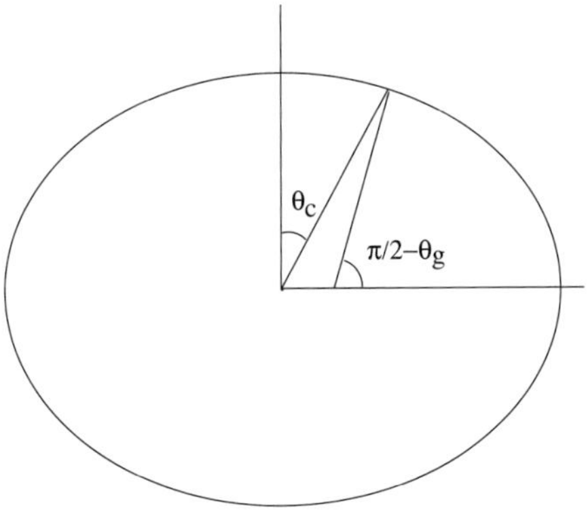


Fig. 13.4. A station at geographical latitude $\pi/2 - \theta_g$ is seen in the spherical Earth to be located at latitude $\pi/2 - \theta_c$ or co-latitude $\theta_c$.

One should be warned that the geographical latitude of a station as listed in common station lists is not the same as the *geocentric* latitude, because geographic latitudes are defined by the angle given by the local vertical to the Earth's ellipsoidal surface.‡ From Figure 13.4, and the value of the Earth's flattening one derives the following relationship between the co-latitudes in a spherical and elliptical Earth:

$$\cot \theta_{\text{geocentric}} = 0.993277 \cot \theta_{\text{geographic}} \, . \tag{13.4}$$

‡ `http://neic.usgs.gov/neis/gis/station_comma_list.asc` lists more than 12 000 seismic station locations with their station codes.

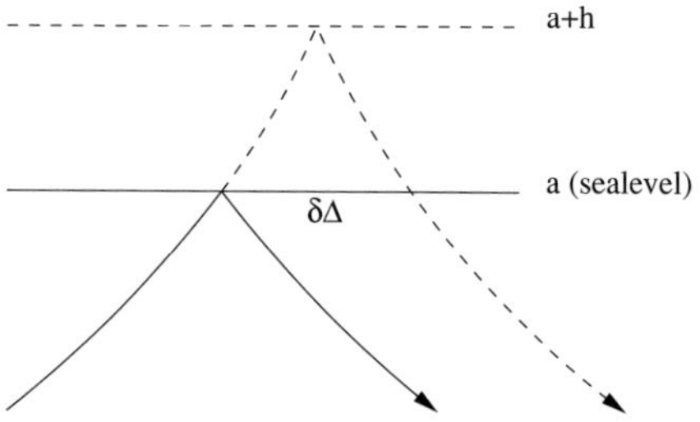


Fig. 13.5. Positive topography of $h$ km adds not only a delay $\delta t$ to the time of a surface-reflected ray (broken line) with respect to the ray in an unperturbed Earth (solid line). A ray with the same slowness also arrives at an epicentral distance that is larger by $\delta\Delta$. Correction for this distance leads to a topographic correction equal to $\delta\tau = \delta t - p\delta\Delta$.

Since all calculations are performed for a spherical model, one should use geocentric latitudes throughout.

## 13.2 Topographic and bathymetric time corrections

Corrections for surface topography are necessary because the seismic stations are usually not exactly located at sealevel (or on the ellipsoid). In the Sun, topographic corrections are important because the Sun's surface – from where Doppler shifts are measured – may vary in height with conditions and may be as much as 600 km below the unperturbed surface in a sunspot (Chou [56]). Although a topographic correction for body waves is different from the ellipticity correction – we are only concerned with the deflection of the end or reflection points – the method is very much the same. We give here a different derivation from the one given in the last section, which does not assume that rays near the surface are straight. In Figure 13.5 we compare a PP or SS reflection in the spherical Earth (solid line) with one in an Earth with (locally) raised topography, keeping the ray parameter $p$ constant. The extra time imposed by the time it takes to reach the reflector follows from (2.36):

$$\delta t_{(p=\text{constant})} = 2\int_a^{a+h} \frac{\mathrm{d}r}{c\sqrt{1-p^2c^2/r^2}}\,.$$

However, this is for a perturbed raypath with the same slowness $p$ as the unperturbed ray. Because of the extra distance travelled in the topography, this ray will not arrive at the original receiver distance, but at a distance $\Delta + \delta\Delta$ (Figure 13.5), where (see 2.37):

$$\delta\Delta = 2\int_a^{a+h} \frac{\mathrm{d}r}{r\sqrt{r^2/p^2c^2-1}}\,.$$

The local slope of the travel time curve $p = \partial T/\partial\Delta$ allows us to correct for the fact that this delay is computed for the wrong ray, one that overshoots the station by an extra epicentral distance $\delta\Delta$. We denote by $\delta\tau$ the perturbation to the travel time keeping $\Delta$ fixed. To first order:

$$\delta t_{(\Delta=\text{constant})} \equiv \delta\tau = \delta t_{(p=\text{constant})} - p\delta\Delta = 2\int_a^{a+h} \frac{1}{c}\sqrt{1-p^2c^2/r^2}\,\mathrm{d}r\,. \quad (13.5)$$

The interpretation of the reduced time $\tau = T - p\Delta$ is that it is the *intercept* time of the $T - \Delta$ curve at $\Delta = 0$ (therefore also called the 'zero-offset' time or 'vertical' travel time). Stork and Clayton [344] have shown that this simple result remains valid in the case of sloping layers. Equation (13.5) is important because it shows that the perturbation in the travel time along a ray of fixed distance due to deflections of interfaces or the surface is given by the perturbation in the vertical travel time of the ray at fixed slowness. Since we already know the ray at fixed slowness, this enables us to quickly compute the perturbation to be expected in a station, because the change in epicentral distance due to the change of the surface height is of second order.

The theory given above applies equally well for pP or sS and is easily adapted for other phases. The factor 2 in (13.5) disappears for P or S arrivals which have only the upcoming ray. For SP or PS reflections the integration needs to be split in separate up- and downgoing parts with appropriate velocities in each leg.

Phases with bounce points below a deep water layer pose special problems. Standard depth determinations such as published by ISC or NEIC use a continental reference model and assume essentially that the pP-wave reflects at the surface, whereas in reality most of this wave's energy is reflected at the ocean bottom and arrives earlier than the pwP phase reflecting from the air/water interface. Standard depths are therefore best interpreted as *depth below the ocean bottom*. But Engdahl et al. [98] correct for the water layer, so that the depths in the 'EHB catalogue' are true depths, i.e. with respect to the Earth's surface.

Neele and de Regt [226] and Dahlen [74] formulate a finite-frequency version of the effect of topography. For a general change $\delta r$ in the radius of an interface or surface:

$$\delta\tau = \int_S K^T_{\delta r}(\boldsymbol{r}_\mathrm{x})\delta r(\boldsymbol{r}_\mathrm{x})\mathrm{d}^2\boldsymbol{r}_\mathrm{x},$$

$$K^T_{\delta r}(\boldsymbol{r}_\mathrm{x}) = \frac{D_{\delta r}}{2\pi V_r}\left(\frac{|\cos i|\mathcal{R}_\mathrm{rs}}{\mathcal{R}_\mathrm{xs}\mathcal{R}_\mathrm{xr}}\right)\frac{\int_0^\infty \omega^3|\dot{m}(\omega)|^2\sin[\omega\Delta T(\boldsymbol{r}_\mathrm{x})-\Delta\Phi(\boldsymbol{r}_\mathrm{x})]\mathrm{d}\omega}{\int_0^\infty \omega^2|\dot{m}(\omega)|^2\mathrm{d}\omega}, \quad (13.6)$$

where the factor $D_{\delta r}$ is the same as given in (8.9–8.11). The factor in brackets (...) is continuous across the interface and can be evaluated on either side. For deep discontinuities such as the phase transitions at 410 and 660 km depth, accurate knowledge of $\delta r$ is not available. Instead, (13.6) may be used to invert for boundary layer topography rather than use it to correct $T$ with prior information.

### *Exercises*

**Exercise 13.1** Use $p \equiv \mathrm{d}T/\mathrm{d}\Delta$ to prove that $\tau$ is a monotonically decreasing function of $p$ by showing that $\mathrm{d}\tau/\mathrm{d}p = -\Delta$.

**Exercise 13.2** In local problems, we may prefer Cartesian coordinates. Show that (for a $z$-axis pointing upward):

$$\delta\tau = 2\int_0^h \frac{1}{c}\sqrt{1-c^2p^2}\,\mathrm{d}z\,.$$

**Exercise 13.3** Verify that the topographic correction for straight rays reduces to the expression we found earlier for the deflection of a discontinuity (13.3).

## 13.3 Crustal time corrections

Crustal thickness and velocity variations may induce travel time anomalies in P- and S-waves that exceed the effect of topography. For global tomography it is generally not possible to resolve crustal variations at anything but the crudest level. Therefore, one often chooses not to try to image those short-wavelength variations but instead to correct for known or inferred crustal structure, such that the delays represent as closely as possible the delays that would have been obtained if the real Earth had the simple laterally homogeneous crust of the background model. As in the case of topography, we can calculate the time correction for a ray with the same slowness $p$ as the unperturbed ray, provided we use the *vertical* travel time $\tau = t - p\Delta$.

If we have a reference level $r = r_d$ below which there are no differences between the crustal (3D) and background (BG) model, we find the correction by subtracting the vertical travel time $\tau_{\mathrm{BG}}$ for the background model from the theoretically predicted travel time and adding the $\tau_{\mathrm{3D}}$ for the crustal structure at the location of reflection points or stations:

$$\delta t_{\mathrm{crust}} = \delta\tau = \tau_{\mathrm{3D}} - \tau_{\mathrm{BG}}\,,$$

where $\tau$ is found from:

$$\tau = \sum_{\mathrm{segments}} \int \frac{1}{c}\sqrt{1-p^2c^2/r^2}\,\mathrm{d}r\,,$$

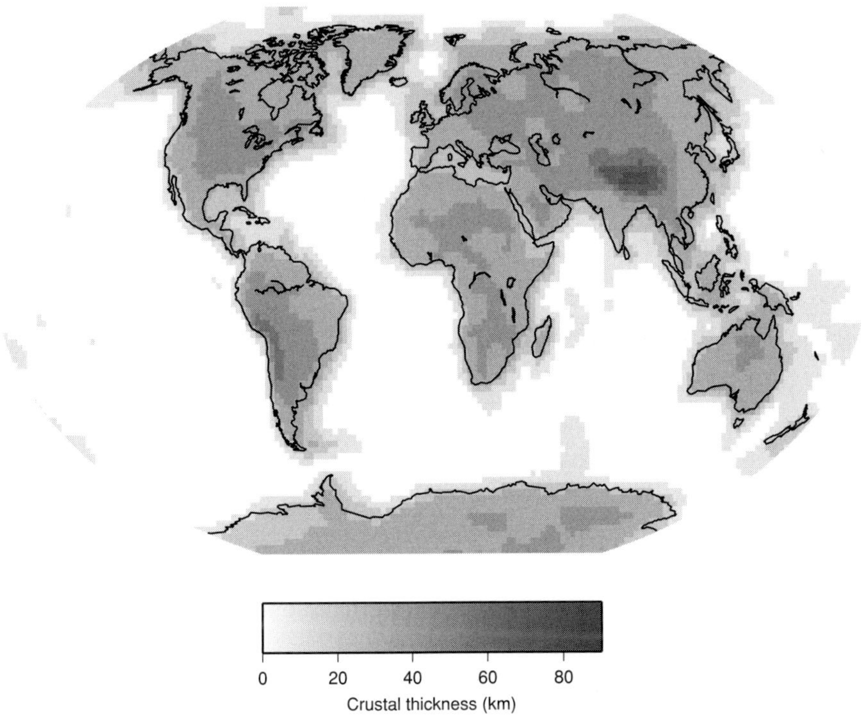


Fig. 13.6. Crustal thickness model CRUST2.0.

where the sum is over all ray segments in $r > r_d$, and the integration limits are $r = r_d$ and the surface $r = a$ unless the ray starts somewhere in the middle of this interval. Crustal models usually include topography $h_{3D}$ averaged over a grid unit, and care should be taken to make sure the topographic correction and the crustal correction are fully compatible. The logical way to do this is to calculate $\tau_{3D}$ for the interval $r_d \leq r \leq a + h_{3D}$ and to add to that the topographic correction for the actual station or reflection point height $h - h_{3D}$. If the background model has a crust that is thicker than that in the crustal model, information about the wave velocity $c$ at the the top of the mantle is missing. Linear extrapolation or simply a constant $c$ is usually adequate to fill the gap.

A convenient crustal model is Gabi Laske's CRUST2.0, an updated version of an earlier model CRUST5.1 (Mooney et al. [216]) and available in the public domain.[†] The most influential crustal parameter, the thickness, is shown in Figure 13.6.

Instead of using a 3D crustal model, one could of course add a 'station correction' term in the tomographic inversion as an unknown for every station (and reflection point where a ray such as PP bounces off the surface). Such an approach ignores

[†] `http://mahi.ucsd.edu/Gabi/rem.dir/crust/crust2.html`

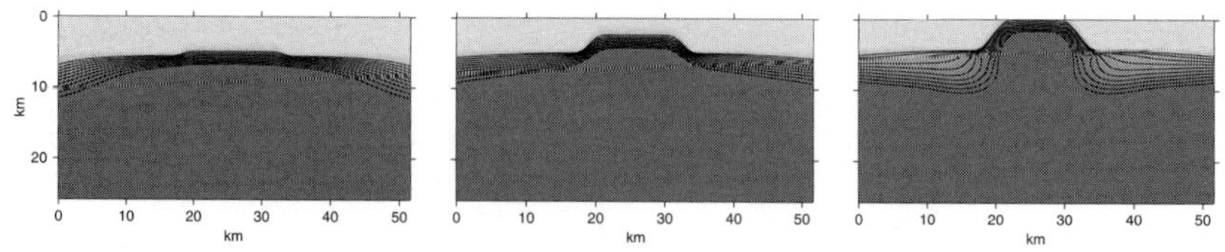


Fig. 13.7. Three snapshots in time of a wavefront entering a 15-km wide island in an ocean that is 5160 m deep, using a 2D finite-difference simulation.

prior knowledge that we have about the Earth from other sources and the danger is that a velocity anomaly under a station – e.g. a deep craton or a mantle plume – will always result in a late or early arrival and become invisible because the time induced by the heterogeneity is instead interpreted as a large crustal correction. Uncertainties in the crustal model can therefore better be dealt with by increasing the error assigned to the corrected time delay.

We have used ray theory to derive the crustal and topographic corrections. Yang and Shen [408] show frequency-dependent effects due to crustal reverberations of up to 0.6 s if strong reverberations arrive early enough to influence the cross-correlation, which is determined by the thickness and layering of the crust. Such frequency-dependent delays can probably best be modelled by incorporating crustal reverberations into the synthetic waveform used to cross-correlate, but this evidently requires prior knowledge of crustal structure, not just its thickness.

For topography at very small length scales, it is also not a-priori obvious that ray theory is sufficiently precise. For very small islands, for example, the acoustic water wave will be significantly (2–3 s) delayed with respect to the P-wave that traverses the island rock, but may diffract into the island basement and influence the signal. Figure 13.7 shows a 2D finite-difference simulation of the effect. Careful analysis of the seismograms in this numerical experiment show that the crustal correction for short periods is accurately predicted by ray theory, but for long periods on the very narrow island (15 km at the base) the correction is 0.2 s fast (Montelli et al., [213]). The effect can be understood if one realizes that a weak water wave signal arrives a few seconds after the onset of the P-wave and influences the cross-correlation maximum. This is an extreme case, though. Most oceanic stations are on islands that are significantly larger, and ray theory seems adequate for such crustal and topographic corrections.

Surface reflections such as for PP-waves occur over a large area defined by the Fresnel zone. Topographic corrections for such waves should therefore use a topography averaged over that area. In practice, this means that the $2^\circ \times 2^\circ$ average crustal structures in model CRUST2.0 contain all or much of the necessary topography and one might be satisfied applying $\delta t_{\text{crust}}$ only.

## 13.4 Surface wave corrections

For surface waves, a simple ellipticity correction is provided by changing the path length across the spherical surface to the actual path length $D_{\text{ellips}}$ along the ellipsoid. The Andoyer–Lambert approximation gives for the distance between points with geocentric colatitudes $\theta_1$ and $\theta_2$ to first order in flattening (Thomas [355]):

$$D_{\text{ellips}} = 6378.2\left[\Delta - 0.847519 \times 10^{-3}\left(\frac{(\Delta - \sin\Delta)(\sin\theta_1 + \sin\theta_2)^2}{1+\cos\Delta} + \frac{(\Delta + \sin\Delta)(\sin\theta_1 - \sin\theta_2)^2}{1-\cos\Delta}\right)\right], \tag{13.7}$$

where $\Delta$ is the epicentral distance calculated for the spherical Earth in radians, and $D_{\text{ellips}}$ is in km.

Topographic corrections for surface wave phase velocities were first derived by Snieder [326], using ray theory; these are a special case of the finite-frequency scattering by topography on boundary layers developed by Zhou et al. [420]. We refer the reader to her paper for the derivation and give here only the final result. If a boundary such as the Moho is deflected by $\delta r$, the phase $\phi$ of a surface wave is perturbed by:

$$\delta\phi = \int_S K_d(\boldsymbol{r})\delta r(\boldsymbol{r})\mathrm{d}^2\boldsymbol{r}\,, \tag{13.8}$$

with

$$K_d = \mathrm{Im}\left(\sum_{\text{mode 1}}\sum_{\text{mode 2}} N[\Omega_d^{(1)} + \Omega_d^{(2)}]_-^+\mathcal{P}\right),$$

where $N$ and $\mathcal{P}$ are given by (11.8) and (11.9) and the interaction coefficients $\Omega_d$ are listed in Table 13.1. The brackets $[.]_-^+$ denote the difference between the quantity above and below the interface; for the free surface there is no contribution from the topside of the boundary, and many terms are zero because of the zero stress boundary condition.

In global tomography, the large variation in crustal thickness between continents and oceans makes the crustal correction problem for surface wave phase velocities nonlinear, and no accurate solution is available, as shown by Montagner and Jobert [210] and Zhou et al. [420]. One should therefore be careful in applying crustal corrections to surface wave phase velocities over very long paths.

Table 13.1. *Boundary interaction coefficients. Subscripts 1 and 2 denote incoming and outgoing surface waves, respectively; $U$, $V$, $W$ are normalized using (10.6). $\eta$ is the scattering angle (Figure 11.2), $\nu = ak$ with $k$ in rad/m if $U$, $V$, $W$ and $r$ are in m. Dots denote differentiation with respect to $r$.*

| | $\Omega_d^{(1)}$ |
|---|---|
| Rayleigh → Rayleigh | $-\rho\omega^2(U_1U_2 + V_1V_2\cos\eta)$ |
| | $+\lambda(\dot{U}_1 + 2r^{-1}U_1 - \nu_1 r^{-1}V_1)(\dot{U}_2 + 2r^{-1}U_2 - \nu_2 r^{-1}V_2)$ |
| | $+\mu[2\dot{U}_1\dot{U}_2 + r^{-2}(2U_1 - \nu_1 V_1)(2U_2 - \nu_2 V_2)]$ |
| | $+\mu(\dot{V}_1 - r^{-1}V_1 + \nu_1 r^{-1}U_1)(\dot{V}_2 - r^{-1}V_2 + \nu_2 r^{-1}U_2)\cos\eta$ |
| | $+\mu\nu_1\nu_2 r^{-2}V_1V_2\cos 2\eta$ |
| Love → Love | $-\rho\omega^2 W_1W_2\cos\eta + \mu(\dot{W}_1 - r^{-1}W_1)(\dot{W}_2 - r^{-1}W_2)\cos\eta$ |
| | $+\mu\nu_1\nu_2 r^{-2}W_1W_2\cos 2\eta$ |
| Love → Rayleigh | $\rho\omega^2 W_1V_2\sin\eta$ |
| | $-\mu(\dot{W}_1 - r^{-1}W_1)(\dot{V}_2 - r^{-1}V_2 + \nu_2 r^{-1}U_2)\sin\eta$ |
| | $-\mu\nu_1\nu_2 r^{-2}W_1V_2\sin 2\eta$ |
| Rayleigh → Love | $-\rho\omega^2 V_1W_2\sin\eta$ |
| | $+\mu(\dot{V}_1 - r^{-1}V_1 + \nu_1 r^{-1}U_1)(\dot{W}_2 - r^{-1}W_2)\sin\eta$ |
| | $+\mu\nu_1\nu_2 r^{-2}V_1W_2\sin 2\eta$ |
| | $\Omega_d^{(2)}$ |
| Rayleigh → Rayleigh | $-4\mu\dot{U}_1\dot{U}_2 - \lambda\dot{U}_1(\dot{U}_2 + 2r^{-1}U_2 - \nu_2 r^{-1}V_2)$ |
| | $-\lambda\dot{U}_2(\dot{U}_1 + 2r^{-1}U_1 - \nu_1 r^{-1}V_1)$ |
| | $-\mu\dot{V}_1(\dot{V}_2 - r^{-1}V_2 + \nu_2 r^{-1}U_2)\cos\eta$ |
| | $-\mu\dot{V}_2(\dot{V}_1 - r^{-1}V_1 + \nu_1 r^{-1}U_1)\cos\eta$ |
| Love → Love | $-\mu[\dot{W}_1(\dot{W}_2 - r^{-1}W_2) + \dot{W}_2(\dot{W}_1 - r^{-1}W_1)]\cos\eta$ |
| Love → Rayleigh | $\mu[\dot{W}_1(\dot{V}_2 - r^{-1}V_2 + \nu_2 r^{-1}U_2) + \dot{V}_2(\dot{W}_1 - r^{-1}W_1)]\sin\eta$ |
| Rayleigh → Love | $-\mu[\dot{V}_1(\dot{W}_2 - r^{-1}W_2) + \dot{W}_2(\dot{V}_1 - r^{-1}V_1 + \nu_1 r^{-1}U_1)]\sin\eta$ |
| | At free surface: |
| | $(\lambda + 2\mu)\dot{U} + \lambda r^{-1}(2U - \nu V) = 0$ |
| | $\mu(\dot{V} - r^{-1}V + \nu r^{-1}U = 0$ |
| | $\mu(\dot{W} - r^{-1}W) = 0$ |

## 13.5 Source corrections

The source location $\boldsymbol{r}_s$ and origin time $T_0$ are usually assumed to be known when we interpret seismic delay times. This represents a classic chicken-and-egg problem: we need to know the Earth's structure if we must pinpoint the source, yet we determine the Earth's structure from observed travel times that require knowledge of the source time and location. Fortunately, source parameters can be accurately determined from stations near the source, where precise knowledge of Earth's structure is not as critical. Many subsequent refinements of the Earth's 1D models have led to ever better source locations. The first model, that of Jeffreys and Bullen [147], dates from the 1930s. Two frequently used 1D models today are IASP91 (Kennett and Engdahl [160]) and AK135 (Kennett et al. [161]). These models are designed to locate earthquakes and their shallow structure is consequently biased toward a typical continental structure since most of the seismic stations are located on land – unlike the Preliminary Reference Earth Model (PREM) which represents an average structure over the whole Earth.

Source locations determined by agencies such as the National Earthquake Information Service (NEIS) in the US or the International Seismological Centre (ISC) in Britain are derived by picking the onset time of seismic waves, and therefore representative for the beginning of the rupture. Source epicentre locations have errors of the order of 10 km in each component (equivalent to 14 km for the horizontal length of the location error vector) though larger errors or bias may exist in some oceanic regions far away from seismic networks (e.g. Kennett and Engdahl [160], Smith and Ekström [323], Shearer [308]). The depth accuracy is more variable, and there is a strong tradeoff between depth and origin time when no nearby stations are available, so that most rays are close to vertical. In that case an upward shift of the hypocentre combined with an earlier origin time leads to the same arrival time. Houser et al. [135] estimate that the effect of the source mislocation in the estimated travel times of long period gives errors in P-wave times of the order of 0.6 s, and 1.6 s for S. The depth/origin time ambiguity can be avoided for deeper earthquakes when reliable pP picks are available. But even then depth determinations may have errors of the order of 10 km.

There are two main strategies to deal with the problem of uncertainties in the source parameters. The first is to render the tomographic system insensitive to source parameter estimates by seeking combinations of data such that the influence of source errors subtracts out. A simple example is to invert for the arrival time *difference* between the PP and the P-wave. This renders the datum insensitive to the source origin time. For teleseismic waves, with the rays for both phases leaving the source in almost the same (vertical) direction, the depth mislocation will also subtract out. More formal approaches to desensitize the linear equations to errors

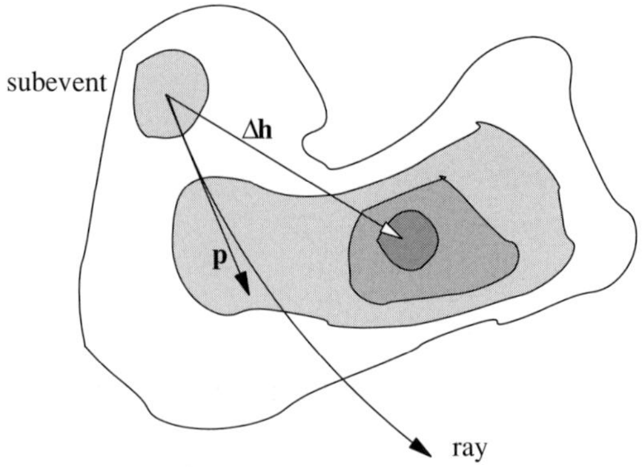


Fig. 13.8. If the hypocentre shifts by $\Delta\boldsymbol{h}$ this leads to a change in raypath length given by the projection of $\Delta\boldsymbol{h}$ onto the unit slowness vector $\hat{\boldsymbol{p}}$. Darker grey indicates larger energy release on the fault plane. This cartoon illustrates the case that the original hypocentre represents a small subevent, whereas a cross-correlation estimate for arrival time will be dominated by the centroid location.

in source parameters are given by Spencer and Gubbins [341] and Masters et al. [200] (see also Chapter 14).

The second option is to incorporate corrections to the published source parameters as additional unknowns into the linear system of equations. This has the advantage that one can incorporate prior knowledge of the uncertainties in the estimates into the inversion, thus retaining at least some of the information provided by independent data – such as P and PP arrival times – that is discarded when we only use the difference between the two observations. This advantage may be small when earthquakes are located with large uncertainties, but the strategy may pay off for earthquakes in areas densely populated with seismic stations such as Japan or the western US, that come with accurate origin times and hypocentral locations.

To derive the formalism of source corrections we use the situation shown in Figure 13.8. If we move the hypocentre by a vector $\Delta\boldsymbol{h}$, the travel time of the wavefront along the ray is given by the projection of $\Delta\boldsymbol{h}$ on the slowness vector $\boldsymbol{p}$. To this end we adopt a local (Cartesian) coordinate system and add the change in the origin time $\Delta T_0$:

$$\delta t = -\Delta T_0 - \boldsymbol{p}\cdot\Delta\boldsymbol{h} = -(\Delta T_0 + p_x\Delta x + p_y\Delta y + p_z\Delta z)\,. \tag{13.9}$$

Thus, for every hypocentre we add four columns to the matrix rows for that event, with entries $-1, -p_x, -p_y$ and $-p_z$, and four unknowns $(\Delta T_0, \Delta x, \Delta y, \Delta z)$ to the vector of unknowns. If, by convention, $x$ points east, $y$ points north, and $z$ points up (i.e. $z$ is negative depth), the slowness vector is expressed in terms of source azimuth $\zeta$ and take-off angle $i_0$ as $\boldsymbol{p} = (\sin\zeta \sin i_0, \cos\zeta \sin i_0, \cos i_0)$.

Unlike the corrections for crustal structure, which we assumed to contain independent information, we add the hypocentral corrections as extra unknowns to the model parameters.

Of course, to make a correction, one needs a starting value for the source parameters in the first place. For the origin time and hypocentre location, different choices are available. The ISC and NEIC catalogues list parameters derived from the onset of the first arriving wave, and can best be interpreted as the time and location of the onset of the rupture. 'Centroid' parameters, determined from long period waveforms (Dziewonski and Woodhouse [94]) are representative of the time-integrated properties of the source. Since an earthquake with moment magnitude $M_w = 6$ ruptures over a distance of about 10 km and has a rupture duration of several seconds, it is clear that the centroid hypocentres and origin times are different from the ISC/NEIC catalogue source parameters even for rather small events. For short-period data, e.g. delay times from the ISC catalogue, one should generally stick to the parameters determined from short-period body waves provided by NEIC or ISC. However, times estimated from cross-correlation will be dominated by the part of the observed wave that has the highest energy and these originate from another location, with a later origin time. This time may still be earlier than the centroid origin time if that time represents lower frequencies than found in typical body waves, so that choosing centroid parameters is no panacea for this mismatch. The situation is sketched in Figure 13.8: if the earthquake is triggered by a smaller subevent, the catalogue origin time may be several seconds earlier than the centroid time, a discrepancy that was noted by Montelli et al. [215]. To first order, one may shift all origin times with the average discrepancy by centering the histograms to zero delay, then use a hypocentral and origin time correction. When combining both catalogue data and data derived from cross-correlations, it is good policy to consider these as coming from two different hypocentres with two different origin times, essentially doubling the number of unknown source corrections. A source time function and a model for the rupture propagation may be needed to correctly estimate the effect of longer ruptures on initial phases for long-period surface waves. Alternatively, waveforms may be clustered in groups with similar source propagation effects (Sigloch and Nolet, [313]).

## 13.6 Amplitude corrections for body waves

Sediment layers with very low impedance (the product of density and velocity $\rho c$) will cause the amplitudes of seismic waves to increase so as to preserve energy. The effect is frequency dependent. This can be understood if we write the seismogram to include crustal reverberations:

$$u(t) = u_f(t) - \mathrm{R}u_f(t - 2\tau_s) + \mathrm{R}^2 u_f(t - 4\tau_s) - \dots ,$$

where $\tau_s$ is the vertical travel time of the wavefront through the sediment layer and R the reflection coefficient at the bottom of the sedimentary layer. The reflection coefficient at the surface has been simplified to be $-1$ (no conversion to S), and $u_f(t)$ is the first arriving P pulse, which is related to the incoming wave $u_0(t)$ by the transmission coefficient $\mathrm{T} = 1 + \mathrm{R}$ at the bottom of the sediment layer:

$$u_f(t) = \mathrm{T}u_0(t) .$$

We define amplitude by the robust observable introduced earlier in Chapter 8:

$$A = \sqrt{\frac{1}{T_\mathrm{p}} \int_0^{T_\mathrm{p}} u^2(t)\mathrm{d}t} \qquad \text{(8.1 again).}$$

where the pulse duration $T_\mathrm{p}$ is defined as the time window where $u_f(t)$ is appreciably different from zero; $u_f(t)$ has an amplitude that will be dominated by the impedance effect of the sedimentary layer. For a pulse duration $T_\mathrm{p} < \tau_s$, the first arriving P-wave will be isolated from the multiples, and the sedimentary amplification has a maximum effect. As $T_\mathrm{p}$ becomes longer, the term $-\mathrm{R}u_f(t - 2\tau_s)$ enters the time window, diminishing the amplitude because of the negative sign. For very thin layers with $\tau_s \ll T_\mathrm{p}$ we can ignore the time delays and

$$u(t) \approx u_f(t)[1 - \mathrm{R} + \mathrm{R}^2 - \dots] = u_f(t)\frac{1}{1 + \mathrm{R}} = u_f(t)\frac{1}{\mathrm{T}} = u_0(t) ,$$

so that the sediment layer has become 'invisible' to the amplitude measurement. Figure 13.9 shows that this stage is reached for $T_\mathrm{p} \approx 9$ s in the case of a very thick layer, but at much shorter periods for thinner layers. In almost all regions in model CRUST2.0 is the layer of unconsolidated sediment thinner than 1 km. Warren and Shearer [391] investigate the effect of similar reverberations in the oceanic layer on PP waves with periods less than 6 s and find that it can be ignored.

Computations for more sophisticated crustal models lead to the same conclusion that measurements of amplitudes for somewhat longer periods are only affected by a frequency-independent amplification effect. Zhou et al. [421] published a map showing that the amplification is approximately frequency independent at $T_\mathrm{p} > 6$ s for most regions on Earth (the exceptions being the world's large sedimentary basins). Tibuleac et al. [363] and Sigloch and Nolet [313] show that actual amplitude measurements from bandpassed seismograms exhibit a much larger scatter at short periods than at periods above 10 s.

Unless an accurate model of the crustal structure is available, we must include the amplitude correction for each station into the set of unknowns to be inverted for.

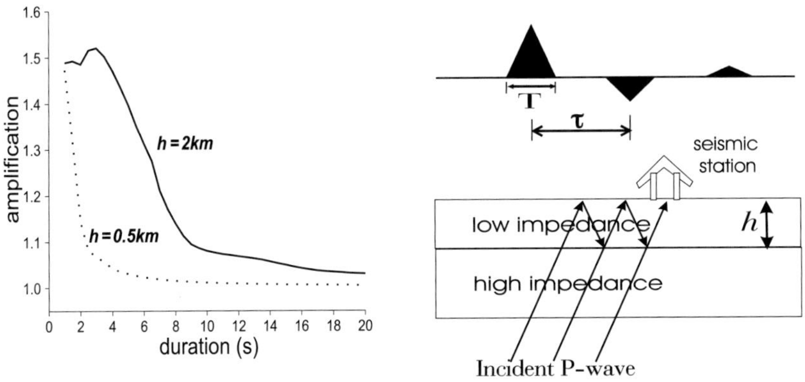


Fig. 13.9. Illustration of the effect of a sedimentary layer on the amplitude of a P-wave. Left: calculated amplification effect of a sedimentary layer ($V_P = 2$ km/s) on top of a consolidated layer ($V_P = 4.5$ km/s) as a function of the duration (dominant period $T$) of a P-wave arriving at a distance of 60°. Both a very thick layer of 2 km and an intermediate thickness of 0.5 km are shown. Right: The frequency dependence of the amplification arises from interference of multiples in the sedimentary layer. Pulses arrive with time spacing $\tau$. Figure courtesy Ying Zhou [421], reproduced with permission from the AGU.

One needs a large number of amplitude measurements from the same station if one wishes to resolve frequency-dependent amplitude corrections at periods shorter than about 6 s. For larger periods one can in most cases use the same correction for different frequency bands.

Similarly, a correction for the scalar moment for each event is needed to avoid a bias in amplitude measurements due to an under- or over-estimate of the earthquake magnitude.

## 13.7 Dispersion corrections

The seismic velocity in an attenuating medium is dependent on the frequency over the width of the absorption band. This relationship is given, to first order, by:

$$c(\omega) = c(\omega_0)\left[1 + \frac{1}{\pi Q}\ln\left(\frac{\omega}{\omega_0}\right)\right]. \qquad \text{(5.16 again)}$$

In general, the tomographic inversion involves data in many different frequency bands, but we invert for velocity anomalies at some reference frequency, since the effect is small and inverting for a model in each frequency band is not practical. We can do this by incorporating dispersion corrections in the theoretical predictions for the data from the background model which is given for that reference frequency. The reference frequency $\omega_0$ must be in the attenuation band if we apply (5.16). It is often defined at a period of 1 second ($\omega_0 = 2\pi$), which may be at the upper end of a major attenuation band for teleseismic waves. If, in fact, this attenuation band ends at a somewhat lower frequency, (5.16) will overestimate the effect when used with $\omega_0 = 2\pi$.

Normal mode programs such as MINEOS (Chapter 9) or the surface wave dispersion programs `love.f` and `rayleigh.f` in the software repository use (5.16) to correct the background model to the appropriate frequency before computing the phase velocity or eigenvector at that frequency, which makes the dispersion correction for these types of data practically automatic. But in the case of finite-frequency delay times, where delays along the same ray path may be determined for many different frequency bands, one should apply a dispersion correction. If the delay times are determined by cross-correlation with a synthetic signal, the proper way to incorporate the effects of attenuation is to convolve the pulse with an operator that incorporates both the dispersive delay $\delta t_{\mathrm{disp}}$ and the attenuation given by $t^*$. If $t_0$ denotes the arrival time at the reference frequency $\omega_0$, the perturbed time is given by

$$t_0 + \delta t_{\mathrm{disp}} = \int \frac{\mathrm{d}s}{c(\omega_0)\left[1 + \frac{1}{\pi Q}\ln\left(\frac{\omega}{\omega_0}\right)\right]} \approx \int \frac{\mathrm{d}s}{c(\omega_0)}\left[1 - \frac{1}{\pi Q}\ln\left(\frac{\omega}{\omega_0}\right)\right],$$

so that

$$\delta t_{\mathrm{disp}} = -\ln\left(\frac{\omega}{\omega_0}\right)\int \frac{\mathrm{d}s}{c(\omega_0)\pi Q} = -\frac{t^*}{\pi}\ln\left(\frac{\omega}{\omega_0}\right), \tag{13.10}$$

and with (2.65) the operator in the frequency domain becomes:

$$u(\omega) \rightarrow u(\omega)\,\mathrm{e}^{-\omega t^*/2}\,\mathrm{e}^{\mathrm{i}\omega\delta t_{\mathrm{disp}}}. \tag{13.11}$$

If we estimate differential travel times, and cross-correlate the same phase in two stations, or two phases in the same station, we must first correct one of the phases for the differential effects of attenuation. This is especially important for multiply reflected phases like SS that cross the highly attenuating asthenosphere more often than S.

## 13.8 Instrument response

Although the Doppler measurements from the solar surface accurately represent the material velocity along the line-of-sight direction, on Earth our seismometers do not record ground velocity, displacement or acceleration exactly. Instead, the instrument records a bandpassed version of the true ground motion, filtered by the properties of the mechanical sensor and associated electronics. When we cross-correlate with a synthetic seismogram, we generally predict pure velocity or

Table 13.2. *WWSSN instrument response*

| | SP | LP |
|---|---|---|
| $T_s$ | 1.0 s | 15.0 s |
| $T_g$ | 0.75 s | 100.0 s |
| $h_s$ | 0.87 | 1.00 |
| $h_g$ | 1.00 | 1.00 |
| $\sigma$ | 0.05 | 0.05 |

displacement, at least within the frequency band of interest. Thus, we must correct the seismogram first for any distortions due to the instrumental filter. Unlike the other corrections described in this chapter, this is a correction that is usually applied directly to the data (the observed seismogram).

To simplify the interpretation of seismograms, seismologists have often tried to standardize the response of instruments employed in networks. An important network that was operational for several decades after 1963 was the World-Wide Standard Seismograph Network. The WWSSN instruments used photographic recording. The response is a function of both the seismometer free period and the period of the galvanometer that picks up a current induced in a coil by the movement of the sensor. A light ray is deflected by a small mirror on the galvanometer and focused on photographic paper on a rotating drum (Figure 6.1). The instrument response of a WWSSN seismograph is given by:

$$V(\omega) = \frac{-\mathrm{i}A\omega^3}{(-\omega^2 - 2\mathrm{i}\epsilon_g\omega + \omega_g^2)(-\omega^2 - 2\mathrm{i}\epsilon_s\omega + \omega_s^2) + 4\epsilon_s\epsilon_g\sigma^2\omega^2}, \tag{13.12}$$

where $V(\omega)$ represents the factor that multiplies a single-frequency signal ($\sin\omega t$) ground displacement to obtain the actual deflection on the photographic record and where $A$ is constant amplification factor, $\omega_s = 2\pi/T_s$ and $\omega_g = 2\pi/T_g$ are the resonance frequencies of the seismometer and galvanometer, respectively, $\sigma$ is a small coupling constant and $\epsilon_s = h_s\omega_s$ and $\epsilon_g = h_g\omega_g$ with $h_s$ and $h_g$ damping constants of the order 1.

The more than 120 WWSSN stations have now been replaced by digital networks. Occasionally, one may wish to use digitized WWSSN records even in modern seismic tomography. I therefore give the characteristics for both the long period (LP) and short period (SP) instruments in Table 13.2. The amplification factor can be read from the seismograms themselves.

Instrument response correction is done by transforming the seismogram $u(t)$ to the spectral domain. Dividing $u(\omega)$ by $V(\omega)$ leads in principle to a corrected seismogram (true ground displacement) when this is transformed back to the time domain. In practice one needs to be careful with such procedures, as $|V| \to 0$ for very low and very high frequencies. Therefore, instrument correction is generally limited to a finite bandwidth of frequencies.

The instrument response of a digital seismograph is often much more complicated than that of an electromagnetic seismograph. To obtain the complete instrument response of a system, one has to multiply the responses of individual filters or other elements such as analogue-to-digital (A/D) converters in the system.

Very generally, the instrument response is described by an amplification factor $A_i$, and a number of 'poles' $p_j$ and 'zeroes' $z_j$:

$$V(\omega) = A_i \frac{\prod_{j=1}^{k}(\mathrm{i}\omega - z_j)}{\prod_{j=1}^{m}(\mathrm{i}\omega - p_j)} . \tag{13.13}$$

Often, one finds $A_i$ split up in two parts: $A_i = S_d A_0$, where $S_d$ is specified as the stage gain at some frequency $\omega_s$, and where $A_0$ normalizes the ratio of the polynomials to 1 at some normalizing frequency $\omega_n$, usually equal to the angular resonance frequency $\omega_s$ of the seismometer.

Even though modern recording instruments allow for large flexibility in defining the instrument response, there are still reasons to let $|V|$ approach zero outside some frequency band. At low frequencies, tides, atmospheric and other disturbances may dominate the signal. At the high frequency end, the sampling rate of the A/D converter determines the bandwidth: if we sample $n$ times per second, the maximum frequency that can still be correctly represented is $n/2$ Hz (one sample for each peak *and* trough). Higher frequencies are mapped back into lower frequency bands, a phenomenon known as aliasing. Aliasing is avoided by a low-pass (or 'anti-alias') filter that blocks high frequencies.

The dimension of $V(\omega)$ for a digital instrument is usually counts per micron per Hz. The term 'counts' refers to the digital number stored by the recorder. When the response is flat to velocity instead of displacement, the response is normally specified in counts per micron/s per Hz. This is the response to a velocity $\mathrm{d}e^{-\mathrm{i}\omega t}/\mathrm{d}t = -\mathrm{i}\omega e^{-\mathrm{i}\omega t}$. Thus if $\tilde{V}$ is the velocity response, we see that it is related to the displacement response by $\tilde{V} = (-\mathrm{i}\omega)^{-1} V$.

For seismograms distributed in SEED format by the IRIS Data Management Centre in Seattle or one of its associated data centres,† standard software is available to correct for the instrument response. One of the easiest ways to perform the

† Standard for the Exchange of Earthquake Data

correction for a variety of common instrument types is through the TRANSFER command in the Seismic Analysis Code (SAC).‡

Note that inversions for the amplitude correction as described in the previous section will absorb a correction for errors in the reported amplification $A_i$ of the instrument. One should, however, be on guard for sudden deterioration of instruments, such as has been reported for a very limited number of STS-1 sensors in the Global Seismograph Network (GSN). Scaling factors have been determined as a function of time for many GSN stations.† Instrument responses of 2 Hz L-22 vertical geophones used in the PASSCAL arrays are within 10% of the specified values, though one should watch out for exceptions. Horizontal geophones require *in situ* calibration because they may deviate by as much as 40% from specification (Pratt et al. [268]).

## 13.9 Clock corrections

Keeping track of the time is as important to seismometry as is recording the correct seismic wave amplitude and phase. For teleseismic events used in global tomography, a timing precision of better than 0.1 s is desirable. In local investigations, we may wish to have a precision better than 10 ms (milliseconds). Thus, clocks in seismic observatories all over the world must be synchronized to exactly the same time. This is not as easy as it may seem. Seismologists have agreed to use Coordinated Universal Time (UTC) as their standard. The start of the day at midnight (0:00h) in UTC is the same time in the UK, 1:00h in Western Europe, 9:00h in Japan, and 19:00h (or 7 p.m.) the previous day in the eastern United States (EST).

There are several other 'Universal Times'. The most important of them is UT1, which is defined by the Earth's rotation. However, the Earth proves to be an unreliable time keeper: not only does the length of day fluctuate because of the influence of climate (and because the mass displacement that accompanies earthquakes or the melting of icecaps slightly change its moment of inertia), the Earth also slows down because it loses energy through tidal friction that dissipates as heat into space. In between the rapid and the secular changes are periodic changes that are as yet poorly understood (McCarthy and Babcock [203]). Length of day variations are of the order of milliseconds. The secular slowing down of the Earth due to tidal dissipation lengthens the day by about 20 $\mu$s per year. The second is officially defined as 9 192 631 770 oscillations of a caesium atom. That corresponded to the

‡ Software to read SEED format and correct for instrument response is publicly available at `www.iris.washington.edu/manuals/`. The SAC program can be downloaded from this site as well.

† `www.ldeo.columbia.edu/~ekstrom/Projects/WQC/SCALING`.

second as defined by the Earth rotation in 1900, but this link is now outdated. To avoid a large discrepancy developing between UT1 and UTC, the two are sometimes synchronized at midnight on December 31, by introducing a leap second into UTC. This practice leads to much inconvenience or even errors and may soon be discontinued.

The precision of modern clocks is more than sufficient, but historical data are affected by systematic errors in time. Röhm et al. [285] found statistical indications for clock errors in ISC data by studying the average station delay as a function of time. Though large deviations are recognizable as outliers and can be removed, more subtle clock errors may influence the amplitude of tomographic inversions but by less than 15% (Röhm et al. [284]). WWSSN seismograms usually have the clock drift corrections written on the photographic records. Such corrections need to be added to the times of the minute marks. Older seismograms generally have timing errors that are too large for modern seismic tomography. Even though clock drift can be assumed linear, errors of one second or more are likely.

Modern, digital instrumentation has clocks synchronized to the very accurate time system maintained by GPS satellites and clock corrections are not needed unless GPS reception is interrupted. But for older digital records, clock errors may still be present. Bolton and Masters [27] recommend constructing summary rays – averaging the travel times from all measurements in one station from groups of closely located events, and removing outliers that turn up in the histograms constructed by differencing the raw data with this average. For a list of known clock errors see Table 1 in [27].

# 14

# Linear inversion

In Chapter 12 we saw how the parametrization of a continuous model allows us to formulate a discrete linear relationship between data $\boldsymbol{d}$ and model $\boldsymbol{m}$. With unknown corrections added to the model vector, this linear relationship remains formally the same if we write the physical model parameters as $\boldsymbol{m}_1$ and the corrections as $\boldsymbol{m}_2$ but combine both in one vector $\boldsymbol{m}$:

$$\boldsymbol{A}_1\boldsymbol{m}_2 + \boldsymbol{A}_2\boldsymbol{m}_2 = \boldsymbol{A}\boldsymbol{m} = \boldsymbol{d} \quad \text{(12.1) again.}$$

Assuming we have $M_1$ model parameters and $M_2$ corrections, this is a system of $N$ equations (data) and $M = M_1 + M_2$ unknowns. For more than one reason the solution of the system is not straightforward:

- Even if we do not include multiple measurements along the same path, many of the $N$ rows will be dependent. Since the data always contain errors, this implies we cannot solve the system exactly, but have to minimize the misfit between $\boldsymbol{A}\boldsymbol{m}$ and $\boldsymbol{d}$. For this misfit we can define different norms, and we face a choice of options.
- Despite the fact that we have (usually) many more data than unknowns (i.e. $N \gg M$), the system is almost certainly ill-posed in the sense that small errors in $\boldsymbol{d}$ can lead to large errors in $\boldsymbol{m}$; a parameter $m_j$ may be completely undetermined ($A_{ij} = 0$ for all $i$) if it represents a node that is far away from any raypath. We cannot escape making a subjective choice among an infinite set of equally satisfactory solutions by imposing a *regularization* strategy.
- For large $M$, the numerical computation of the solution has to be done with an iterative matrix solver which is often halted when a satisfactory fit is obtained. Such efficient shortcuts interfere with the regularization strategy.

We shall deal with each of these aspects in succession. Appendix D introduces some concepts of probability theory that are needed in this chapter.

## 14.1 Maximum likelihood estimation and least squares

In experimental sciences, the most commonly used misfit criterion is the criterion of least squares, in which we minimize $\chi^2$ ('chi square') as a function of the model:

$$\chi^2(\boldsymbol{m}) = \sum_{i=1}^{N} \left( \frac{|\sum_{j=1}^{M} A_{ij} m_j - d_i|^2}{\sigma_i^2} \right) = \min, \tag{14.1}$$

where $\sigma_i$ is the standard deviation in datum $i$; $\chi^2$ is a direct measure of the data misfit, in which we weigh the misfits inversely with their standard errors $\sigma_i$.

For uncorrelated and normally distributed errors, the principle of maximum likelihood leads naturally to the least squares definition of misfit. If there are no sources of bias, the expected value $E(d_i)$ of $d_i$ (the average of infinitely many observations of the same observable) is equal to the 'correct' or error-free value. In practice, we have only one observation for each datum, but we usually have an educated guess at the magnitude of the errors. We almost always use a normal distribution for errors, and assume errors to be uncorrelated, such that the probability density is given by a Gaussian or 'normal' distribution of the form:

$$P(d_i) = \frac{1}{\sigma_i \sqrt{2\pi}} \exp\left( -\frac{|d_i - E(d_i)|^2}{2\sigma_i^2} \right). \tag{14.2}$$

The joint probability density for the observation of an $N$-tuple of data with independent errors $\boldsymbol{d} = (d_1, d_2, ..., d_N)$ is found by multiplying the individual probability densities for each datum:

$$P(\boldsymbol{d}) = \prod_{i=1}^{N} \frac{1}{\sigma_i \sqrt{2\pi}} \exp\left( -\frac{|d_i - E(d_i)|^2}{2\sigma_i^2} \right). \tag{14.3}$$

If we replace the expected values in (14.3) with the predicted values from the model parameters, we obtain again a probability, but now one that is conditional on the model parameters taking the values $m_j$:

$$P(\boldsymbol{d}|\boldsymbol{m}) = \prod_{i=1}^{N} \frac{1}{\sigma_i \sqrt{2\pi}} \exp\left( -\frac{|d_i - \sum_j A_{ij} m_j|^2}{2\sigma_i^2} \right). \tag{14.4}$$

We usually assume that there are no extra errors introduced by the modelling (e.g. we ignore the approximation errors introduced by linearizations, neglect of anisotropy, or the shortcomings of ray theory etc.). In fact, if such modelling errors are also uncorrelated, unbiased and normally distributed, we can take them into account by including them in $\sigma_i$ – but this is a big 'if'.†

† See Tarantola [351] for a much more comprehensive discussion of this issue.

Clearly, one would like to have a model that is associated with a high probability for its predicted data vector. This leads to the definition of the likelihood function $\mathcal{L}$ for the model $\boldsymbol{m}$ given the observation of the data $\boldsymbol{d}$:

$$\mathcal{L}(\boldsymbol{m}|\boldsymbol{d}) = P(\boldsymbol{d}|\boldsymbol{m}) \propto \exp\left(-\frac{1}{2}\chi^2(\boldsymbol{m})\right).$$

Thus, maximizing the likelihood for a model involves minimizing $\chi^2$. Since this involves minimizing the sum of squares of data misfit, the method is more generally known as the *method of least squares*. The strong point of the method of least squares is that it leads to very efficient methods of solving (12.1). Its major weakness is the reliance on a normal distribution of the errors, which may not always be the case. Because of the quadratic dependence on the misfit, outliers – misfits of several standard deviations – have an influence on the solution that may be out of proportion, which means that errors may dominate in the solution. For a truly normal distribution, large errors have such a low probability of occurrence that we would not worry about this. In practice however, many data do suffer from outliers. For picked arrival times Jeffreys [146] has already observed that the data have a tail-like distribution that deviates from the Gaussian for large deviations from the mean $t_m$, mainly because a later arrival is misidentified as P or S:

$$P(t) = \frac{1-\epsilon}{\sigma\sqrt{2\pi}} \mathrm{e}^{-(t-t_m)^2/2\sigma^2} + \epsilon g(t),$$

where the probability density $g(t)$ varies slowly and where $\epsilon \ll 1$. A simple method to bring the data distribution close to normal is to reject outliers with a delay that exceeds the largest delay time to be expected from reasonable effects of lateral heterogeneity. This decision can be made after a first trial inversion: for example, one may reject all data that leave a residual in excess of $3\sigma$ after a first inversion attempt.

If we divide all data – and the corresponding row of $\boldsymbol{A}$ – by their standard deviations, we end up with a data vector that is univariant, i.e. all standard deviations are equal to 1. Thus, without loss of generality, we may assume that the data are univariant, in which case we see from (14.1) that $\chi^2$ is simply the squared length of the residual vector $|\boldsymbol{r}| = |\boldsymbol{d} - \boldsymbol{Am}|$. From Figure 14.1 we see that $\boldsymbol{r}$ is then perpendicular to the subspace spanned by all vectors $\boldsymbol{Ay}$ (the 'range' $R(\boldsymbol{A})$ of $\boldsymbol{A}$). For if it was not, we could add a $\delta\boldsymbol{m}$ to $\boldsymbol{m}$ such that $\boldsymbol{A}\delta\boldsymbol{m}$ reduces the length of $\boldsymbol{r}$. Thus, for all $\boldsymbol{y}$ the dot product between $\boldsymbol{r}$ and $\boldsymbol{Ay}$ must be zero:

$$\boldsymbol{r} \cdot \boldsymbol{Ay} = \boldsymbol{A}^T\boldsymbol{r} \cdot \boldsymbol{y} = \boldsymbol{A}^T(\boldsymbol{d} - \boldsymbol{Am}) \cdot \boldsymbol{y} = 0,$$

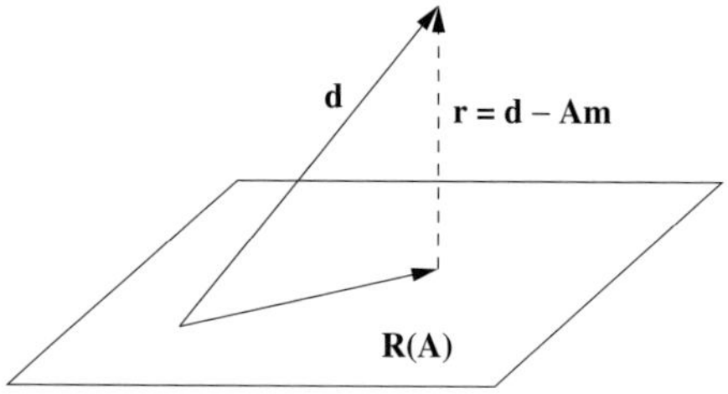


Fig. 14.1. If the data vector $\boldsymbol{d}$ does not lie in the range of $\boldsymbol{A}$, the best we can do is to minimize the length of the residual vector $\boldsymbol{r}$. This implies that $\boldsymbol{r}$ must be perpendicular to any possible vector $\boldsymbol{Ay}$.

where $\boldsymbol{A}^T$ is the transpose of $\boldsymbol{A}$ (i.e. $A^T_{ij} = A_{ji}$). Since this dot product is 0 for *all* $\boldsymbol{y}$, clearly $\boldsymbol{A}^T(\boldsymbol{d} - \boldsymbol{Am}) = 0$, or:

$$\boldsymbol{A}^T\boldsymbol{Am} = \boldsymbol{A}^T\boldsymbol{d}, \tag{14.5}$$

which is known as the set of 'normal equations' to solve the least-squares problem.

Chi square is an essential statistical measure of the goodness of fit. In the hypothetical case that we satisfy every datum with a misfit of one standard deviation we find $\chi^2 = N$; clearly values much higher than $N$ are unwanted because the misfit is higher than could be expected from the knowledge of data errors, and values much lower than $N$ indicate that the model is trying to fit the data errors rather than the general trend in the data. For example, if two very close rays have travel time anomalies differing by only 0.5 s and the standard deviation is estimated to be 0.7 s, we should accept that a smooth model predicts the same anomaly for each, rather than introducing a steep velocity gradient in the 3D model to try to satisfy the difference. Because we want $\chi^2 \approx N$, it is often convenient to work with the reduced $\chi^2$ or $\chi^2_{\text{red}}$, which is defined as $\chi^2/N$, so that the optimum solution is found for $\chi^2_{\text{red}} \approx 1$.

But how close should $\chi^2$ be to $N$? Statistical theory shows that $\chi^2$ itself has a variance of $2N$, or a standard deviation of $\sqrt{2N}$. Thus, for 1 000 000 data the true model would with 67% confidence be found in the interval $\chi^2 = 1\,000\,000 \pm 1414$. Such theoretical bounds are almost certainly too narrow because our estimates of the standard deviations $\sigma_i$ are themselves uncertain. For example, if the true $\sigma_i$ are equal to 0.9 but we used 1.0 to compute $\chi^2$, our computed $\chi^2$ itself is in error (i.e. too low) by almost 20%, and a model satisfying this level of misfit is probably not good enough. It is therefore important to obtain accurate estimates of the standard errors, e.g. using (6.2) or (6.12). Provided one is confident that the estimated standard errors are unbiased, one should still aim for a model that brings $\chi^2$ very close to $N$, say to within 20 or 30%.

An additional help in deciding how close one wishes to be to a model that fits at a level given by $\chi^2 = N$ is to plot the tradeoff between the model norm and $\chi^2$

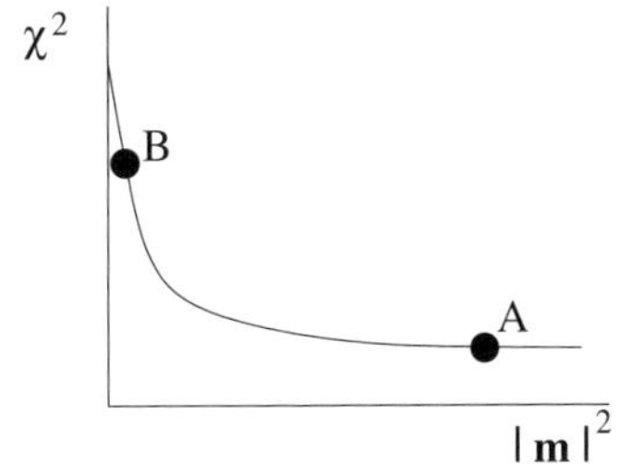


Fig. 14.2. The L- or tradeoff curve between $\chi^2$ and model norm $|\boldsymbol{m}|^2$.

(sometimes called the L-curve), shown schematically in Figure 14.2. If the tradeoff curve shows that one could significantly reduce the norm of the model while paying only a small price in terms of an increase in $\chi^2$ (point A in Figure 14.2), this is an indication that the standard errors in the data have been underestimated. For common data errors do not correlate between nearby stations, but the true delays should correlate – even if the Earth's properties vary erratically (because of the overlap in finite-frequency sensitivity). The badly correlating data can only be fit by significantly increasing the norm and complexity of the model, which is what we see happening on the horizontal part of the tradeoff curve. Conversely, if we notice that a significant decrease in $\chi^2$ can be obtained at the cost of only a minor increase in model norm (point B), this indicates an overestimate of data errors and tells us we may wish to accept a model with $\chi^2 < N$. If the deviations required are unexpectedly large, this is an indication that the error estimation for the data may need to be revisited.

Depending on where on the L-curve we find that $\chi^2 = N$, we find that we do or do not have a strong constraint on the norm of the model. If the optimal data fit is obtained close to point B where the L-curve is steep, even large changes in $\chi^2$ have little effect on the model norm. On the other hand, near point A even large changes in the model give only a small improvement of the data fit. Both A and B represent unwanted situations, since at A we are trying to fit data errors, which leads to erratic features in the model, whereas at B we are damping too strongly. In a well designed tomography experiment, $\chi^2 \approx N$ near the bend in the L-curve.

We used the term 'model norm' here in a very general sense – one may wish to inspect the Euclidean $|\boldsymbol{m}|^2$ as well as more complicated norms that we shall encounter in Section 14.5.

In many cases one inverts different data groups that have uncorrelated errors. For example Montelli et al. [215] combine travel times from the ISC catalogues with cross-correlation travel times from broadband seismometers. The ISC set, with about $10^6$ data was an order of magnitude larger than the second data set ($10^5$), and a brute force least-squares inversion would give preference to the short period ISC data in cases where there are systematic incompatibilities. This is easily diagnosed by computing $\chi^2$ for the individual data groups. One would wish to weigh the data sets such that each group individually satisfies the optimal $\chi^2$ criterion, i.e. if $\chi_i^2$ designates the misfit for data set $i$ with $N_i$ data, one imposes $\chi_i^2 \approx N_i$ for each data set. This may be accomplished by giving each data set equal weight and minimizing a weighted penalty function:

$$\mathcal{P} = \sum_i \frac{1}{N_i} \chi_i^2 .$$

Note that this gives a solution that deviates from the maximum likelihood solution, and we should only resort to weighting if we suspect that important conditions are violated, especially those of zero mean, uncorrelated and normally distributed errors. More often, an imbalance for individual $\chi_i^2$ simply reflects an over- or underestimation of the standard deviations for one particular group of data, and may prompt us to revisit our estimates for prior data errors.

Early tomographic studies often ignored a formal statistical appraisal of the goodness of fit, and merely quoted how much better a 3D tomographic model satisfies the data when compared to a 1D (layered or spherically symmetric) background or 'starting' model, using a quantity named 'variance reduction', essentially the reduction in the Euclidean norm of the misfit vector. This reduction is as much a function of the fit of the 1D starting model as of the data fit itself – i.e. the same 3D model can have different variance reductions depending on the starting model – and is therefore useless as a statistical measure of quality for the tomographic model.

## Exercises

**Exercise 14.1** Derive the normal equations by differentiating the expression for $\chi^2$ with respect to $m_k$ for $k = 1, ..., M$. Assume univariant data ($\sigma_i = 1$).

**Exercise 14.2** Why can we not conclude from (14.5) that $\boldsymbol{Am} \equiv \boldsymbol{d}$?

## 14.2 Alternatives to least squares

In the parlance of mathematics, the squared Euclidean norm $\sum_i |r_i|^2$ is one of a class of Lebesgue norms defined by the power $p$ used in the sum: $(\sum_i |r_i|^p)^{1/p}$. Thus, the Euclidean norm is also known as the '$L_2$' norm because $p = 2$. Of special interest are the $L_1$ norm ($p = 1$) and the case $p \to \infty$ which leads to minimizing the maximum among all $|r_i|$.

Instead of simply rejecting outliers, which always requires the choice of a hard bound for acceptance, we may downweight data that show a large misfit in a previous inversion attempt, and repeat the process until it converges. In 1898, the

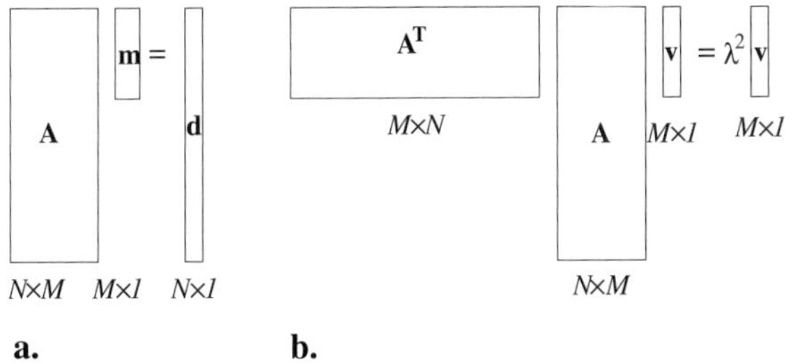


Fig. 14.3. (a) The original matrix system $\boldsymbol{Am} = \boldsymbol{d}$. (b) The eigenvalue problem for the least-squares matrix $\boldsymbol{A}^T\boldsymbol{A}$.

Belgian mathematician Charles Lagrange proposed such an 'iteratively weighted' least-squares solution, by iteratively solving:

$$|\boldsymbol{W}_p\boldsymbol{Am} - \boldsymbol{W}_p\boldsymbol{d}|^2 = \min,$$

where $\boldsymbol{W}_p$ is a diagonal matrix with elements $|r_i|^{p-2}$ and $0 \leq p < 2$, which are determined from the misfits $r_i$ in datum $i$ after the previous iteration. We can start with an unweighted inversion to find the first $\boldsymbol{r}$. The choice $p = 1$ leads to the minimization of the $L_1$ norm if it converges. The elements of the residual vector $r_i$ vary with each iteration, and convergence is not assured, but the advantage is that the inversion makes use of the very efficient numerical tools available for linear least-squares problems. The method was introduced in geophysics by Scales et al. [304].

## 14.3 Singular value decomposition

Though the least squares formalism handles the incompatibility problem of data in an overdetermined system, we usually find that $\boldsymbol{A}^T\boldsymbol{A}$ has a determinant equal to zero, i.e. eigenvalues equal to zero, and its inverse does not exist. Even though in tomographic applications $\boldsymbol{A}^T\boldsymbol{A}$ is often too large to be diagonalized, we shall analyse the inverse problem using singular values ('eigenvalues' of a non-square matrix), since this formalism gives considerable insight.

Let $\boldsymbol{v}_i$ be an eigenvector of $\boldsymbol{A}^T\boldsymbol{A}$ with eigenvalue $\lambda_i^2$, so that $\boldsymbol{A}^T\boldsymbol{A}\boldsymbol{v} = \lambda_i^2\boldsymbol{v}$. We may use squared eigenvalues because $\boldsymbol{A}^T\boldsymbol{A}$ is symmetric and has only non-negative, real eigenvalues. Its eigenvectors are orthogonal. The choice of $\lambda^2$ instead of $\lambda$ as eigenvalue is for convenience: the notation $\lambda_i^2$ avoids the occurrence of $\sqrt{\lambda_i}$ later in the development. We can arrange all $M$ eigenvectors as columns in an $M \times M$ matrix $\boldsymbol{V}$ and write (see Figure 14.3):

$$\boldsymbol{A}^T\boldsymbol{A}\boldsymbol{V} = \boldsymbol{V}\boldsymbol{\Lambda}^2. \tag{14.6}$$

The eigenvectors are normalized such that $\boldsymbol{V}^T\boldsymbol{V} = \boldsymbol{V}\boldsymbol{V}^T = \boldsymbol{I}$.

With (14.6) we can study the underdetermined nature of the problem $\boldsymbol{Am} = \boldsymbol{d}$, of which the least-squares solution is given by the system $\boldsymbol{A}^T\boldsymbol{Am} = \boldsymbol{A}^T\boldsymbol{d}$. The eigenvectors $\boldsymbol{v}_i$ span the $M$-dimensional model space so $\boldsymbol{m}$ can be written as a linear combination of eigenvectors: $\boldsymbol{m} = \boldsymbol{V}\boldsymbol{y}$. Since $\boldsymbol{V}$ is orthonormal, $|\boldsymbol{m}| = |\boldsymbol{y}|$ and we can work with $\boldsymbol{y}$ instead of $\boldsymbol{m}$ if we wish to restrict the norm of the model. Using this:

$$\boldsymbol{A}^T\boldsymbol{A}\boldsymbol{V}\boldsymbol{y} = \boldsymbol{V}\boldsymbol{\Lambda}^2\boldsymbol{y} = \boldsymbol{A}^T\boldsymbol{d}\,,$$

or, multiplying both on the left with $\boldsymbol{V}^T$ and using the orthogonality of $\boldsymbol{V}$:

$$\boldsymbol{\Lambda}^2\boldsymbol{y} = \boldsymbol{V}^T\boldsymbol{A}^T\boldsymbol{d}\,.$$

Since $\boldsymbol{\Lambda}$ is diagonal, this gives $y_i$ (and with that $\boldsymbol{m} = \boldsymbol{V}\boldsymbol{y}$) simply by dividing the $i$-th component of the vector on the right by $\lambda_i^2$. But clearly, any $y_i$ which is multiplied by a zero eigenvalue can take any value without affecting the data fit! We find the *minimum norm solution*, the solution with the smallest $|\boldsymbol{y}|^2$, by setting such components of $\boldsymbol{y}$ to 0. If we rank the eigenvalues $\lambda_1^2 \geq \lambda_2^2 \geq ...\lambda_K^2 > 0, 0, ..., 0$, then the last $M-K$ columns of $\boldsymbol{V}$ belong to the nullspace of $\boldsymbol{A}^T\boldsymbol{A}$. We truncate the matrices $\boldsymbol{V}$ and $\boldsymbol{\Lambda}$ to an $M \times K$ matrix $\boldsymbol{V}_K$ and a $K \times K$ diagonal matrix $\boldsymbol{\Lambda}$ to obtain the minimum norm estimate:

$$\hat{\boldsymbol{m}}_{\text{min norm}} = \boldsymbol{V}_K\boldsymbol{\Lambda}_K^{-2}\boldsymbol{V}_K^T\boldsymbol{A}^T\boldsymbol{d}\,. \tag{14.7}$$

Note that the inverse of $\boldsymbol{\Lambda}_K$ exists because we have removed the zero eigenvalues. The orthogonality of the eigenvectors still guarantees $\boldsymbol{V}_K^T\boldsymbol{V}_K = \boldsymbol{I}_K$, but now $\boldsymbol{V}_K\boldsymbol{V}_K^T \neq \boldsymbol{I}_M$.

To see how errors in the data propagate into the model, we use the fact that (14.7) represents a linear transformation of data with a covariance matrix $\boldsymbol{C}_d$. The posteriori covariance of transformed data $\boldsymbol{Td}$ is equal to $\boldsymbol{T}\boldsymbol{C}_d\boldsymbol{T}^T$ (see Equation 14.39 in Appendix D). In our case we have scaled the data such that $\boldsymbol{C}_d = \boldsymbol{I}$ so that the posteriori model covariance is:

$$\begin{aligned}\boldsymbol{C}_{\hat{m}} &= \boldsymbol{V}_K\boldsymbol{\Lambda}_K^{-2}\boldsymbol{V}_K^T\boldsymbol{A}^T\boldsymbol{I}\boldsymbol{A}\boldsymbol{V}_K\boldsymbol{\Lambda}_K^{-2}\boldsymbol{V}_K^T \\ &= \boldsymbol{V}_K\boldsymbol{\Lambda}_K^{-2}\boldsymbol{\Lambda}_K^2\boldsymbol{\Lambda}_K^{-2}\boldsymbol{V}_K^T \\ &= \boldsymbol{V}_K\boldsymbol{\Lambda}_K^{-2}\boldsymbol{V}_K^T\,.\end{aligned} \tag{14.8}$$

Thus the posteriori variance of the estimate for parameter $m_i$ is given by:[†]

$$\sigma_{m_i}^2 = \sum_{j=1}^{K}\frac{V_{ij}^2}{\lambda_j^2}\,. \tag{14.9}$$

[†] To distinguish data uncertainty from model uncertainty we denote the model standard deviation as $\sigma_{m_i}$ and the data standard deviation as $\sigma_i$.

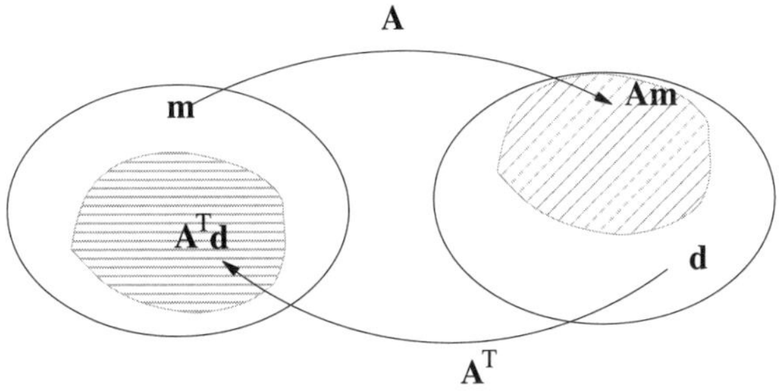


Fig. 14.4. Mappings between the model space (left) and the data space (right). The range of $\boldsymbol{A}$ is indicated by the grey area within the data space. The range of the backprojection $\boldsymbol{A}^T$ is indicated by the grey area in the model space.

This equation makes it clear that removing zero singular values is not sufficient, since the errors blow up as $\lambda_j^{-2}$, rendering the incorporation of small $\lambda_j$ very dangerous. Dealing with small eigenvalues is known as *regularization* of the problem. Before we discuss this in more detail, we need to show the connection between the development given here and the theory of singular value decomposition which is more commonly found in the literature.

One way of looking at the system $\boldsymbol{Am} = \boldsymbol{d}$ is to see the components $m_i$ as weights in a summation of the columns of $\boldsymbol{A}$ to fit the data vector $\boldsymbol{d}$. The columns make up the range of $\boldsymbol{A}$ in the data space (Figure 14.4). Similarly, the rows of $\boldsymbol{A}$ – the columns of $\boldsymbol{A}^T$ – make up the range of the backprojection $\boldsymbol{A}^T$ in the model space. The rest of the model space is the nullspace: if $\boldsymbol{m}$ is in the nullspace, $\boldsymbol{Am} = 0$. Components in the nullspace do not contribute to the data fit, but add to the norm of $\boldsymbol{m}$. We find the minimum norm solution by avoiding any components in the nullspace, in other words by selecting a model in the range of $\boldsymbol{A}^T$:

$$\hat{\boldsymbol{m}} = \boldsymbol{A}^T \boldsymbol{y}$$

and find $\boldsymbol{y}$ by solving for:

$$\boldsymbol{A}\boldsymbol{A}^T \boldsymbol{y} = \boldsymbol{d} .$$

The determinant of $\boldsymbol{AA}^T$ is likely to be zero, so just as in the case of least squares we shall wish to eliminate zero eigenvalues. Let the eigenvectors of $\boldsymbol{AA}^T$ be $\boldsymbol{u}_i$ with eigenvalues $\tilde{\lambda}_i^2$:

$$\boldsymbol{AA}^T\boldsymbol{U} = \boldsymbol{U}\tilde{\boldsymbol{\Lambda}}^2 . \tag{14.10}$$

Since $\boldsymbol{AA}^T$ is symmetric, the eigenvectors are orthogonal and we can scale them to be orthonormal, such that $\boldsymbol{U}^T\boldsymbol{U} = \boldsymbol{UU}^T = \boldsymbol{I}$. Multiplying (14.10) on the left by

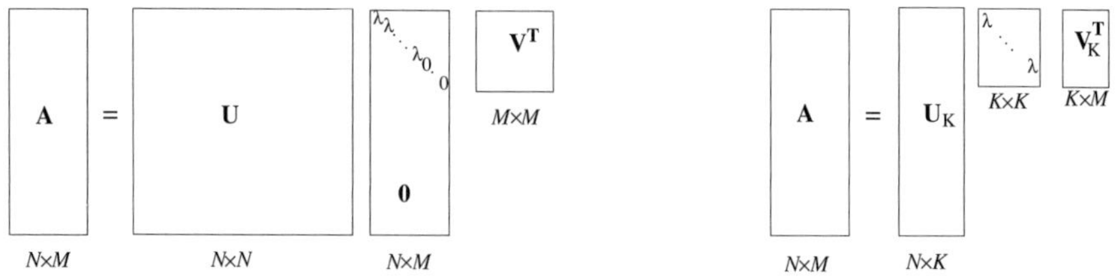


**a.** **b.**

Fig. 14.5. (a) The full eigenvalue problem for $\boldsymbol{A}\boldsymbol{A}^T$ leads to a matrix with small or zero eigenvalues on the diagonal. (b) removing zero eigenvalues has no effect on $\boldsymbol{A}$.

$\boldsymbol{A}^T$ and grouping $\boldsymbol{A}^T\boldsymbol{U}$ we see that $\boldsymbol{A}^T\boldsymbol{U}$ is an eigenvector of $\boldsymbol{A}^T\boldsymbol{A}$:

$$\boldsymbol{A}^T\boldsymbol{A}(\boldsymbol{A}^T\boldsymbol{U}) = (\boldsymbol{A}^T\boldsymbol{U})\tilde{\boldsymbol{\Lambda}}^2$$

and comparison with (14.6) shows that $\boldsymbol{A}^T\boldsymbol{u}_i$ must be a constant$\times\boldsymbol{v}_i$, and $\tilde{\lambda}_i = \lambda_i$. We choose the constant to be $\lambda_i$, so that

$$\boldsymbol{A}^T\boldsymbol{U} = \boldsymbol{V}\boldsymbol{\Lambda}. \tag{14.11}$$

Multiplying this on the left by $\boldsymbol{A}$ we obtain:

$$\boldsymbol{A}\boldsymbol{A}^T\boldsymbol{U} = \boldsymbol{U}\boldsymbol{\Lambda}^2 = \boldsymbol{A}\boldsymbol{V}\boldsymbol{\Lambda}\,,$$

or, dividing on the right by $\lambda_i$ for all $\lambda_i \neq 0$, and defining $\boldsymbol{u}_i\lambda_i = \boldsymbol{A}\boldsymbol{v}_i$ with a nullspace eigenvector $\boldsymbol{v}_i$ in case $\lambda_i = 0$:

$$\boldsymbol{A}\boldsymbol{V} = \boldsymbol{U}\boldsymbol{\Lambda}\,. \tag{14.12}$$

In the same way, by multiplying (14.12) on the right by $\boldsymbol{V}^T$ we find:

$$\boldsymbol{A} = \boldsymbol{U}\boldsymbol{\Lambda}\boldsymbol{V}^T \tag{14.13}$$

which is the *singular value decomposition* of $\boldsymbol{A}$. Note that in this development we have carefully avoided using the inverse of $\boldsymbol{\Lambda}$, so there is no need to truncate it to exclude zero singular values. However, because the tail of the diagonal matrix $\boldsymbol{\Lambda}$ contains only zeroes, (14.13) is equivalent to the truncated version (Figure 14.5):

$$\boldsymbol{A} = \boldsymbol{U}\boldsymbol{\Lambda}\boldsymbol{V}^T = \boldsymbol{U}_K\boldsymbol{\Lambda}_K\boldsymbol{V}_K^T\,. \tag{14.14}$$

## *Exercises*

**Exercise 14.3** Show that the choice (14.11) indeed implies that $\boldsymbol{U}^T\boldsymbol{U} = \boldsymbol{I}$. Hint: use (14.12).

**Exercise 14.4** Show that $\hat{\boldsymbol{m}} = \boldsymbol{V}_K\boldsymbol{\Lambda}_K^{-1}\boldsymbol{U}_K^T\boldsymbol{d}$ is equivalent to $\hat{\boldsymbol{m}}_{\text{min norm}}$.

## 14.4 Tikhonov regularization

The truncation to include only nonzero singular values is an example of regularization of the inverse problem. Removing zero $\lambda_i$ is not sufficient however, since small singular values may give rise to large modelling errors, as shown by (14.9). This equation tells us that small errors in the data vector may cause very large excursions in model space in the direction of $\boldsymbol{v}_k$ if $\lambda_k \ll 1$. It thus seems wise to truncate $\boldsymbol{V}$ in (14.13) even further, and exclude eigenvectors belonging to small singular values. The price we pay is a small increase in $\chi^2$, but we are rewarded by a significant reduction in the modelling error. We could apply a sharp cut-off by choosing $K$ at some nonzero threshold level for the singular values. Less critical to the choice of threshold is a tapered cut-off. We show that the latter approach is equivalent to adding $M$ equations of the form $\epsilon_n m_i = 0$, with $\epsilon_n$ small, to the tomographic system. Such equations act as artificial 'data' that bias the model parameters towards zero:

$$\begin{pmatrix} \boldsymbol{A} \\ \epsilon_n \boldsymbol{I} \end{pmatrix} \boldsymbol{m} = \begin{pmatrix} \boldsymbol{d} \\ \boldsymbol{0} \end{pmatrix}. \tag{14.15}$$

If the $j$-th column of $\boldsymbol{A}$ – associated with parameter $m_j$ – has large elements, the addition of one additional constraint $\epsilon_n m_j = 0$ will have very little influence. But the more $m_j$ is underdetermined by the undamped system, the more the damping will push $m_j$ towards zero. The least squares solution of (14.15) is:

$$(\boldsymbol{A}^T \boldsymbol{A} + \epsilon_n^2 \boldsymbol{I})\boldsymbol{m} = \boldsymbol{A}^T \boldsymbol{d}\,. \tag{14.16}$$

The advantage of the formulation (14.15) is that it can easily be solved iteratively, without a need for singular value decomposition. But the solution of (14.15) does have a simple representation in terms of singular values, and it is instructive to analyse it with SVD. If $\boldsymbol{v}_k$ is an eigenvector of $\boldsymbol{A}^T \boldsymbol{A}$ with eigenvalue $\lambda_k^2$, then the damped matrix gives:

$$(\boldsymbol{A}^T \boldsymbol{A} + \epsilon_n^2 \boldsymbol{I})\boldsymbol{v}_k = (\lambda_k^2 + \epsilon_n^2)\boldsymbol{v}_k\,, \tag{14.17}$$

and we see that the damped system has the same eigenvectors but with raised eigenvalues $\lambda_k^2 + \epsilon_n^2 > 0$. The minimum norm solution (14.7) is therefore replaced by:

$$\hat{\boldsymbol{m}}_{\text{damped}} = \boldsymbol{V}_K (\boldsymbol{\Lambda}_K^2 + \epsilon_n^2 \boldsymbol{I})^{-1} \boldsymbol{V}_K^T \boldsymbol{A}^T \boldsymbol{d} \tag{14.18}$$

with the posteriori model variance given by:

$$\sigma_{m_i}^2 = \sum_{j=1}^{K} \frac{V_{ij}^2}{\lambda_j^2 + \epsilon_n^2}. \tag{14.19}$$

Since there are no zero eigenvalues, we may set $K = N$, but of course this maximizes the variance and some truncation may still be needed. For simplicity, we assumed a damping with the same $\epsilon_n$ everywhere on the diagonal. The method is often referred to as Tikhonov regularization, after its original discoverer [364]. Because one adds $\epsilon^2$ to the diagonal of $\boldsymbol{A}^T\boldsymbol{A}$ it is also known as 'ridge regression'.

Spakman and Nolet [338] vary the damping factor $\epsilon_n$ along the diagonal. When corrections are part of the model, one should vary damping factors such that damping results in corrections that are reasonable in view of the prior uncertainty (for example, one would judge corrections as large as 100 km for hypocentral parameters usually unacceptable and increase $\epsilon_n$ for those corrections).

A comparison of (14.19) with (14.9) shows that damped model errors blow up at most by a factor $\epsilon_n^{-1}$. Thus, damping reduces the variance of the solution. This comes at a price however: by discarding eigenvectors, we reduce our ability to shape the model. The small eigenvalues are usually associated with vectors that are strongly oscillating in space: the positive and negative parts cancel upon integration and the resulting integral (12.12) is small. Damping small eigenvalues is thus expected to lead to smoother models. However, even long-wavelength features of the model may be biased towards zero because of regularization.

The fact that biased estimations produce smaller variances is a well known phenomenon in statistical estimation, and it is easily misunderstood: one can obtain a very small model parameter $m_i$ with a very small posteriori variance $\sigma_i^2$, yet learn nothing about the model because the bias is of the order of the true $m_i$. We shall come back to this in the section on resolution, but first investigate a more powerful regularization method, based on Bayesian statistics.

### *Exercises*

**Exercise 14.5** Show that the minimization of $|\boldsymbol{Am} - \boldsymbol{d}|^2 + \epsilon^2|\boldsymbol{m}|^2$ leads to (14.16).

**Exercise 14.6** In the L-curve for (14.18), indicate where $\epsilon = 0$ and where $\epsilon \to \infty$.

## 14.5 Bayesian inference

The simple Tikhonov regularization by norm damping we introduced in the previous section, while reducing the danger of excessive error propagation, is usually not satisfactory from a geophysical point of view. At first sight, this may seem surprising: for, when the $m_i$ represent perturbations with respect to a background model, the damping towards 0 is defensible if we prefer the model values given by the background model in the absence of any other information. However, if the information given by the data is unequally distributed, some parts of the model may be damped more than others, introducing an apparent structure in $\boldsymbol{m}$ that may be very misleading. The error estimate (14.19) does not represent the full modelling error because it neglects the bias. In general, we would like the model to have a minimum of unwarranted *structure*, or detail. Jackson [145] and Tarantola [349], significantly extending earlier work by Franklin [105], introduced the Bayesian method into geophysical inversion to deal with this problem, named after the Reverend Thomas Bayes (1702–1761), a British mathematician whose theorem on joint probabilities is a cornerstone of this inference method.

We shall give a brief exposé of Bayesian estimation for the case of $N$ observations in a data vector $\boldsymbol{d}^{\text{obs}}$. Let $P(\boldsymbol{m})$ be the prior probability density for the model $\boldsymbol{m} = (m_1, m_2, ..., m_M)$, e.g. a Gaussian probability of the form:

$$P(\boldsymbol{m}) = \frac{1}{(2\pi)^{M/2}} \frac{1}{|\det \boldsymbol{C}_m|^{1/2}} \exp\left(-\frac{1}{2}\boldsymbol{m} \cdot \boldsymbol{C}_m^{-1}\boldsymbol{m}\right). \tag{14.20}$$

Here, $\boldsymbol{C}_m$ is the *prior* covariance matrix for the model parameters. By 'prior' we mean that we generally have an idea of the allowable variations in the model values, e.g. how much the 3D Earth may differ from a 1D background model without violating more general laws of physics. We may express such knowledge as a prior probability density for the model values. The diagonal elements of $\boldsymbol{C}_m$ are the variances of that prior distribution. The off-diagonal elements reflect the correlation of model parameters – often it helps to think of them as describing the likely 'smoothness' of the model.

In a strict Bayesian philosophy such constraints may be 'subjective'. This, however, is not to say that we may impose constraints following the whim of an arbitrary person. An experienced geophysicist may often develop a very good intuition of the prior uncertainty of model parameters, perhaps because he has done experiments in the laboratory on analogue materials, or because he has experience with tomographic inversions in similar geological provinces. We shall classify such defensible subjective notions to be 'objective' after all.

The random errors in our observations make that the observed data vector $\boldsymbol{d}^{\text{obs}}$ deviates from the true (i.e. error-free) data $\boldsymbol{d}$. For the data we assume the normal distribution (14.2). Assuming the linear relationship $\boldsymbol{Am} = \boldsymbol{d}$ has no errors (or incorporating those errors into $\sigma_i$ as discussed before), we find the conditional probability density for the observed data, *given a model* $\boldsymbol{m}$:

$$P(\boldsymbol{d}|\boldsymbol{m}) = \frac{1}{(2\pi)^{N/2}} \frac{1}{|\det \boldsymbol{C}_d|^{1/2}} \exp\left(-\frac{1}{2}(\boldsymbol{Am} - \boldsymbol{d}^{\text{obs}}) \cdot \boldsymbol{C}_d^{-1}(\boldsymbol{Am} - \boldsymbol{d}^{\text{obs}})\right), \tag{14.21}$$

where $\boldsymbol{C}_d$ is the matrix with data covariance, usually taken to be diagonal with entries $\sigma_i^2$ because we have little knowledge about data correlations.

Though we have an expression for the data probability $P(\boldsymbol{d}|\boldsymbol{m})$, for solution of the inverse problem we are more interested in the probability of the model, *given the observed data* $\boldsymbol{d}^{\text{obs}}$. This is where Bayes' theorem is useful. It starts from the recognition that the joint probability can be split up in a conditional and marginal probability in two ways, assuming the probabilities for model and data are independent:

$$P(\boldsymbol{m}, \boldsymbol{d}^{\text{obs}}) = P(\boldsymbol{m}|\boldsymbol{d}^{\text{obs}})P(\boldsymbol{d}^{\text{obs}}) = P(\boldsymbol{d}^{\text{obs}}|\boldsymbol{m})P(\boldsymbol{m}),$$

from which we find Bayes' theorem:

$$P(\boldsymbol{m}|\boldsymbol{d}^{\text{obs}}) = \frac{P(\boldsymbol{d}^{\text{obs}}|\boldsymbol{m})P(\boldsymbol{m})}{P(\boldsymbol{d}^{\text{obs}})}. \tag{14.22}$$

Using (14.20) and (14.21):

$$P(\boldsymbol{m}|\boldsymbol{d}^{\text{obs}}) \propto \exp\left[-\frac{1}{2}(\boldsymbol{Am} - \boldsymbol{d}^{\text{obs}}) \cdot \boldsymbol{C}_d^{-1}(\boldsymbol{Am} - \boldsymbol{d}^{\text{obs}}) - \frac{1}{2}\boldsymbol{m} \cdot \boldsymbol{C}_m^{-1}\boldsymbol{m}\right].$$

Thus, we obtain the maximum likelihood solution by minimizing:

$$(\boldsymbol{Am} - \boldsymbol{d}^{\text{obs}}) \cdot \boldsymbol{C}_d^{-1}(\boldsymbol{Am} - \boldsymbol{d}^{\text{obs}}) + \boldsymbol{m} \cdot \boldsymbol{C}_m^{-1}\boldsymbol{m} = \chi^2(\boldsymbol{m}) + \boldsymbol{m} \cdot \boldsymbol{C}_m^{-1}\boldsymbol{m} = \min,$$

or, differentiating with respect to $m_i$:

$$\boldsymbol{A}^T\boldsymbol{C}_d^{-1}(\boldsymbol{Am} - \boldsymbol{d}^{\text{obs}}) + \boldsymbol{C}_m^{-1}\boldsymbol{m} = 0.$$

One sees that this is – again – a system of normal equations belonging to the 'damped' system:

$$\begin{pmatrix} \boldsymbol{C}_d^{-\frac{1}{2}}\boldsymbol{A} \\ \boldsymbol{C}_m^{-\frac{1}{2}} \end{pmatrix} \boldsymbol{m} = \begin{pmatrix} \boldsymbol{C}_d^{-\frac{1}{2}}\boldsymbol{d} \\ \boldsymbol{0} \end{pmatrix}. \tag{14.23}$$

Of course, if we have already scaled the data to be univariant the data covariance matrix is $\boldsymbol{C}_d = \boldsymbol{I}$. This simply shows that we are sooner or later obliged to scale the system with the data uncertainty. The prior smoothness constraint is unlikely to be a 'hard' constraint, and in practice we face again a tradeoff between the data fit and the damping of the model, much as in Figure 14.2. We obtain a manageable flexibility in the tradeoff between smoothness of the model and $\chi^2$ by scaling $\boldsymbol{C}_d^{-\frac{1}{2}}$ with a scaling factor $\epsilon$. Varying $\epsilon$ allows us to tweak the model damping until $\chi^2 \approx N$. Equation (14.23) is thus usually encountered in the equivalent, simplified form:

$$\begin{pmatrix} \boldsymbol{A} \\ \epsilon\boldsymbol{C}_m^{-\frac{1}{2}} \end{pmatrix} \boldsymbol{m} = \begin{pmatrix} \boldsymbol{d} \\ \boldsymbol{0} \end{pmatrix}. \tag{14.24}$$

How should one specify $\boldsymbol{C}_m$? The model covariance essentially tells us how model parameters are correlated. Usually, such correlations are only high for nearby parameters. Thus, $\boldsymbol{C}_m$ smoothes the model when operating on $\boldsymbol{m}$. Conversely, $\boldsymbol{C}_m^{-1}$ roughens the model, and (14.24) expresses the penalization of those model elements that dominate after the roughening operation. The simplest roughening operator is the Laplacian $\nabla^2$, which is zero when a model parameter is exactly the average of its neighbours. If we parametrize the model with tetrahedra or blocks, so that every node has well-defined nearest neighbours, we can minimize the difference between parameter $m_i$ and the average of its neighbours (Nolet [235]):

$$\frac{1}{2}\sum_i \frac{1}{N_i}\sum_{j\in\mathcal{N}_i}(m_i - m_j)^2 = \min,$$

where $\mathcal{N}_i$ is the set of $N_i$ nearest neighbours of mode $i$. Differentiating with respect to $m_k$ gives $M$ equations:

$$m_k - \frac{1}{N_k}\sum_{j\in\mathcal{N}_k} m_j = 0, \tag{14.25}$$

in which we recognize the $k$-th row of $\boldsymbol{C}_m^{-\frac{1}{2}}\boldsymbol{m}$ in (14.24).

One disadvantage of the system (14.24) is that it often converges much more slowly than the Tikhonov system (14.15) in iterative matrix solvers (VanDecar and Snieder [381]). The reason is that we are simultaneously solving a system arising from a set of integral equations, and the regularization system which involves finite-differencing. Without sacrificing the Bayesian philosophy, it is possible to transform (14.24) to a simple norm damping. Spakman and Nolet [338] introduce $\boldsymbol{m} = \boldsymbol{C}_m^{\frac{1}{2}}\boldsymbol{m}'$. Inserting this into (14.24) we find:

$$\begin{pmatrix} \boldsymbol{A}\boldsymbol{C}_m^{\frac{1}{2}} \\ \epsilon \boldsymbol{I} \end{pmatrix} \boldsymbol{m}' = \begin{pmatrix} \boldsymbol{d} \\ \boldsymbol{0} \end{pmatrix}. \tag{14.26}$$

Though it is not practical to invert the matrix $\boldsymbol{C}_m^{-\frac{1}{2}}$ that is implicit in (14.25) to find an exact expression for $\boldsymbol{C}_m^{\frac{1}{2}}$, many explicit smoothers of $\boldsymbol{m}$ may act as an appropriate 'correlation' matrix $\boldsymbol{C}_m^{\frac{1}{2}}$ for regularization purposes. After inversion for $\boldsymbol{m}'$, the tomographic model is obtained from the smoothing operation $\boldsymbol{m} = \boldsymbol{C}_m^{\frac{1}{2}}\boldsymbol{m}'$. The system (14.26) has the same form as the Tikhonov regularization (14.15). Despite this resemblance, in my own experience the acceleration of convergence is only modest compared to inverting (14.24) directly.

## 14.6 Information theory

Given the lack of resolution, geophysicists are condemned to accept the fact that there are infinitely many models that all satisfy the data within the error bounds. The Earth is a laboratory, but one that is very different from those in experimental physics, where we are taught to carefully design an experiment so that we have full control. Understandably, we feel unhappy with a wide choice of regularizations, resulting in our inability to come up with a unique outcome of the experiment. The temptation is always to resort to some 'higher' – if not metaphysical – principle that allows us to choose the 'best' model among the infinite set before we start plotting tomographic cross-sections. It should be recognized that this simply replaces one subjective choice (that of a model) with another (that of a criterion). Though some tomographers religiously adhere to such metaphysical considerations, I readily confess to being an atheist. In my view, such external criteria are simply a matter of taste. As an example, the methods of regularization are related to concepts known from the field of information theory, notably to the concept of information entropy. We shall briefly look into this, but warn the reader that, in the end, there is no panacea for our fall from Paradise.

We start with a simple application of the concept of information entropy: suppose we have only one datum, a delay measured along a ray of length $L$. We then have a $1 \times M$ system, or just one equation:

$$d_1 = \int \boldsymbol{m}(\boldsymbol{r})\mathrm{d}s = \sum_i m_i \mathrm{d}s,$$

As a thought experiment, assume that the segments of $\mathrm{d}s_i$ are of equal length $\mathrm{d}s$, and that we allow only one of them to cause the travel time anomaly. Which one? Information theory looks at this problem in the following way: let $P_i$ be the probability that $m_i \neq 0$. By the law of probabilities, $\sum P_i = 1$. Intuitively, we judge that in the absence of any other information, all $P_i$ should be equal – if not this would constitute additional information on the $m_i$. Formally, we may get to this conclusion by defining the information entropy:

$$I = \sum_i P_i \ln P_i, \tag{14.27}$$

which can be understood if we consider that any $P_i = 0$ will yield $I = -\infty$, thus minimizing the 'disorder' in the solution (note that if any $P_i = 1$, all others must be 0, again minimizing disorder). We express our desire to have a solution with minimum unwarranted information as the desire to maximize $I$, while still satisfying $\sum P_i = 1$. Such problems are solved with the method of Lagrange multipliers. This method recognizes that the absolute maximum of $I$ – zero for all probabilities equal to 1 – does not satisfy the constraint that $\sum P_i = 1$. So we relax the maximum condition by adding $\lambda(\sum P_i - 1)$ to $I$ and require:

$$I + \lambda\left(\sum P_i - 1\right) = \text{Max}\,.$$

Since the added factor is required to be zero, the function to maximize has not really changed as long as we satisfy that constraint. All we have done is add another dimension, or dependent variable, the Lagrange multiplier $\lambda$. We recover the original constraint by maximizing with respect to $\lambda$. Taking the derivative with respect to $P_i$ now gives an equation that involves $\lambda$:

$$\frac{\partial}{\partial P_i}\left(\sum_i P_i \ln P_i + \lambda \sum_i P_i\right) = 0,$$

or

$$\ln P_i = -(1+\lambda) \rightarrow P_i = \mathrm{e}^{-1-\lambda}.$$

We find the Lagrange multiplier from the constraint:

$$\sum_i P_i = N\mathrm{e}^{-1-\lambda} = 1 \rightarrow \lambda = \ln N - 1,$$

or

$$P_i = \mathrm{e}^{\ln(1/N)} = \frac{1}{N}.$$

Thus, if we impose the criterion of maximum entropy for the 'information' in our model, all $m_i$ are equally likely to contribute. The reasoning does not change much if we allow every $m_i$ to contribute to the anomaly and again maximize (14.27). In that case, all $m_i$ are equally likely to contribute. In the absence of further information, there is no reason to assume that one would contribute more than any other, and all are equal: $m_i = d_1/\sum \mathrm{d}s = d_1/L$. The smoothest model is the model with the highest information entropy. Such reasoning provides a 'higher principle' to justify the damping towards smooth models.

Constable et al. [64] named the construction of the smoothest model that satisfies the data with the prescribed tolerance *Occam's inversion*, after the fourteenth century philosopher William of Occam, or Ockham, who advocated the principle that simple explanations are more likely than complicated ones and who applied what came to be known as *Occam's razor* to eliminate unnecessary presuppositions.

However, one should not assume that smooth models are free of presuppositions: in fact, if we apply (14.25) in (14.24) we arbitrarily impose that smooth structures are more 'likely' than others. Artefacts may be suppressed, but so will sharp boundaries, e.g. the top of a subduction zone. Loris et al. [188], who invert for models that can be expanded with the fewest wavelets of a given wavelet basis, provide a variant on Occam's razor that is in principle able to preserve sharp features while eliminating unwarranted detail.

An interesting connection arises if we assume that sparse model parametrizations are a priori more probable than parametrizations with many basis functions. Assume that the prior model probability $P(\boldsymbol{m})$ is inversely proportional to the number of basis functions with nonzero coefficients in an exponential fashion:

$$P(\boldsymbol{m}) \propto \mathrm{e}^{-K},$$

where $K$ is the number of basis functions. If we insert this into Bayes' equation, we find that the maximum likelihood equation becomes:

$$\ln \chi^2(\boldsymbol{m}) - K = \min,$$

which is Akaike's [1] criterion for the optimum selection of the number of parameters $K$, used in seismic tomography by Zollo et al. [422]. Note, however, that this criterion lacks a crucial element: it does not impose any restrictions on the shape of the basis functions. Presumably one could use it by ranking independently defined basis functions in order of increasing roughness, again appealing to William of Occam for his blessing.

## 14.7 Numerical considerations

With $N$ often of the order of $10^5 - 10^7$ data, and $M$ only one order of magnitude smaller than $N$, the matrix system $\boldsymbol{Am} = \boldsymbol{d}$ is gigantic in size. Some reduction in the number of rows $N$ can be obtained by combining (almost) coincident raypaths into summary rays (see Section 6.1). The correct way to do this is to sum the rows of all $N_{\mathcal{S}}$ data belonging to a summary ray group $\mathcal{S}$ into one new averaged row that replaces them in the matrix:

$$\sum_{j=1}^{M} \frac{1}{N_{\mathcal{S}}} \left( \sum_{i \in \mathcal{S}} A_{ij} \right) m_j = \frac{1}{N_{\mathcal{S}}} \sum_{i \in \mathcal{S}} d_i \pm \sigma_{\mathcal{S}}, \tag{14.28}$$

with the variance $\sigma_{\mathcal{S}}^2$ equal to

$$\sigma_{\mathcal{S}}^2 = \frac{1}{N_{\mathcal{S}}^2} \sum_{i \in \mathcal{S}} \sigma_i^2 + \sigma_0^2 .$$

Here, $\sigma_0^2$ is added to account for lateral variations within the summary ray that affect the variance of the sum. Gudmundsson et al. [126] analysed the relationship between the width of a bundle and the variance in teleseismic P delay times from the ISC catalogue.

Care must be taken in defining the volume that defines the members of the summary ray. Events with a common epicentre but different depth provide important vertical resolution in the earthquake region and should often be treated separately. When using ray theory and large cells to parametrize the model we do not lose much information if we average over large volumes with size comparable to the model cells. But the Fréchet kernels of finite-frequency theory show that the sensitivity narrows down near source and receiver, and summarizing may undo some of the benefits of a finite-frequency approach.

Summary rays are sometimes applied to counteract the effect of dominant ray trajectories on the model – which may lead to strong parameter correlations along the prevailing ray direction – by ignoring the reduction of the error in the average. However, this violates statistical theory if we seek the maximum likelihood solution for normally distributed errors. The uneven distribution of sensitivity is better fought using unstructured grids with adapted resolution, and smoothness damping using a correlation matrix $\boldsymbol{C}_m$ that promotes equal parameter correlation in all directions.

If the parametrization is local, many elements of $\boldsymbol{A}$ are zero. For a least-squares solution, $\boldsymbol{A}^T\boldsymbol{A}$ has lost much of this sparseness, though, so we shall wish to avoid constructing $\boldsymbol{A}^T\boldsymbol{A}$ explicitly.† We can obtain a large savings in memory space by only storing the nonzero elements of $\boldsymbol{A}$. We do this row-wise – surprisingly the multiplications $\boldsymbol{Am}$ and $\boldsymbol{A}^T\boldsymbol{d}$ can both be done in row-order, using the following 'row-action' algorithms:

$\boldsymbol{p} = \boldsymbol{Am}$:
```
for i = 1, N
    for j = 1, M
        p_i ← p_i + A_ij m_j
```

$\boldsymbol{q} = \boldsymbol{A}^T\boldsymbol{d}$:
```
for i = 1, N
    for j = 1, M
        q_j ← q_j + A_ij d_i
```

where only nonzero elements of $A_{ij}$ should take part. This often leads to complicated bookkeeping. Claerbout's [60] dot-product test: $\boldsymbol{q}\cdot\boldsymbol{Ap} = \boldsymbol{A}^T\boldsymbol{q}\cdot\boldsymbol{p}$ – for random vectors $\boldsymbol{p}$ and $\boldsymbol{q}$ – can be used as a first (though not conclusive) test to validate the coding.

Early tomographic efforts in the medical and biological sciences led to a rediscovery of row-action methods (Censor [44]). The early methods, however, had the disadvantage that they introduced an unwanted scaling into the problem that interferes with the optimal regularization one wishes to impose (see van der Sluis and van der Vorst [377] for a detailed analysis).

† The explicit computation and use of $\boldsymbol{A}^T\boldsymbol{A}$ is also unwise from the point of view of numerical stability since its condition number – the measure of the sensitivity of the solution to data errors – is the square of that of $\boldsymbol{A}$ itself. For a discussion of this issue see *Numerical Recipes* [269].

Conjugate gradient methods work without implicit scaling. The stablest algorithm known today is LSQR, developed by Paige and Saunders [249] and introduced into seismic tomography by the author [233, 234]. We give a short derivation of LSQR. The main idea of the algorithm is to develop orthonormal bases $\boldsymbol{\mu}_k$ in model space, and $\boldsymbol{\rho}_k$ in data space. The first basis vector in data space, $\boldsymbol{\rho}_1$, is simply in the direction of the data vector: $\beta_1\boldsymbol{\rho}_1 = \boldsymbol{d}$, and $\boldsymbol{\mu}_1$ is the backprojection of $\boldsymbol{\rho}_1$: $\alpha_1\boldsymbol{\mu}_1 = \boldsymbol{A}^T\boldsymbol{\rho}_1$. Coefficients $\alpha_i$ and $\beta_i$ are normalization factors such that $|\boldsymbol{\rho}_i| = |\boldsymbol{\mu}_i| = 1$. We find the second basis vector in data space by mapping $\boldsymbol{\mu}_1$ into data space, and orthogonalize to $\boldsymbol{\rho}_1$:

$$\beta_2\boldsymbol{\rho}_2 = \boldsymbol{A}\boldsymbol{\mu}_1 - (\boldsymbol{A}\boldsymbol{\mu}_1 \cdot \boldsymbol{\rho}_1)\boldsymbol{\rho}_1 = \boldsymbol{A}\boldsymbol{\mu}_1 - \alpha_1\boldsymbol{\rho}_1,$$

where we use $\boldsymbol{A}\boldsymbol{\mu}_1 \cdot \boldsymbol{\rho}_1 = \boldsymbol{\mu}_1 \cdot \boldsymbol{A}^T\boldsymbol{\rho}_1 = \boldsymbol{\mu}_1 \cdot \alpha_1\boldsymbol{\mu}_1$. Similarly:

$$\alpha_2\boldsymbol{\mu}_2 = \boldsymbol{A}^T\boldsymbol{\rho}_2 - \beta_2\boldsymbol{\mu}_1.$$

Although it would seem that we have to go through more and lengthier orthogonalizations as the basis grows, it turns out that – at least in theory, ignoring roundoff errors – the orthogonalization to the previous basis function only is sufficient. For example, for $\boldsymbol{\rho}_3$ we find $\beta_3\boldsymbol{\rho}_3 = \boldsymbol{A}\boldsymbol{\mu}_2 - \alpha_2\boldsymbol{\rho}_2$. Taking the dot product with $\boldsymbol{\rho}_1$, we find:

$$\beta_3\boldsymbol{\rho}_3 \cdot \boldsymbol{\rho}_1 = \boldsymbol{A}\boldsymbol{\mu}_2 \cdot \boldsymbol{\rho}_1 - \alpha_2\boldsymbol{\rho}_2 \cdot \boldsymbol{\rho}_1 = \boldsymbol{\mu}_2 \cdot \boldsymbol{A}^T\boldsymbol{\rho}_1 = \boldsymbol{\mu}_2 \cdot (\alpha_1\boldsymbol{\mu}_1) = 0\,,$$

and $\boldsymbol{\rho}_3$ is perpendicular to $\boldsymbol{\rho}_1$. A similar proof by induction can be made for all $\boldsymbol{\rho}_k$ and $\boldsymbol{\mu}_k$ in the iterative sequence:

$$\beta_{k+1}\boldsymbol{\rho}_{k+1} = \boldsymbol{A}\boldsymbol{\mu}_k - \alpha_k\boldsymbol{\rho}_k \tag{14.29}$$

$$\alpha_{k+1}\boldsymbol{\mu}_{k+1} = \boldsymbol{A}^T\boldsymbol{\rho}_{k+1} - \beta_{k+1}\boldsymbol{\mu}_k. \tag{14.30}$$

If we expand the solution after $k$ iterations:

$$\boldsymbol{m}_k = \sum_{j=1}^{k} \gamma_j\boldsymbol{\mu}_j,$$

$$\sum_{j=1}^{k} \gamma_j\boldsymbol{A}\boldsymbol{\mu}_j = \boldsymbol{d},$$

and with (14.29):

$$\sum_{j=1}^{k} \gamma_j(\beta_{j+1}\boldsymbol{\rho}_{j+1} + \alpha_j\boldsymbol{\rho}_j) = \beta_1\boldsymbol{\rho}_1.$$

Taking the dot product of this with $\boldsymbol{\rho}_1$ yields $\gamma_1 = \beta_1/\alpha_1$, whereas subsequent factors are found by taking the product with $\boldsymbol{\rho}_k$ to give $\gamma_k = -\beta_k\gamma_{k-1}/\alpha_k$.

## 14.8 Appendix D: Some concepts of probability theory and statistics

I assume the reader is familiar with discrete probabilities, such as the probability that a flipped coin will come up with head or tail. If added up for all possible outcomes, the sum of all probabilities is 1.

This concept of probability cannot directly be applied to variables that can take any value within prescribed bounds. For such variables we use probability *density*. The probability density $P(X_0)$ for a random variable $X$ at $X_0$ is equal to the probability that $X$ is within the interval $X_0 \leq X \leq X_0 + \mathrm{d}X$, divided by $\mathrm{d}X$.

This can be extended to multiple variables. If $P(\boldsymbol{d})$ is the probability density for the data in vector $\boldsymbol{d}$, then the probability that we find the data within a small $N$-dimensional volume $\Delta\boldsymbol{d}$ in data space is given by $0 \leq P(\boldsymbol{d})\Delta\boldsymbol{d} \leq 1$. We only deal with normalized probability densities, i.e. the integral over all data:

$$\int P(\boldsymbol{d})\mathrm{d}^N\boldsymbol{d} = 1\,. \tag{14.31}$$

Joint probability densities give the probability that two or more random variables take a particular value, e.g. $P(\boldsymbol{m}, \boldsymbol{d})$. If the distributions for the two variables are independent, the *joint* probability density is the product of the individual densities:

$$P(\boldsymbol{m}, \boldsymbol{d}) = P(\boldsymbol{m})P(\boldsymbol{d}). \tag{14.32}$$

Conversely, one finds the *marginal* probability density of one of the variables by integrating out the second variable:

$$P(\boldsymbol{m}) = \int P(\boldsymbol{m}, \boldsymbol{d})\mathrm{d}^N\boldsymbol{d}. \tag{14.33}$$

The *conditional* probability density gives the probability of the first variable under the condition that the second variable has a given value, e.g. $P(\boldsymbol{m}|\boldsymbol{d}^{\mathrm{obs}})$ gives the probability density for model $\boldsymbol{m}$ given an observed set of data in $\boldsymbol{d}^{\mathrm{obs}}$.

The *expectation* or expected value $E(X)$ of $X$ is defined as the average over all values of $X$ weighted by the probability density:

$$\bar{X} \equiv E(X) = \int P(X)X\,\mathrm{d}X. \tag{14.34}$$

The expectation is a linear functional:

$$E(aX + bY) = aE(X) + bE(Y)\,, \tag{14.35}$$

and for independent variables it is separable:

$$E(XY) = E(X)E(Y)\,. \tag{14.36}$$

The *variance* is a measure of the spread of $X$ around its expected value:

$$\sigma_X^2 = E[(X - \bar{X})^2]\,, \tag{14.37}$$

where $\sigma_X$ itself is known as the standard deviation. The covariance between two random variables $X$ and $Y$ is defined as

$$\mathrm{Cov}(X, Y) = E[(X - \bar{X})(Y - \bar{Y})]\,. \tag{14.38}$$

In the case of an $N$-tuple of variables this defines an $N \times N$ covariance matrix, with the variance on the diagonal. The covariance matrix of a linear combination of variables is found by applying the linearity (14.35). Consider a linear transformation $\boldsymbol{x} = \boldsymbol{T}\boldsymbol{y}$. Since the spread of a variable does not change if we redefine the average as zero, we can assume that $E(x_i) = 0$ without loss of generality. Then:

$$\begin{aligned}\mathrm{Cov}(x_i, x_j) &= E\left(\sum_k T_{ij} y_k \sum_l T_{jl} y_l\right) = \sum_{kl} T_{ij} T_{jl} E(y_k y_l)\\ &= \sum_{kl} T_{ij} T_{jl} \mathrm{Cov}(y_k, y_l),\end{aligned}$$

or, in matrix notation:

$$\boldsymbol{C}_x = \boldsymbol{T}\boldsymbol{C}_y\boldsymbol{T}^T\,. \tag{14.39}$$

# 15

# Resolution and error analysis

One of the most important tasks of the seismic tomographer is to make sure he or she knows the limitations of the final model, and is able to convey that knowledge to others in a digestible form. This is not an easy problem: even within a narrow band of acceptable $\chi^2$ values, there will be infinitely many models that satisfy the data at this misfit level. Yet some features will change little among those models. Such features are 'resolved' if the change is less than some pre-specified variance. Of course, one cannot calculate infinitely many models and usually resigns oneself to present one possible inversion outcome with an assessment of its resolution and uncertainty.

To estimate resolution and uncertainty is a major task that will usually consume far more time than the actual inversion. As we shall see, all of the methods we currently know have shortcomings. Our means to present the results in an accessible form are equally poorly developed. There exists also some confusion about the meaning of damping parameters and their role in resolution and sensitivity tests. Many tomographers do not distinguish clearly between Bayesian constraints (damping parameters based on somewhat objective information) and damping parameters used to obtain a smooth model, which are inherently subjective if not based on prior information.

## 15.1 Resolution matrix

We restrict ourselves to the resolution and error analysis of linear problems of the form $\boldsymbol{A}\boldsymbol{m} = \boldsymbol{d}$. Whatever method we use to find a solution $\hat{\boldsymbol{m}}$, the estimate can be formally expressed as a linear combination of the data:

$$\hat{\boldsymbol{m}} = \boldsymbol{A}^{-}\boldsymbol{d}\,. \tag{15.1}$$

In virtually all cases, $\boldsymbol{A}^{-}$ is not a true inverse and therefore called a 'generalized inverse'. This means that neither $\boldsymbol{A}^{-}\boldsymbol{A}$ nor $\boldsymbol{A}\boldsymbol{A}^{-}$ necessarily equals the identity matrix, but a well-designed solver will make sure that both expressions are at least close to $\boldsymbol{I}$. For the second product this implies that we impose that $\boldsymbol{A}\hat{\boldsymbol{m}} = \boldsymbol{A}\boldsymbol{A}^{-}\boldsymbol{d} \approx \boldsymbol{d}$ (good data fit). The consequences of the reverse product $\boldsymbol{A}^{-}\boldsymbol{A}$ not being equal to $\boldsymbol{I}$ become visible when we formally analyse how close $\hat{\boldsymbol{m}}$ is to the true Earth model $\boldsymbol{m}^{\text{true}}$. We write the observed data $\boldsymbol{d}$ as the sum of the 'true', i.e. error-free data and an error term $\boldsymbol{e}$:

$$\boldsymbol{d} = \boldsymbol{d}^{\text{true}} + \boldsymbol{e} = \boldsymbol{A}\boldsymbol{m}^{\text{true}} + \boldsymbol{e}.$$

For simplicity we assume that the linear mapping $\boldsymbol{A}\boldsymbol{m}$ is adequate to predict the error-free data and does not introduce errors of its own, e.g. because the model parametrization is inadequate. From this we find the error in the solution $\hat{\boldsymbol{m}}$:

$$\hat{\boldsymbol{m}} - \boldsymbol{m}^{\text{true}} = \boldsymbol{A}^{-}\boldsymbol{d} - \boldsymbol{m}^{\text{true}} = (\boldsymbol{A}^{-}\boldsymbol{A} - \boldsymbol{I})\boldsymbol{m}^{\text{true}} + \boldsymbol{A}^{-}\boldsymbol{e}\,, \qquad (15.2)$$

and it is clear that the total model error has two components: lack of resolution because $\boldsymbol{A}^{-}\boldsymbol{A} \neq \boldsymbol{I}$ and propagated data errors ($\boldsymbol{A}^{-}\boldsymbol{e}$).

Another way to look at the effect of resolution on model error is to recognize that in the error-free case we have:

$$\hat{\boldsymbol{m}} = \boldsymbol{A}^{-}\boldsymbol{A}\boldsymbol{m}^{\text{true}} = \boldsymbol{R}\boldsymbol{m}^{\text{true}}\,, \qquad (15.3)$$

where $\boldsymbol{R} = \boldsymbol{A}^{-}\boldsymbol{A}$ is known as the resolution matrix, a kind of blurring window through which we can observe the true Earth. If the problem at hand is small enough for singular value decomposition, $\boldsymbol{R}$ can be constructed from the eigenvectors $\boldsymbol{v}_i$ of $\boldsymbol{A}^T\boldsymbol{A}$. Substituting the truncated SVD solution (see Exercise 14.4) $\boldsymbol{A}^{-} = \boldsymbol{V}_K\boldsymbol{\Lambda}_K^{-1}\boldsymbol{U}_K^T$, we find:

$$\boldsymbol{R} = \boldsymbol{A}^{-}\boldsymbol{A} = \boldsymbol{V}_K\boldsymbol{\Lambda}_K^{-1}\boldsymbol{U}_K^T\boldsymbol{U}_K\boldsymbol{\Lambda}_K\boldsymbol{V}_K^T = \boldsymbol{V}_K\boldsymbol{V}_K^T\,. \qquad (15.4)$$

The model estimate $\hat{m}_i$ is called unbiased if (15.3) yields true averages, i.e. if

$$\sum_{j=1}^{M} R_{ij} = 1\,. \qquad (15.5)$$

It is clear that (15.4) does not guarantee bias-free estimates because the truncation to $K$ eigenvectors causes $\boldsymbol{V}_K\boldsymbol{V}_K^T \neq \boldsymbol{I}$ and in fact, minimum-norm solutions are usually heavily biased. The problem is that $\boldsymbol{R}$ represents both the resolution and the bias, i.e. $\sum_j R_{ij} < 1$ (and sometimes much smaller than 1) so that $\sum_j R_{ij}m_j^{\text{true}}$ is not the true average over model parameters, but over the damped ones. This reveals a shortcoming of resolution estimations using $\boldsymbol{R}$ that is serious, even though it is widely used for smaller tomographic experiments. Another shortcoming of the method is that most global tomographic problems are too large for singular value analysis, so that $\boldsymbol{R}$ cannot even be computed. However, both Vasco et al. [382] and

Boschi et al. [28] have been able to compute the resolution matrix even for large systems, and for those with access to powerful parallel machines this problem may be less serious.

When calculating the posteriori model covariance $\boldsymbol{C}_{\hat{m}}$ with (14.8), we must be careful with the choice of the damping. From a Bayesian point of view, *we should only use objective constraints* on the model values itself or on their correlations (off-diagonal elements in the prior model covariance). In theory, there are no physical constraints on the correlation lengths, since the Earth may contain very sharp transitions between different compositions of the rock. This implies that we should assume the prior $\boldsymbol{C}_{\hat{m}}$ to be diagonal in (14.24) when computing $\boldsymbol{R}$ with a matrix that contains prior constraints. Thus, even if we obtain a preferred solution using smoothness constraints, it would give a falsely optimistic estimate of resolution if we include those constraints in the system when computing the resolution matrix. This confines us to simple ridge regression as in (14.15), in which the regression coefficient $\epsilon_n^2$ assures us that the parameters of the solution remain within physically acceptable bounds. Usually, such physical constraints require an $\epsilon_n$ that is less than the one we used to obtain the preferred solution, i.e. it is less than the coefficient used to obtain a $\chi^2$ close to $N$ and it may not be a preferred 'smooth' solution.

The two roles of $\epsilon_n$ – the role of regularization parameter and of an objective prior constraint to the model – result in two options for the formulation of the resolution matrix. We write (14.15) in the form:

$$\boldsymbol{A}_\epsilon \boldsymbol{m} = \boldsymbol{d}_0 \,, \qquad \text{with } \boldsymbol{A}_\epsilon = \begin{pmatrix} \boldsymbol{A} \\ \epsilon_n \boldsymbol{I} \end{pmatrix}, \qquad \boldsymbol{d}_0 = \begin{pmatrix} \boldsymbol{d} \\ \boldsymbol{0} \end{pmatrix}.$$

If the role of $\epsilon_n$ is only that of a damping parameter, with no prior connection to the true Earth, the trailing zeros of $\boldsymbol{d}_0$ are not considered as true 'data', and $\boldsymbol{d} = \boldsymbol{A}\boldsymbol{m}^{\text{true}}$. With that

$$\begin{aligned}
\hat{\boldsymbol{m}} &= \boldsymbol{V}(\boldsymbol{\Lambda}^2 + \epsilon_n^2 \boldsymbol{I})^{-1} \boldsymbol{V}^T \boldsymbol{A}_\epsilon^T \boldsymbol{d}_0 \\
&= \boldsymbol{V}(\boldsymbol{\Lambda}^2 + \epsilon_n^2 \boldsymbol{I})^{-1} \boldsymbol{V}^T \boldsymbol{A}^T \boldsymbol{d} \\
&= \boldsymbol{V}(\boldsymbol{\Lambda}^2 + \epsilon_n^2 \boldsymbol{I})^{-1} \boldsymbol{V}^T \boldsymbol{A}^T \boldsymbol{A} \boldsymbol{m}^{\text{true}} \\
&= \boldsymbol{V}(\boldsymbol{\Lambda}^2 + \epsilon_n^2 \boldsymbol{I})^{-1} \boldsymbol{V}^T \boldsymbol{V} \boldsymbol{\Lambda}^2 \boldsymbol{V}^T \boldsymbol{m}^{\text{true}} \\
&= \boldsymbol{V}(\boldsymbol{\Lambda}^2 + \epsilon_n^2 \boldsymbol{I})^{-1} \boldsymbol{\Lambda}^2 \boldsymbol{V}^T \boldsymbol{m}^{\text{true}} \,,
\end{aligned}$$

so that the resolution matrix for the damped solution is defined as

$$\boldsymbol{R}^{\text{damped}} = \boldsymbol{V}(\boldsymbol{\Lambda}^2 + \epsilon_n^2 \boldsymbol{I})^{-1} \boldsymbol{\Lambda}^2 \boldsymbol{V}^T \,. \tag{15.6}$$

On the other hand, if we assume that the zeros at the trailing end of $\boldsymbol{d}_0$ and their standard deviations $\epsilon_n^{-1}$ contain real prior information about the allowed prior variance of the model parameters, then we must set $\boldsymbol{d}_0 = \boldsymbol{A}_\epsilon \boldsymbol{m}^{\text{true}}$. This gives:

$$\begin{aligned}\hat{\boldsymbol{m}} &= \boldsymbol{V}(\boldsymbol{\Lambda}^2 + \epsilon_n^2 \boldsymbol{I})^{-1}\boldsymbol{V}^T \boldsymbol{A}_\epsilon^T \boldsymbol{d}_0 \\ &= \boldsymbol{V}(\boldsymbol{\Lambda}^2 + \epsilon_n^2 \boldsymbol{I})^{-1}\boldsymbol{V}^T \boldsymbol{A}_\epsilon^T \boldsymbol{A}_\epsilon \boldsymbol{m}^{\text{true}} \\ &= \boldsymbol{V}(\boldsymbol{\Lambda}^2 + \epsilon_n^2 \boldsymbol{I})^{-1}\boldsymbol{V}^T \boldsymbol{V}(\boldsymbol{\Lambda}^2 + \epsilon^2 \boldsymbol{I})\boldsymbol{V}^T \boldsymbol{m}^{\text{true}} \\ &= \boldsymbol{V}\boldsymbol{V}^T \boldsymbol{m}^{\text{true}} = \boldsymbol{m}^{\text{true}},\end{aligned}$$

with the expected outcome that the resolution matrix for a model estimate from Bayesian inference is the unit matrix:

$$\boldsymbol{R}^{\text{Bayes}} = \boldsymbol{V}\boldsymbol{V}^T = \boldsymbol{I}\,. \tag{15.7}$$

In this case we resolve every component of the model completely. Most likely, however, the posterior model variance given by (14.19) is now uncomfortably high, and we shall wish to truncate the number of eigenvectors, again introducing a bias.

One could conceivably defend a more permissive philosophy and argue that heavily oscillating solutions with strong gradients are 'unphysical', so that the regularization itself tells us what limits to impose on parameter derivatives; this would equate subjective damping parameters with objective ones and allow us to include smoothness damping even when analysing resolution. However, necessity may be the mother of invention in this case: Shearer and Earle [309], for example, analyse scattered wave energy and conclude that variations of $V_P$ can be 3–4 per cent over distances of 4 km in the upper mantle, or 0.5 per cent over 8 km in the lower mantle. Even if such variations cannot be maintained over distances comparable to the model grid separation, this does not bode well for efforts to find objective constraints on model gradients.

It is a fact of life that inversions without smoothing constraints easily degrade into heavily oscillating tomograms, in which one cannot distinguish the wood from the trees when searching for larger structures that can be understood in terms of dynamical processes. Therefore, the damping values that we actually use in the inversion usually reflect the values that yield the maximum smoothness for the solution while still getting an acceptable data fit, rather than independent constraints on model parameters or their derivatives. But generally, prior variances based on physical limitations for the model are too large to lead to an acceptable posteriori variance. What this is then telling us is that there is no visually pleasing solution if we strive for a resolution equal to that allowed by the parametrization.

Unfortunately, many tomographers choose to analyse the resolving power using the same damping coefficients that led to a lower but acceptable data fit and a lower variance. However, if we lower $K$ to obtain an acceptable model variance with (14.8), the resolution matrix loses much of its physical significance, because we bias our model towards zero. This may lead to meaningless results, e.g. one may damp an ill-resolved parameter to a value equal or close to zero. This parameter will have a very small variance, giving the impression we have 'resolved' it with a very small posteriori error. But the actual error in the estimate is large, being governed by the systematic error introduced by damping – the bias – rather than by random data errors.

The posteriori covariance matrix $\boldsymbol{C}_{\hat{m}}$ can also be computed from the singular value decomposition, using (14.8). Again, it is imperative that we use the full rank $N$ when doing this. Unresolved parameters will then show up with a variance dominated by the physics-based prior uncertainty assigned to them in the Bayesian regularization.

### *Exercise*

**Exercise 15.1** Someone suggests you can undo the fact that $\sum_j R_{ij} \neq 1$ by dividing every $\hat{m}_i$ by $\sum_j R_{ij}$ and multiplying the variance by $(\sum_j R_{ij})^{-2}$. Why would this not work?

## 15.2 Backus–Gilbert theory

An alternative to specifying the inverse mapping $\boldsymbol{A}^-$ is to specify the resolution $\boldsymbol{R} = \boldsymbol{A}^-\boldsymbol{A}$ instead, leading to a class of techniques labelled as 'optimally localized averaging' (OLA) methods in helioseismology, and sometimes referred to in the mathematical literature as 'mollifying', but really dating back to the groundbreaking work of Backus and Gilbert [14, 13, 12] named 'quelling' by Backus [10].† Backus–Gilbert estimates of resolution and uncertainty do not suffer from the shortcomings listed in the previous section: they are for unbiased estimates of a local average. If the data statistics are known, then the statistics of the errors in the solution – i.e. in the local averages – can be calculated easily. A reduction in the variance of the estimate is obtained by enlarging the volume over which the model parameter is averaged, not by increasing the bias. Thus, Backus–Gilbert theory provides an important tool for validation of the tomographic model. Unfortunately, it has its own drawback: every estimate requires a full inversion of the system. Backus–Gilbert estimation is therefore usually done only for selected model parameters or regions of special interest. That should not deter us from applying it – certainly not for smaller local tomography experiments, or when powerful computational resources are available even for global inversions. Though the method receives much interest in helioseismology, it is woefully underused in terrestrial applications. Notable exceptions are the estimation of lower mantle density by Ishii and Tromp [142], of

† See also Parker [255, 256] for an easily accessible introduction to the original Backus–Gilbert theory which was formulated for continuous functions $m(r)$ rather than for parametrized models as we do here.

tomographic imaging of the African superplume and the transition zone by Ritsema et al. [280, 281] and of azimuthal anisotropy in the upper mantle transition zone by Trampert and van der Heijst [366].

The starting point of Backus–Gilbert theory is the recognition that a meaningful variance can never be obtained for a point estimate of the model value. For example, we could always introduce a speckle with a very high $V_P$ that is small enough to leave $\chi^2$ practically unaffected. Meaningful estimates of model values spread out over a volume that is large enough to influence the data. In practice, such a volume is often provided by the model parametrization, which in turn reflects the data resolution we expect. But this is not flexible enough, since we would like to know how the size of the volume affects the variance of the model estimate.

We assume a local parametrization. This is not absolutely necessary, but it keeps the mathematics a little simpler. Instead of solving for individual parameters $m_i$, we may solve for an estimate $\hat{m}_k$ that is a local average of several parameters around node $k$. Since our original inverse problem is linear, and since averaging is also a linear operation, it follows that $\hat{m}_k$ must be a linear function of the data:

$$\hat{m}_k = \sum_i a_i^{(k)} d_i \,. \tag{15.8}$$

Note the similarity with (15.1). In fact, the $a_i^{(k)}$ form the row of a generalized inverse, but since Backus–Gilbert analysis is usually only done for selected parameters, we prefer the notation introduced in (15.8) for this section. From (15.8) we immediately obtain an expression for the uncertainty in the estimate; assuming univariant data for which $\boldsymbol{C}_d = \boldsymbol{I}$:

$$\sigma_{\hat{m}}^{(k)\,2} = \sum_{i,j} a_i^{(k)} a_j^{(k)} C_{ij} = \sum_i (a_i^{(k)})^2 \,. \tag{15.9}$$

How does one choose the weights $a_i^{(k)}$? Substituting for $d_i$:

$$\hat{m}_k = \sum_i a_i^{(k)} \sum_j A_{ij} m_j = \sum_j R_{kj} m_j ,$$

where we define the resolution matrix as:

$$R_{kj} = \sum_{i=1}^{N} a_i^{(k)} A_{ij} \,. \tag{15.10}$$

We see that the $a_i^{(k)}$ determine the $k$-th row of a resolution matrix, and obviously we wish this matrix to be close to the unit matrix, or to a localized average – meaning that $R_{kj} \approx 0$ if parameters $m_j$ and $m_k$ are separated beyond a certain distance (the 'resolving length' $L$) in space. In addition, we shall wish to specify that the rows

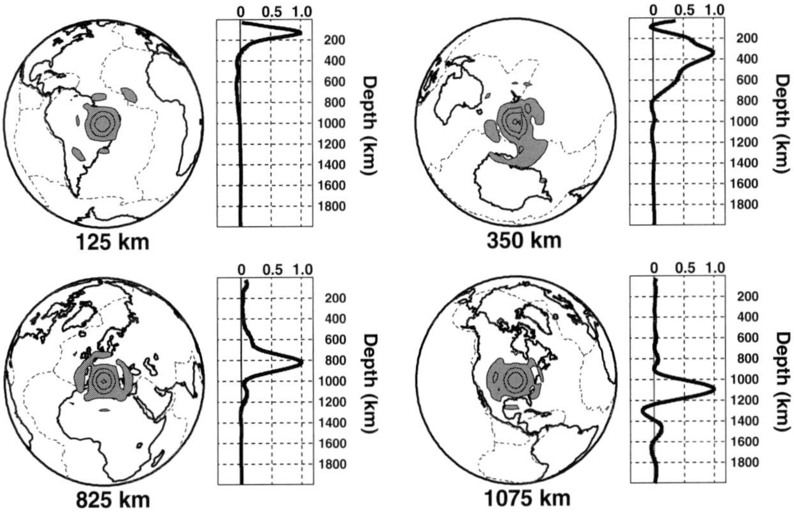


Fig. 15.1. Example of Backus–Gilbert averaging kernels. The target depth is indicated beneath each location. The map shows a horizontal cross-section of $\boldsymbol{R}$ in greyscale for kernel values in excess of 0.1; The curves on the right of each globe show the behaviour of $\boldsymbol{R}$ in the vertical direction. This is representative for the resolution that can be obtained when surface wave phase velocities, body wave travel times, and normal mode splitting parameters are jointly inverted for $V_S$. Figure courtesy Jeroen Ritsema.

of $\boldsymbol{R}$ sum to 1, so that $\boldsymbol{Rm}$ represents a true average. For a target node $k$, we can plot $R_{kj}$ at the locations of nodes $j$. An example is shown in Figure 15.1.

We may define various measures for the quality of the averaging kernel, e.g.:

$$\text{Min}\, L^2 = \alpha \sum_j R_{kj}^2 L_{kj}^2 ,$$

where $L_{kj}$ is the distance between nodes $k$ and $j$ of the model grid, the same as the $L_{kj}$ defined earlier in (12.10). The constant $\alpha$ is arbitrary and can be 1, or chosen such that $L$ can be physically interpreted as the resolving length, the width of the region over which $R_{kj}$ is significantly different from zero. We use this to minimize:

$$L^2 + \epsilon^2 \sigma_{\hat{m}}^{(k)\,2} = \alpha \sum_j R_{kj}^2 L_{kj}^2 + \epsilon^2 \sum_j (a_j^{(k)})^2 = \text{min} . \tag{15.11}$$

The weight $\epsilon^2$ is added to allow for a tradeoff between the size of the resolving length and the variance (15.9) of the estimate. Another measure, termed SOLA by Pijpers and Thompson [261], can be used to mollify the resolution to a pre-specified averaging kernel, say $D_{kj}$:

$$\sum_j [R_{kj} - D_{kj}]^2 + \epsilon^2 \sum_j a_j^{(k)\,2} = \text{min} . \tag{15.12}$$

Suitable choices for $D$ are a spherical volume of radius $\ell$ ($D_{kj} = 1$ for $L_{kj} < \ell$ and 0 elsewhere), or a Gaussian-shaped function. Both have an easy interpretation in terms of resolving length. We find the coefficients $a_i^{(k)}$ by minimizing either (15.11) or (15.12) under the condition that $\hat{m}_k$ is a true average of $m_k$, i.e.:

$$\sum_j R_{kj} = (\sum_j A_{ij}) \sum_i a_i^{(k)} = \sum c_i a_i^{(k)} = 1, \tag{15.13}$$

where we introduce the row-sum vector $\boldsymbol{c}$ with elements $c_i = \sum_j A_{ij}$. We introduce a similar row-sum vector $\boldsymbol{b}$ with elements $b_j = R_{kj}$, and from now on omit the superscript $k$, with the understanding that this is a resolution analysis for element $m_k$ only. Differentiating with respect to $a_i$ leads to:

$$(\boldsymbol{A}\boldsymbol{A}^T + \epsilon^2 \boldsymbol{I})\boldsymbol{a} = \boldsymbol{A}\boldsymbol{b}.$$

These are the normal equations for the following least-squares problem:

$$\begin{pmatrix} \boldsymbol{A}^T \\ \epsilon \boldsymbol{I} \end{pmatrix} \boldsymbol{a} = \begin{pmatrix} \boldsymbol{b} \\ \boldsymbol{0} \end{pmatrix}, \tag{15.14}$$

which has to be solved under the condition:

$$\boldsymbol{c} \cdot \boldsymbol{a} = 1 \qquad (c_i = \sum_j A_{ij}). \tag{15.15}$$

Normally, such problems are solved using the method of Lagrange multipliers. This, however, leads to an ill-conditioned system of linear equations. Nolet [234] obtained a somewhat more stable system by expressing one of the $a_i$ in terms of the others using (15.15). Assuming $c_1 \neq 0$ we take $a_1$ and solve for a truncated vector $\boldsymbol{a}'$. Defining:

$$\begin{aligned} \boldsymbol{a}' &= (a_2, a_3, ...a_N) \\ c_1\boldsymbol{c}' &= (c_2, c_3, ..., c_N) \\ \boldsymbol{B} &= \begin{pmatrix} (\boldsymbol{c}')^T \\ \epsilon \boldsymbol{I}_{N-1} \end{pmatrix}, \end{aligned}$$

where $\boldsymbol{I}_{N-1}$ the unit matrix of order $N-1$. With $\hat{\boldsymbol{e}}_1 = (1, 0, ..., 0)$:

$$a_1 = c_1^{-1} - \boldsymbol{c}' \cdot \boldsymbol{a}', \; \boldsymbol{a} = \boldsymbol{B}\boldsymbol{a}' + c_1^{-1}\hat{\boldsymbol{e}}_1,$$

and we obtain the following system:

$$\begin{pmatrix} \boldsymbol{A}^T\boldsymbol{B} \\ \epsilon \boldsymbol{B} \end{pmatrix} \boldsymbol{a}' = \begin{pmatrix} \boldsymbol{b} - c_1^{-1}\boldsymbol{A}^T\hat{\boldsymbol{e}}_1 \\ -\epsilon c_1^{-1}\hat{\boldsymbol{e}}_1 \end{pmatrix}. \tag{15.16}$$

The kernel needs to be re-calculated for every target location, which makes the method computationally expensive. In helioseismology the data are obtained on a regular grid such as a CCD of 1024 by 1024 points, and one can use translation invariance to compute one $\boldsymbol{a}$ that can be used for all targets that are not influenced by the boundaries of the data domain.

In cases where the science requires a good estimate of both the resolution and the uncertainty of an estimate – e.g. the average velocity anomaly in a well delineated unit such as a downgoing slab – Backus–Gilbert estimation is clearly the method of choice. Criterion (15.12) is in that case more flexible than (15.11) since it allows us to specify the averaging volume.

## 15.3 Sensitivity tests

To obtain a 'quick-and-dirty' estimate of the influence of damping on the solution we can apply sensitivity tests. For very large tomographic systems, such tests are often the only feasible method for getting some idea of the resolving power without taxing available computing resources beyond reasonable limits.

The idea of a sensitivity test is very simple. Suppose we wish to know how a particular feature of the solution is influenced by the regularization, for example a spike-like feature at a certain depth. We design a model with just this feature, $\boldsymbol{m}_\delta$, and compute an artificial ('synthetic') data vector $\boldsymbol{d}_\delta = \boldsymbol{A}\boldsymbol{m}_\delta$. We then invert the data to get a model estimate, using the same damping parameters we used to obtain the original solution: $\hat{\boldsymbol{m}} = \boldsymbol{A}^- \boldsymbol{d}_\delta$, and we inspect the result by comparing the test output $\hat{\boldsymbol{m}}$ with the input $\boldsymbol{m}_\delta$. Often, a sharp feature will have spread out horizontally ('smearing'), or vertically ('leaking'). By adding random errors with a given distribution (usually Gaussian) we can study the effect of data errors on the regularized solution. An example of a sensitivity test is shown in Figure 15.2.

To save effort, we may combine more than one feature in the synthetic model. As long as the smeared solutions do not overlap, this will still give us independent estimates of the effects of regularization. If the input features are point-like, we are testing the regularization effects for a wide spectrum of wavelengths (recall that the Fourier transform of the delta function is equal to 1). In an effort to study the resolving power globally across the model, one often distributes many spikes in a regular pattern, or uses a checkerboard-like synthetic model of alternating positive and negative anomalies. However, the gain in efficiency is somewhat offset by the fact that the regularity of the pattern introduces a dominant wavelength into the model. Leveque et al. [183] warn against a simplistic interpretation of such narrow-band tests. The panacea is to perform the test for a range of patterns covering a wide band of wavelengths. But the interpretation is now effectively in the spectral domain, and usually more difficult to judge than the space domain interpretation of simple spikes.

Note that, by allowing the damping parameters to regularize the solution beyond what is imposed by objective physical constraints, sensitivity tests are no replacement for a proper analysis of resolution and variance in the final model.

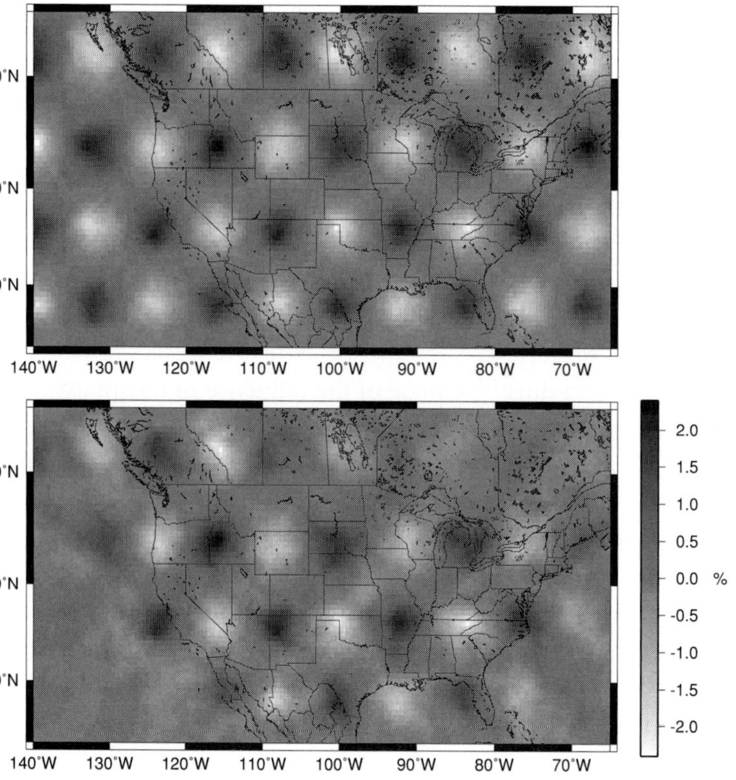


Fig. 15.2. Example of a sensitivity test, using a checkerboard of Gaussian spikes with a width of 400 km, at a depth of 600 km beneath North America. Top: input model ($V_P$ in per cent). Bottom: Result from inversion of the synthetic data generated by the input model. Figure courtesy Karin Sigloch.

Clearly, both have their roles in judging the 'preferred' model, but one should always keep in mind that the results of sensitivity tests reflect the influence of the subjective regularization of the tomographer, whereas those of a true resolution analysis are independent of the choice of final model.

I conclude this chapter by discussing various other tests that allow one to judge the reliability of the result or the adequacy of the regularization.

*Backprojection of data misfits.*

The data misfit vector for the solution $\boldsymbol{m}$ is $\boldsymbol{r} = \boldsymbol{d} - \boldsymbol{A}\boldsymbol{m}$. A histogram of its elements should be inspected to check for anomalies far away from a Gaussian distribution. If there are strong 'tails' this may be an indication that outliers are still present in the data and need to be removed. Once a proper histogram has been obtained with $\chi^2 \approx N$, one should backproject the residual vector to the model space ($\boldsymbol{A}^T\boldsymbol{r}$) and inspect maps of backprojected misfits. These should be uncorrelated in space. If they are not randomly distributed, this may indicate an improper weighting of model parameters (through the prior covariance matrix $\boldsymbol{C}_m$ used in the regularization) or an imbalance in the spatial parametrization of the model.

*Monte Carlo tests of error propagation*

One may generate a data vector consisting of random noise only and invert this to judge the propagation of data errors. An obvious variant on this is to add the noise to the original data vector and inspect the variability of the resulting models. Both tests can be used to get an estimate of the model covariance by repeating them for many different random outcomes of the data vector. Such 'Monte Carlo' experiments need always be done many times to cover a range of random realizations, but limitations in the compute power often pose restrictions. For example, Houser et al. [135] generate 100 global models in a global tomography experiment.

If we generate $K$ models $\boldsymbol{m}^{(k)}$ in this way, an unbiased estimate of the posteriori covariance matrix is:

$$\tilde{\boldsymbol{C}}_{\hat{m}} = \frac{1}{K-1} \sum_{k=1}^{K} (\boldsymbol{m}^{(k)} - \bar{\boldsymbol{m}})(\boldsymbol{m}^{(k)} - \bar{\boldsymbol{m}})^T , \qquad (15.17)$$

where $T$ indicates transpose and where $\bar{\boldsymbol{m}}$ is the average over all $K$ models generated this way. For Monte Carlo tests, a Gaussian error distribution is acceptable provided one has removed outliers from the data set. Note again that $\tilde{\boldsymbol{C}}_m$ is only a proper statistical estimate of covariance if the Monte Carlo inversions are done with objective damping parameters.

In the absence of more detailed information on the distribution of errors, one usually assumes that these are normally distributed after the removal of outliers. However, for bad data with a very low signal-to-noise ratio (certainly less than 1) one can also simply scramble the elements of the data vector to obtain a noise vector with a distribution close to the actual distribution.

*Cross-validation, jackknifing and bootstrapping*

Resampling techniques such as cross-validation or bootstrapping should only be applied in seismic tomography with extreme care because they require a strictly overdetermined system. They must therefore be applied to the damped system and the damping parameter should, again, reflect true physical information on the range of model values or its derivatives (model smoothness), rather than the subjective damping that was applied to obtain a solution that pleases the eye of the geophysicist. However, if you are lucky enough to have useful prior information, these techniques may help you get robust estimates of model error and resolution without any prior assumptions about the distribution of data errors.

In the classical variant of cross-validation, one leaves out one datum, inverts the data set, then compares how well the omitted datum is predicted. One repeats this for all or a large number of data and computes the root-mean-square misfit of the predictions. This allows one to compare different models, or different damping factors. We have already encountered an example of cross-validation in Section 11.2.

Clearly, the method is very expensive to apply. A more efficient variant is to remove a fraction of the data (say 10%) and apply the test to the remaining data, a technique referred to as 'jackknifing'. The larger the fraction one chooses, the faster the computations are performed but the obvious tradeoff is that the prediction suffers from the fact that too many of the data may be missing, and the resulting linear system may not be fully representative of the complete data set. Bootstrapping is like jackknifing, but the removed data are replaced by duplicating data randomly selected from the surviving data set (Efron [95]).

Because of the overdetermined nature of the problem, the data fit also contains information about the data errors (if most of the misfit is due to data errors and not to the inadequate nature of the parametrization!), and in fact there are optimum estimators for the damping parameters that can be obtained this way without first analysing the data to determine the standard deviations (Golub et al. [119]).

### *Hypothesis testing*

Occasionally one wishes to test whether a particular model arising from a hypothesis, say $\boldsymbol{m}_h$, satisfies the data. Simply calculating the misfit $\boldsymbol{r}_h = \boldsymbol{d} - \boldsymbol{A}\boldsymbol{m}_h$ is in this case not sufficient, because it could be that, even if $\chi^2$ for this particular model is too high, a small adjustment of $\boldsymbol{m}_h$ brings $\chi^2$ within an acceptable range without invalidating the hypothesis itself. Deal and Nolet [80] introduced a test to see whether $\boldsymbol{m}_h$ is within the subspace of models that satisfy the data to an acceptable precision. There are various ways to do this. The simplest is to merely subtract $\boldsymbol{A}\boldsymbol{m}_h$ from the data and invert for the minimum norm model change to be added to $\boldsymbol{m}_h$ that yields an acceptable $\chi^2$:

$$\boldsymbol{A}\delta\boldsymbol{m} = \boldsymbol{d} - \boldsymbol{A}\boldsymbol{m}_h$$

$$\hat{\boldsymbol{m}} = \delta\hat{\boldsymbol{m}} + \boldsymbol{m}_h \,.$$

For example, Deal et al. [81, 82] invert for a regular tomographic model containing subduction zones in the west Pacific. To this they fit an analytical model $\boldsymbol{m}_h$ of slab temperature anomalies $\Delta T$, using velocity perturbations $\delta V_P = (\partial V_P/\partial T)\Delta T$. The procedure is equivalent to adding components of the nullspace of $\boldsymbol{A}$ to the original solution $\hat{\boldsymbol{m}}$, and the operator that does this is named the 'nullspace shuttle'.

# 16
# Anisotropy

Until now, we have considered the Earth to be isotropic, even though minerals are anisotropic at the scale of a single crystal. For a planet to behave like an anisotropic solid these crystals must 'line up' and be oriented in the same direction over length scales comparable to that of the Fresnel zone of a seismic wave, i.e. over tens or hundreds of kilometres. Surprisingly, there is now ample evidence that such lattice-preferred orientation (LPO) occurs in nature and that the Earth is at least weakly anisotropic near the surface and in its inner core. Since the magnetic field influences the acoustic wave speed in the Sun's convection zone, and the magnetic field has a distinct direction, anisotropy must affect helioseismic observations as well. However, the strong magnetic field associated with sunspots couples acoustic and magneto-acoustic waves and still poses significant problems of interpretation. Away from such anomalies, the averaging of Doppler measurements over annuli (see Chapter 6) destroys any azimuthal anisotropy, however, and magnetic anisotropy plays no role in the interpretation of solar travel times.

The first indication that anisotropy measurably affects terrestrial seismic waves came in 1964, when Hess [132] discovered that the horizontally travelling Pn-waves in the oceans travel with a velocity that depends on direction, indicating that the fast direction of olivine crystals is aligned in the direction of spreading. It was not until 1993 that the complementary observation was made: using ocean-bottom seismometers on the mid-Atlantic ridge, Blackman et al. [25] showed that vertically travelling P-waves are fast under the ridge, again indicating a preferred orientation along the flow axis, which is presumed to be vertical.

The most convincing observations for anisotropy are those of S-wave birefringence, or splitting due to different velocities for different polarity. In 1982, Ando and Ishikawa [8] first observed such splitting of S-waves into different polarizations. Irrefutable evidence that the splitting is due to anisotropy came from observation of the splitting of the SKS-waves into two arrivals with different polarity by Silver and Chan [314] and Vinnik et al. [384], using the earliest digital recordings from the RTSN, NARS and GEOSCOPE arrays.

For surface waves, it has long been known that Love and Rayleigh wave phase velocities over oceanic regions are incompatible but can be explained by assuming a faster S-velocity for horizontally polarized waves. Such 'radial anisotropy' (with symmetry axis in the direction of the radius vector, also referred to as 'transverse isotropy') was for the first time globally mapped by Nataf et al. [225]. More recent models by Ekström and Dziewonski [96] and Panning and Romanowicz [251] still show disagreements in both the amplitude and the depth extent of radial anisotropy, indicating the difficulty in resolving the exact structure of the anisotropic perturbations.

But radial anisotropy, while sufficient to explain much of the average Love–Rayleigh discrepancy in many regions, is unable to fully explain the observations for surface waves travelling in different directions. The existence of such 'azimuthal anisotropy' had already been observed for surface waves in the oceans by Forsyth [103]. Crampin [67] recognized that an important indicator of azimuthal anisotropy is provided by the polarization of fundamental mode Love waves. In an isotropic Earth, or even in an Earth with radial isotropy along a vertical axis, the particle motion of Love waves is purely transverse. The presence of azimuthal anisotropy is revealed if Love waves have a vertical or radial component of motion that is out of phase with the transverse component, such that the particle motion is elliptical. Park and Yu [254] clearly observed this type of particle motion on long period seismograms. An early effort to map azimuthal anisotropy on a global scale was reported by Montagner and Tanimoto [212].

The inverse problem is severely hampered by a large degree of underdeterminacy. In fact, the major problem with anisotropy is how to isolate resolvable parameters, and this question will be the dominating theme of this chapter.

## 16.1 The elasticity tensor

The full elasticity tensor $c_{ijkl}$ has $3^4 = 81$ elements. The symmetry relationships $c_{ijkl} = c_{klij} = c_{jikl} = c_{ijlk}$ reduce the number of independent elements to 21, but even this cannot hide the huge impediment facing anisotropic tomography: the number of unknowns is an order of magnitude larger than that for the isotropic situation in which we only deal with Lamé's parameters $\lambda$ and $\mu$ (or with $V_P$ and $V_S$). Therefore we cannot avoid making drastic simplifications to reduce the number of parameters. Often, a tradeoff exists between invoking anisotropy or resolving short-wavelength heterogeneities, as was shown for Pn-waves by Hearn [130], and for surface waves by Laske et al. [172]. But an even greater economy of model parameters can be obtained if the number of anisotropic parameters is significantly reduced. Symmetries in the nature of the anisotropy do indeed allow for such reductions, though not without a price: the unknown orientation of the symmetry axis may make the problem nonlinear.

Babuska and Cara [9] list the number of independent elastic coefficients for various crystal symmetries, which range from 13 for a monoclinic crystal such as hornblende, to three for a cubic crystal (garnets). Fortunately, the Earth need not be composed of a single aligned mineral to exhibit such higher symmetries. In fact, the random orientations of individual crystals would render the Earth isotropic for seismic wavelengths if it was not for the preferential alignment in response to stresses and strains. At low pressures, large cracks may develop a preferred orientation and result in seismic anisotropy with a high degree of symmetry. Below the crust most of these cracks are closed, and the occurrence of seismic anisotropy is often proposed to be indicative of LPO, i.e. a large-scale alignment of the crystal lattice. Such alignment is interpreted to be linked to the flow pattern of the rock and is an important observable since it allows for direct geodynamic interpretations. However, alternative interpretations exist. Horizontal layers with a thickness much smaller than the wavelength of seismic waves will cause an apparent transverse isotropy (e.g. Helbig and Schoenberg [131]), and so can fluid-filled cracks that exhibit 'shape-preferred orientation' or SPO. The latter mechanism has been invoked by Moore et al. [218] to explain the occurrence of transverse isotropy near the core–mantle boundary, and Crampin and Peacock [68] propose that anisotropy is caused by stress-aligned fluid-saturated grain boundary cracks over a wide range of depths, including the upper mantle.

Experimental results indicate that olivine crystals, which have orthorhombic symmetry with nine independent elastic coefficients, line up such that the $a$-axis with a fast P-wave velocity orients itself in the flow direction. For a good review of LPO mechanisms see Savage [303] who observes that the orientation of the $a$-axis with flow 'appears to be valid for many, but not all cases'. One such exception is created by the presence of water, as shown by Jung and Karato [154]. But the observation of seismic anisotropy in itself indicates that some form of mineral or structural alignment must be happening, whatever the exact cause and interpretation. If only one axis lines up, the random orientation of the other two axes results in a hexagonal symmetry, with five independent coefficients of elasticity, and this is often accepted as the mode of anisotropy that dominates in the Earth. Perhaps this is merely wishful thinking, but in any case, even with a dense coverage of stations, five elastic parameters plus the two angles that define the axis of symmetry already constitute more unknowns than one can realistically hope to resolve.

The coefficients $c_{ijkl}$ are often given as the coefficients of the Voigt matrix $C_{IJ}$, rather than in the original fourth-order tensor notation $c_{ijkl}$, by grouping the index combinations $ij$ and $kl$ into a single index describing the six independent combinations of the Cartesian indices:

$$11 \rightarrow 1, \quad 22 \rightarrow 2, \quad 33 \rightarrow 3, \quad 23 \rightarrow 4, \quad 13 \rightarrow 5, \quad 12 \rightarrow 6,$$

or

$$xx \rightarrow 1, \quad yy \rightarrow 2, \quad zz \rightarrow 3, \quad yz \rightarrow 4, \quad xz \rightarrow 5, \quad xy \rightarrow 6.$$

The two most important forms for the Voigt matrix are the isotropic case, with Lamé's parameters $\lambda$ and $\mu$:

$$\begin{pmatrix} \lambda + 2\mu & \lambda & \lambda & 0 & 0 & 0 \\ \lambda & \lambda + 2\mu & \lambda & 0 & 0 & 0 \\ \lambda & \lambda & \lambda + 2\mu & 0 & 0 & 0 \\ 0 & 0 & 0 & \mu & 0 & 0 \\ 0 & 0 & 0 & 0 & \mu & 0 \\ 0 & 0 & 0 & 0 & 0 & \mu \end{pmatrix}, \tag{16.1}$$

and hexagonal symmetry, with Love's parameters $A, C, F, L, N$; for a vertical symmetry axis (the case of 'radial anisotropy'):

$$\begin{pmatrix} A & A - 2N & F & 0 & 0 & 0 \\ A - 2N & A & F & 0 & 0 & 0 \\ F & F & C & 0 & 0 & 0 \\ 0 & 0 & 0 & L & 0 & 0 \\ 0 & 0 & 0 & 0 & L & 0 \\ 0 & 0 & 0 & 0 & 0 & N \end{pmatrix}.$$

Four of Love's parameters can be directly associated with the velocities of vertically or horizontally polarized P- and S-waves:

$$A = \rho V_{\mathrm{PH}}^2$$
$$C = \rho V_{\mathrm{PV}}^2$$
$$L = \rho V_{\mathrm{SV}}^2$$
$$N = \rho V_{\mathrm{SH}}^2$$

though the fifth one, $F$, does not have such a clear interpretation – it is equal to $\lambda$ in the case of an isotropic medium. Often, the radial anisotropy is expressed in terms of the ratios:

$$\xi = \frac{N}{L} = \frac{V_{\rm SH}^2}{V_{\rm SV}^2}$$
$$\phi = \frac{C}{A} = \frac{V_{\rm PV}^2}{V_{\rm PH}^2}$$
$$\eta = \frac{F}{A-2L} .$$

It is important to note that the simplicity of the hexagonal Voigt matrix is only obtained because the $z$-axis was chosen as the axis of symmetry. An arbitrary orientation of the axis would have ruined its sparsity. One way around this is to assume a priori that the orientation is known and adopt this as one of the coordinate axes, as when one assumes radial anisotropy.

For the case of fundamental mode surface waves propagating in an Earth with radial anisotropy, one often assumes that Love waves represent SH, whereas Rayleigh waves represent SV velocities. Such a simplistic interpretation breaks down for higher modes. This is easily understood when one considers that shear waves – which can be synthesized by summing higher modes (see Figure 10.4) – change propagation and polarization angle with epicentral distance, or even for the same ray with distance along the trajectory. The modes that build up these body waves thus have to be sensitive to both SH and SV energy. Panning and Romanowicz [251] compute Fréchet kernels in 2D using the NACT approximation (see Appendix C) for S and ScS waveforms. Some examples for the transverse component are shown in Figure 16.1. The change in sensitivity with ray angle is clearly visible.

The Voigt matrix provides a convenient notational framework for anisotropy. Alternatives exist that preserve the tensorial nature of elasticity (see Browaeys and Chevrot [33]). A particularly useful notation for the generally oriented hexagonal case is that of Chevrot [53], which brings out the part of the tensor that depends on the orientation of the symmetry axis $\hat{\mathbf{s}}$ and the isotropic part:

$$\begin{aligned} c_{ijkl} = &(C_{11} - 2C_{66})\delta_{ij}\delta_{kl} + C_{66}(\delta_{ik}\delta_{jl} + \delta_{il}\delta_{jk}) \\ &+ (C_{13} - C_{11} + 2C_{66})(\delta_{ij}s_k s_l + \delta_{kl}s_i s_j) \\ &+ (C_{44} - C_{66})(\delta_{ik}s_j s_l + \delta_{il}s_j s_k + \delta_{jk}s_i s_l + \delta_{jl}s_i s_k) \\ &+ (C_{11} + C_{33} - 2C_{13} - 4C_{44})s_i s_j s_k s_l . \end{aligned} \quad (16.2)$$

Comparison with (2.39) shows that the first line in this equation represents the isotropic part, all other terms depend on the orientation of the symmetry axis $\hat{\mathbf{s}}$. If the medium is fully isotropic $C_{44} = C_{55} = C_{66}$ and $C_{11} = C_{22} = C_{33}$ (see 16.1). In an anisotropic Earth, the isotropic moduli are not uniquely defined, because they depend on the mode of averaging among different directions (see, e.g. Panning and

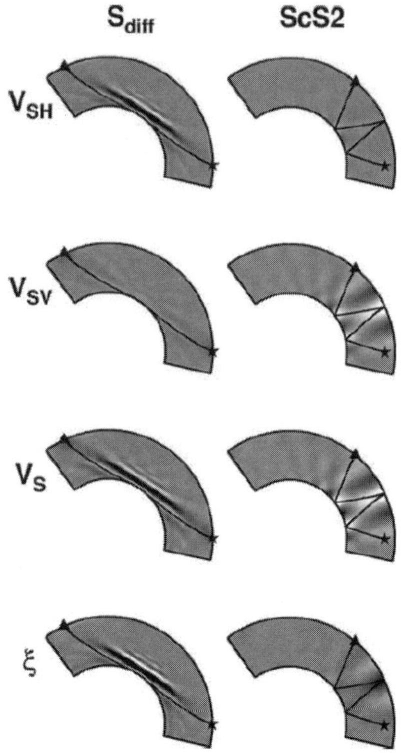


Fig. 16.1. Two-dimensional Fréchet kernels for the waveforms on the transverse component for diffracted S (left) and twice reflected ScS (right). Figure courtesy Mark Panning [251], reproduced with permission from Blackwell Publishing.

Nolet [250]). For most purposes it is sufficient to invert for isotropic velocities and a few constants that characterize the anisotropy, as we do in this section.

We can further recombine the elastic coefficients. A convenient parametrization is then in terms of the P- and S-velocities $V_P = \sqrt{C_{11}/\rho}$ and $V_S = \sqrt{C_{44}/\rho}$ along the symmetry axis, a parameter $\epsilon = (C_{11} - C_{33})/2\rho V_P^2$ denoting the difference in P-wave speed for a P-wave travelling perpendicular to the symmetry axis, a similar parameter $\gamma = (C_{66} - C_{44})/2\rho V_S^2$ for S-waves, and $\delta = (C_{13} - C_{33} + 2C_{44})/2\rho V_P^2$ which governs the transition between slow and fast P-waves at intermediate angles.† With these definitions, the elasticity tensor for such a transversely isotropic medium can be written concisely in the form:

$$c_{ijkl} = \lambda\delta_{ij}\delta_{kl} + \mu(\delta_{ik}\delta_{jl} + \delta_{il}\delta_{jk}) + \delta c_{ijkl} \,,$$

† The parameter $\delta$ as defined here is not the same as the parameter $\delta$ originally introduced by Thomsen [356], which is widely used in exploration seismics, though $\epsilon$ and $\gamma$ agree with Thomsen's original definitions.

where the anisotropic perturbation $\delta c_{ijkl}$ is given by:

$$\begin{aligned}\delta c_{ijkl} &= 2\rho V_{\mathrm{P}}^2 \epsilon (s_i s_j s_k s_l - \delta_{ij} s_k s_l - \delta_{kl} s_i s_j) \\ &\quad + 2\rho V_{\mathrm{S}}^2 \gamma (2\delta_{ij} s_k s_l + 2\delta_{kl} s_i s_j - \delta_{ik} s_j s_l - \delta_{il} s_j s_k - \delta_{jk} s_i s_l - \delta_{jl} s_i s_k) \\ &\quad + \rho V_{\mathrm{P}}^2 \delta (\delta_{ij} s_k s_l + \delta_{kl} s_i s_j - 2 s_i s_j s_k s_l) \,. \qquad (16.3)\end{aligned}$$

## Exercises

**Exercise 16.1** Verify that the Voigt matrix $C_{IJ}$ for isotropic solids is invariant to any interchange of the $x$-, $y$- and $z$-axes.

**Exercise 16.2** Verify that the Voigt matrix for hexagonal solids with the $z$-axis as axis of symmetry is invariant to an interchange of the $x$- and $y$-axes.

## 16.2 Waves in homogeneous anisotropic media

To analyse the effect of anisotropy on wave propagation we assume the medium to be homogeneous. Starting from the elastodynamic equation (2.8), after transformation to the frequency domain and outside of the source region:

$$-\rho\omega^2 u_i = \sum_{jml} \frac{\partial}{\partial x_j}\left(c_{ijml}\frac{\partial u_l}{\partial x_m}\right) = \sum_{jml} c_{ijml}\frac{\partial^2 u_l}{\partial x_j x_m} \,, \qquad (16.4)$$

we assume that the wavefield is a plane wave of the form:

$$u_i = A_i \exp(\mathrm{i}\boldsymbol{k}\cdot\boldsymbol{x}) \,,$$

that travels in a direction given by the unit vector $\hat{\boldsymbol{k}}$ such that $\boldsymbol{k} = k\hat{\boldsymbol{k}}$. Substitution in (16.4) gives:

$$\rho\omega^2 u_i = \sum_{jml} k^2 \hat{k}_m \hat{k}_j c_{ijml} u_l \,,$$

or, in matrix form:

$$\boldsymbol{\Gamma u} = C^{-2}\boldsymbol{u} \qquad \text{with} \quad \Gamma_{il} = \sum_{jm} \frac{c_{ijml}\hat{k}_m\hat{k}_j}{\rho} \,, \quad C = \frac{\omega}{k} \,, \qquad (16.5)$$

which defines eigenvalues $C^{-2}$ and eigenvectors that determine the polarization and the slowness of the different plane waves that solve (16.4). For a minor degree of anisotropy, these solutions will still have polarizations close to those of P- and S-waves, i.e. close to and almost perpendicular to the direction of wave propagation. The solutions are therefore often named 'quasi-P' or qP and 'quasi-S' or qS. In contrast to the isotropic case, there is not one single P or S velocity: the

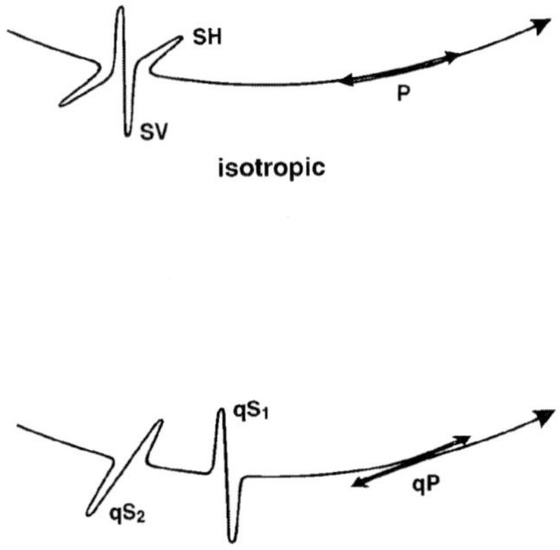


Fig. 16.2. An S-wave with a polarization such that it has components in both the vertical (SV) and horizontal (SH) direction is split into two after traversing an anisotropic layer. The $qS_1$ wave has a polarization equal to the fast velocity for rays travelling under this angle, the $qS_2$ wave is perpendicular to $qS_1$ and has a slower velocity. The polarization of the P-wave is slightly disturbed. From Savage [303], reproduced with permission from the AGU.

phase velocity $C$ depends on the wave direction $\hat{\boldsymbol{k}}$; the two eigenvalues for qS-waves will generally have different velocities, which we shall denote as 'fast' and 'slow'. Thus, an S-wave traversing an anisotropic layer will 'split' into two polarizations separated in time. The P-wave will simply change polarization direction. Figure 16.2 shows the effect of a weakly anisotropic layer on such waves.

## 16.3 S-wave splitting

Imagine an S-wave travelling in the vertical direction through an anisotropic layer. If the wave converts from a P-wave at the core–mantle boundary (e.g. SKS), the polarization is known to be in the ray plane, which we name the radial ($R$) direction (not to be confused with the meaning of radial in 'radial anisotropy'!). The displacement is zero in the transverse ($T$) direction before the wave hits the anisotropic region (see Figure 16.3):

$$u_T(t) = 0, \qquad u_R(t) = f(t)\,.$$

As we saw in the previous section, in the anisotropic region the velocity of the S-wave depends on the propagation direction and polarization. Assume that the fast qS velocity is for waves polarized at an angle $\phi$ with the $R$-direction; the slow qS-wave is then polarized perpendicular to the fast one. Upon exiting the anisotropic region, the S-wave is split into a fast pulse, and a slow one delayed

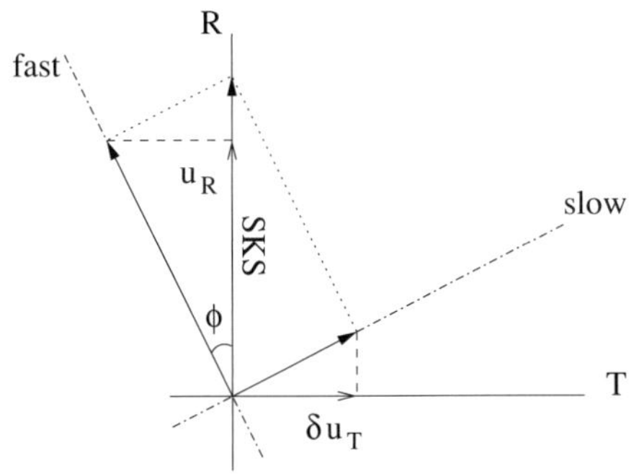


Fig. 16.3. The radial and transverse component of an S-wave with particle motion in the ray plane.

by $\tau$ seconds.

$$u_\phi(t) = f(t)\cos\phi, \qquad u_{\phi-\pi/2}(t) = f(t-\tau)\sin\phi\,.$$

If we measure, or project, the seismogram along the predicted $R$ and $T$ directions, we observe:

$$u_R(t) = f(t)\cos^2\phi + f(t-\tau)\sin^2\phi \tag{16.6}$$

$$\delta u_T(t) = -[f(t) - f(t-\tau)]\cos\phi\sin\phi \approx -\frac{\mathrm{d}f}{\mathrm{d}t}\tau\cos\phi\sin\phi\,. \tag{16.7}$$

We see that – in this simple case, assuming ray theory – the anisotropy is characterized by two parameters: delay $\tau$ and direction $\phi$. Unless the thickness of the layer can be determined independently, $\tau$ cannot be directly translated into an anisotropic velocity deviation. Because we assume only a minor effect of anisotropy, $\tau$ is small; this justifies the use of the notation $\delta u_T$ for the small transverse component.

The presence of lateral heterogeneity, or of multiple regions of anisotropy, may complicate this simple picture; when present, such complications may quickly reduce our ability to interpret observations of anisotropy using ray theory. Recent progress, using a finite-frequency interpretation, has been made by Favier and Chevrot [100] and Chevrot [53]. This section closely follows the latter paper because it adheres to our view that one needs to minimize the number of model parameters in the inversion. Sieminski et al. [312] study the sensitivity for all 21 constants separately.

If we abandon the assumption that the splitting is caused by a simple layer, the S-wave may split up into a distribution of later arrivals, and we should analyse a time window rather than attempt the identification of two split arrivals from the S-wave polarization. To this end, we define the splitting intensity as twice the normalized dot product of the transverse component with the time derivative $\dot{u}_R$ of the radial component. Using Parseval's theorem (2.68):

$$S = 2\frac{\int \delta u_T(t)\dot{u}_R(t)\mathrm{d}t}{\int \dot{u}_R(t)^2\mathrm{d}t} = 2\frac{\mathrm{Re}\int_0^\infty \mathrm{i}\omega\delta u_T(\omega)u_R(\omega)^*\mathrm{d}\omega}{\int_0^\infty |\omega^2 u_R(\omega)|^2\mathrm{d}\omega}. \tag{16.8}$$

The derivation of Fréchet kernels proceeds along the same lines as those for the isotropic case given in Chapter 7, using first-order (Born) theory. We perturb $c_{ijkl}$ in the elastodynamic equations:

$$-\rho\omega^2[u_i + \delta u_i] = \sum_{jkl} \frac{\partial}{\partial x_j}\left([c_{ijkl} + \delta c_{ijkl}]\frac{\partial}{\partial x_k}[u_l + \delta u_l]\right),$$

and retain first-order terms only:

$$-\rho\omega^2\delta u_i - \sum_{jkl}\frac{\partial}{\partial x_j}\left(c_{ijkl}\frac{\partial}{\partial x_k}\delta u_l\right) = \sum_{jkl}\frac{\partial}{\partial x_j}\left(\delta c_{ijkl}\frac{\partial}{\partial x_k}u_l\right).$$

The right-hand side is a force in the $i$-direction. We now introduce the important assumption that the unperturbed tensor $c_{ijkl}$ is for a homogeneous, isotropic medium, and that we can ignore interactions with the free surface, so that the far-field response is given by the Green's function $G^i_j$ given by (4.10). Integrating over a volume $V$ that encloses the anisotropic regions:

$$\begin{aligned}
\delta u_i &= \int \sum_{jkl} G^i_j \frac{\partial}{\partial x_j}\left(\delta c_{ijkl}\frac{\partial u_l}{\partial x_k}\right)\mathrm{d}^3\boldsymbol{x} \\
&= \int \sum_{jkl} \frac{\partial}{\partial x_j}\left(G^i_j\delta c_{ijkl}\frac{\partial u_l}{\partial x_k}\right)\mathrm{d}^3\boldsymbol{x} - \int \sum_{jkl}\frac{\partial G^i_j}{\partial x_j}\frac{\partial u_l}{\partial x_k}\delta c_{ijkl}\mathrm{d}^3\boldsymbol{x} \\
&= \int \sum_{jkl}\left(G^i_j\delta c_{ijkl}\frac{\partial u_l}{\partial x_k}\right)n_j\mathrm{d}^2\boldsymbol{x} - \int \sum_{jkl}\frac{\partial G^i_j}{\partial x_j}\frac{\partial u_l}{\partial x_k}\delta c_{ijkl}\mathrm{d}^3\boldsymbol{x} \\
&= -\int \sum_{jkl}\frac{\partial G^i_j}{\partial x_j}\frac{\partial u_l}{\partial x_k}\delta c_{ijkl}\mathrm{d}^3\boldsymbol{x}
\end{aligned}$$

where we have applied Gauss' theorem. The surface integral is zero even if the anisotropy extends to the free surface because the summation $\sum_{jkl}\delta c_{ijkl}\partial_k u_l n_j$ represents the change in normal stress at the surface which has to remain stress-free. Since the anisotropy is generally located close to the surface, Favier et al. [101] advocate retaining the near-field terms into the expression for the Green's function. For the zero-order field $\boldsymbol{u}$ we adopt an SKS-wave polarized in the radial direction; projection of $\delta\boldsymbol{u}$ on the transverse component yields the expression for $\delta u_T$ needed in (16.8). We skip the details of the considerable algebra and refer readers to the original paper by Chevrot [53]. Here we only summarize the results.

The interaction coefficients for $\epsilon$ and $\delta$ are the same but of opposite sign, so that only the difference $\eta = \epsilon - \delta$ can be resolved. The splitting is also insensitive to isotropic velocity variations since the corresponding Fréchet kernel is zero. The splitting intensity is then expressed as:

$$S = \int [K_\gamma \gamma + K_\eta \eta] \mathrm{d}^3 \boldsymbol{r}_\mathrm{x} \,, \tag{16.9}$$

with the Fréchet kernel given as a sum of local field ($p=0$), near-field ($p=1$), intermediate field ($p=2$) and far-field ($p=3$) contributions according to the dependence on the distance $r^{p-4}$ between scatterer location $\boldsymbol{r}_\mathrm{x}$ and the receiver:

$$K_{\gamma,\eta} = \sum_{p=0}^{3} \Omega_{\gamma,\eta,p} \frac{V_\mathrm{S}^{1-p} r^{p-4}}{2\pi} \frac{\int_0^\infty \omega^p |\dot{m}(\omega)|^2 \cos(\omega \Delta T - p\pi/2) \mathrm{d}\omega}{\int_0^\infty \omega^2 |\dot{m}(\omega)|^2 \mathrm{d}\omega} \,. \tag{16.10}$$

The interaction coefficient $\Omega_X$ is given in Table 16.1. As in Chapter 7, $\Delta T$ represents the detour time, which to first order can be computed from the homogeneous, isotropic background model, and $\dot{m}(\omega)$ represents the spectrum of the moment rate tensor. The expressions for the interaction coefficients depend on the ray direction as well as on the polarization of the incoming wave 16.1.[†] Chevrot gives analytical expressions for the ratio of the frequency integrals in (16.10) if the spectrum is Gaussian.

With this parametrization, the inverse problem has been reduced to only four parameters $\eta$, $\gamma$ and the angles $\theta_s$ and $\phi_s$ which determine the direction $\hat{\boldsymbol{s}}$ of the symmetry axis. Admittedly, this has introduced a nonlinearity into the inverse problem if the symmetry axis is far away from the initially assumed direction, but this is a small price to pay for the gain in economy of parameters. The inversion for $\theta_s$ and $\phi_s$ can be handled iteratively using:

$$\left(\frac{\partial S}{\partial \theta_s}\right) = \gamma \left(\frac{\partial K_\gamma}{\partial s_i}\right)\left(\frac{\partial s_i}{\partial \theta_s}\right) + \eta \left(\frac{\partial K_\eta}{\partial s_i}\right)\left(\frac{\partial s_i}{\partial \theta_s}\right) , \tag{16.11}$$

$$\left(\frac{\partial S}{\partial \phi_s}\right) = \gamma \left(\frac{\partial K_\gamma}{\partial s_i}\right)\left(\frac{\partial s_i}{\partial \phi_s}\right) + \eta \left(\frac{\partial K_\eta}{\partial s_i}\right)\left(\frac{\partial s_i}{\partial \phi_s}\right) . \tag{16.12}$$

The theory of plate tectonics favours a horizontal flow direction under the plates, certainly below the oceans, so it often makes sense to start out with a horizontal $\hat{\boldsymbol{s}}$. Exceptions occur where the flow may be dominantly vertical, such as below ocean ridges, or be more complicated, as in accretionary wedges.

[†] Note that our definition of $\Omega_{\gamma,\eta,p}$ for the far-field terms ($p=3$) differs in sign from that in Chevrot [53] to enable the compact notation (16.10).

Table 16.1. *Interaction coefficients for SKS splitting intensity. Unit vectors $\hat{\boldsymbol{\gamma}}_1$ and $\hat{\boldsymbol{\gamma}}_2$ give the propagation direction of the incoming and outgoing wave, $\hat{\boldsymbol{q}}_1$ denotes the polarization of the incoming wave. $\hat{\boldsymbol{s}}$ is the direction of hexagonal symmetry, $\hat{\boldsymbol{t}}$ points in the direction of the transverse component. Kernels for $\Omega_\epsilon$ are one half of the corresponding kernels $\Omega_\eta$.*

| | |
|---|---|
| | Local field $p = 0$ |
| $\Omega_{\eta,0}$ | $12\frac{V_P^2}{V_S^2}(\hat{\gamma}_1\cdot\hat{s})(\hat{q}_1\cdot\hat{s})[(\hat{\gamma}_2\cdot\hat{t}) - 5(\hat{\gamma}_2\cdot\hat{t})(\hat{\gamma}_2\cdot\hat{s})^2 + 2(\hat{\gamma}_2\cdot\hat{s})(\hat{t}\cdot\hat{s})]$ |
| $\Omega_{\gamma,0}$ | $[(\hat{q}_1\cdot\hat{s})(\hat{\gamma}_2\cdot\hat{\gamma}_1) + (\hat{\gamma}_1\cdot\hat{s})(\hat{\gamma}_2\cdot\hat{q}_1)][60(\hat{\gamma}_2\cdot\hat{t})(\hat{\gamma}_2\cdot\hat{s}) - 12(\hat{t}\cdot\hat{s})]$ $-24(\hat{\gamma}_2\cdot\hat{t})(\hat{\gamma}_1\cdot\hat{s})(\hat{q}_1\cdot\hat{s})$ |
| | Near field $p = 1$ |
| | $\Omega_{\eta,1} = \Omega_{\eta,0} \qquad \Omega_{\gamma,1} = \Omega_{\gamma,0}$ |
| | Intermediate field $p = 2$ |
| $\Omega_{\eta,2}$ | $4\frac{V_P^2}{V_S^2}(\hat{\gamma}_1\cdot\hat{s})(\hat{q}_1\cdot\hat{s})[(\hat{\gamma}_2\cdot\hat{t}) - 6(\hat{\gamma}_2\cdot\hat{t})(\hat{\gamma}_2\cdot\hat{s})^2 + 3(\hat{\gamma}_2\cdot\hat{s})(\hat{t}\cdot\hat{s})]$ |
| $\Omega_{\gamma,2}$ | $[(\hat{q}_1\cdot\hat{s})(\hat{\gamma}_2\cdot\hat{\gamma}_1) + (\hat{\gamma}_1\cdot\hat{s})(\hat{\gamma}_2\cdot\hat{q}_1)][24(\hat{\gamma}_2\cdot\hat{t})(\hat{\gamma}_2\cdot\hat{s}) - 6(\hat{t}\cdot\hat{s})$ $-8(\hat{\gamma}_2\cdot\hat{t})(\hat{\gamma}_1\cdot\hat{s})(\hat{q}_1\cdot\hat{s})]$ |
| | Far field $p = 3$ |
| $\Omega_{\eta,3}$ | $-4\frac{V_P^2}{V_S^2}[(\hat{t}\cdot\hat{s}) - (\hat{\gamma}_2\cdot\hat{s})(\hat{\gamma}_2\cdot\hat{t})](\hat{\gamma}_2\cdot\hat{s})(\hat{\gamma}_1\cdot\hat{s})(\hat{q}_1\cdot\hat{s})$ |
| $\Omega_{\gamma,3}$ | $2[(\hat{q}_1\cdot\hat{s})(\hat{\gamma}_2\cdot\hat{\gamma}_1) + (\hat{\gamma}_1\cdot\hat{s})(\hat{\gamma}_2\cdot\hat{q}_1)][(\hat{t}\cdot\hat{s}) - 2(\hat{\gamma}_2\cdot\hat{s})(\hat{\gamma}_2\cdot\hat{t})]$ |

A further reduction in the number of parameters can be acquired using empirical correlations between the anisotropic parameters. Becker et al. [18] found for weakly anisotropic mantle xenoliths:

$$
\begin{aligned}
\gamma &= -0.023(\pm 0.0120) + 0.233(\pm 0.07)\epsilon, \quad (\epsilon \le -0.03)\\
&= -0.003(\pm 0.0006) + 0.900(\pm 0.10)\epsilon, \quad (\epsilon > -0.03)\\
\delta &= 0.027(\pm 0.003) + 1.857(\pm 0.04)\epsilon \quad (\epsilon \le -0.03)\\
&= 0.003(\pm 0.010) + 0.667(\pm 0.06)\epsilon \quad (\epsilon > -0.03)\\
\eta &= -0.027(\pm 0.003) - 0.857(\pm 0.04)\epsilon \quad (\epsilon \le -0.03)\\
&= -0.003(\pm 0.010) + 0.333(\pm 0.06)\epsilon \quad (\epsilon > -0.03)\,.
\end{aligned}
\tag{16.13}
$$

## 16.4 Surface wave anisotropy

For surface waves, handling azimuthal anisotropy is fairly straightforward if we can assume that ray theory is valid, so that we may assign a local phase velocity to every surface location $(\theta, \phi)$. We introduce anisotropy by making the phase velocity angle-dependent. For a weakly anisotropic medium, Smith and Dahlen [324] show that the phase velocity satisfies a harmonic dependence on azimuth $\psi$ of the form

$$C(\psi, \omega) = C_0(\omega) + a_1(\omega)\cos 2\psi + a_2(\omega)\sin 2\psi + a_3(\omega)\cos 4\psi + a_4(\omega)\sin 4\psi \,. \tag{16.14}$$

Again, the problem is that of a proliferation of new parameters, and all our attention should be focused on reducing the number of unknowns. Even the simple parametrization (16.14) adds four new parameters $a_i$ to the isotropic phase velocity at every location in the model. Since this implies a fivefold increase in the number of parameters of an inverse problem that is usually already ill-constrained, it is worthwhile to consider if some of the $a_i$ can be ignored altogether.

In a global inversion for azimuthal anisotropy, Trampert and Woodhouse [367] find that Love waves are insensitive to the $a_1$ and $a_2$ terms. Montagner and Tanimoto [212] expect a weak dependence of Rayleigh waves on $a_3$ and $a_4$ for plausible petrological models. In fact, Simons et al. [318] incorporated only $a_1$ and $a_2$ terms into a PWI analysis of Rayleigh waves and were able to reduce $\chi^2$ by 20% with respect to an isotropic inversion. Laske and Masters [172] ignore azimuthal anisotropy for Love waves; even for the Rayleigh wave anisotropy they conclude that the improvement in $\chi^2$ is small and could also be achieved by reducing the smoothness of the isotropic solution. Debayle and Kennett [83] force the anisotropy to be constant in a small number of layers in an effort to render the inverse problem more manageable.

Since surface waves average in a horizontal direction and are sensitive over a depth extent that depends on frequency, the information obtained from them is complementary to that obtained from S-wave splitting, where the ability to recover the depth of splitting is minimal. Ideally, one would invert both SKS- and surface-wave data at the same time. There are various ways to use this. Simons et al. [318] assume that for Rayleigh waves, $a_1$ and $a_2$ are dominated by only two elastic constants, $G_c$ and $G_s$, respectively, where $G_c$ and $G_s$ are defined in terms of the Voigt coefficients:

$$G_c = \frac{1}{2}(C_{55} - C_{44}), \qquad G_s = C_{54}.$$

Table 16.2. *Interaction coefficients for surface waves in an isotropic medium with anisotropic perturbations of hexagonal symmetry; $U$, $V$, $W$ are normalized using (10.6), $\nu = ak$ with $k$ in rad/m if $U$, $V$, $W$ and $r$ are in m. Subscript 1 denotes the incoming wave, subscript 2 the outgoing wave. $\hat{s}$ is a unit vector in the direction of the axis of symmetry. Fortran code implementing complex conjugate expressions is given in the electronic supplement to Panning and Nolet [250].*

We use the following abbreviations:

$$
\begin{aligned}
\bar{\nu}^2 &= \nu_1 \nu_2 \\
s_r &= (\hat{s} \cdot \hat{r}) \\
s'_\theta &= (\hat{s} \cdot \hat{\theta}_1) \\
s'_\phi &= (\hat{s} \cdot \hat{\phi}_1) \\
s''_\theta &= (\hat{s} \cdot \hat{\theta}_2) \\
s''_\phi &= (\hat{s} \cdot \hat{\phi}_2) \\
A_1 &= s_r U_1 + \mathrm{i} s'_\theta V_1 \\
A_2 &= s_r U_2 - \mathrm{i} s''_\theta V_2
\end{aligned}
$$

$$\Omega_\epsilon = 2\rho\alpha^2(\Omega_\epsilon^{(1)} + \Omega_\epsilon^{(2)} + \Omega_\epsilon^{(3)})$$

For $\Omega_\epsilon^{(1)}$:

| | |
|---|---|
| Rayleigh → Rayleigh | $s_r^2 \dot{A}_2 \dot{A}_1 - \mathrm{i} r^{-1} s_r [s''_\theta \nu_2 A_2 \dot{A}_1 - s'_\theta \nu_1 \dot{A}_2 A_1] + \bar{\nu}^2 r^{-2} s'_\theta s''_\theta A_2 A_1$ |
| Rayleigh → Love | $\mathrm{i} s_r^2 s''_\phi \dot{A}_1 \dot{W}_2 + r^{-1}[s_r s''_\theta s''_\phi \nu_2 \dot{A}_1 W_2 - s_r s'_\theta s''_\phi \nu_1 A_1 \dot{W}_2] + \mathrm{i} \bar{\nu}^2 r^{-2} s''_\theta s'_\theta s''_\phi A_1 W_2$ |
| Love → Rayleigh | $-\mathrm{i} s_r^2 s'_\phi \dot{W}_1 \dot{A}_2 + r^{-1} s_r s'_\phi [s'_\theta \nu_1 W_1 \dot{A}_2 - s''_\theta \nu_2 \dot{W}_1 A_2] - \mathrm{i} \bar{\nu}^2 r^{-2} s'_\theta s'_\phi s''_\theta W_1 A_2$ |
| Love → Love | $s_r^2 s'_\phi s''_\phi \dot{W}_1 \dot{W}_2 - \mathrm{i} r^{-1} s_r s'_\phi s''_\phi [s''_\theta \nu_2 \dot{W}_1 W_2 - s'_\theta \nu_1 W_1 \dot{W}_2] + \bar{\nu}^2 r^{-2} s'_\phi s''_\phi s'_\theta s''_\theta W_1 W_2$ |

For $\Omega_\epsilon^{(2)}$:

| | |
|---|---|
| Rayleigh → Rayleigh | $-(\dot{U}_2 - \nu_2 r^{-1} V_2)(s_r \dot{A}_1 + \mathrm{i} \nu_1 r^{-1} s'_\theta A_1)$ |
| Rayleigh → Love | 0 |
| Love → Rayleigh | $-(\dot{U}_2 - \nu_2 r^{-1} V_2)(-\mathrm{i} s'_\phi s_r \dot{W}_1 + \nu_1 r^{-1} s'_\theta s'_\phi W_1)$ |
| Love → Love | 0 |

For $\Omega_\epsilon^{(3)}$:

| | |
|---|---|
| Rayleigh → Rayleigh | $-(\dot{U}_1 - \nu_1 r^{-1} V_1)(s_r \dot{A}_2 - \mathrm{i} \nu_2 r^{-1} s''_\theta A_2)$ |
| Rayleigh → Love | $-(\dot{U}_1 - \nu_1 r^{-1} V_1)(\mathrm{i} s''_\phi s_r \dot{W}_2 + \nu_2 r^{-1} s''_\phi s''_\theta W_2)$ |
| Love → Rayleigh | 0 |
| Love → Love | 0 |

Table 16.2. *(continued)*

| | |
|---|---|
| | $\Omega_\delta = 2\rho\alpha^2(\Omega_\delta^{(1)} + \Omega_\delta^{(2)} + \Omega_\delta^{(3)})$ |
| | $\Omega_\delta^{(1)} = -\frac{1}{2}\Omega_\epsilon^{(2)}$ |
| | $\Omega_\delta^{(2)} = -\frac{1}{2}\Omega_\epsilon^{(3)}$ |
| | $\Omega_\delta^{(3)} = -\Omega_\epsilon^{(1)}$ |
| | $\Omega_\gamma = -2\rho\beta^2 \sum_{i=1}^{6} \Omega_\gamma^{(i)}$ |
| | $\Omega_\gamma^{(1)} = 2\Omega_\epsilon^{(2)}$ |
| | $\Omega_\gamma^{(2)} = 2\Omega_\epsilon^{(3)}$ |
| | For $\Omega_\gamma^{(3)}$ |
| Rayleigh → Rayleigh | $s_r^2[\dot{U}_2\dot{U}_1 + \cos\eta\,\dot{V}_2\dot{V}_1] + i\nu_1 r^{-1}s_r s_\theta'[\dot{U}_2U_1 + \dot{V}_2V_1\cos\eta]$ $- i\nu_2 r^{-1}s_r s_\theta''[U_2\dot{U}_1 + V_2\dot{V}_1\cos\eta]$ $+ \bar{\nu}^2 r^{-2}s_\theta' s_\theta''[U_2U_1 + V_2V_1\cos\eta]$ |
| Rayleigh → Love | $\sin\eta[s_r^2\dot{W}_2\dot{V}_1 + i\nu_1 r^{-1}s_r s_\theta'\dot{W}_2V_1 - i\nu_2 r^{-1}s_r s_\theta''W_2\dot{V}_1$ $+ \bar{\nu}^2 r^{-2}s_\theta' s_\theta''W_2V_1]$ |
| Love → Rayleigh | $-\sin\eta[s_r^2\dot{V}_2\dot{W}_1 + i\nu_1 r^{-1}s_r s_\theta'\dot{V}_2W_1 - i\nu_2 r^{-1}s_r s_\theta''V_2\dot{W}_1$ $+ \bar{\nu}^2 r^{-2}s_\theta' s_\theta''V_2W_1]$ |
| Love → Love | $\cos\eta[s_r^2\dot{W}_2\dot{W}_1 + i\nu_1 r^{-1}s_r s_\theta'\dot{W}_2W_1 - i\nu_2 r^{-1}s_r s_\theta''W_2\dot{W}_1$ $+ \bar{\nu}^2 r^{-2}s_\theta' s_\theta''W_2W_1]$ |
| | For $\Omega_\gamma^{(4)}$ |
| Rayleigh → Rayleigh | $s_r^2[\dot{U}_2\dot{U}_1 + \bar{\nu}^2 r^{-2}\cos\eta\,U_2U_1] + is_r s_\theta'[\dot{U}_2\dot{V}_1 + \bar{\nu}^2 r^{-2}\cos\eta\,U_2V_1]$ $+ s_\theta' s_\theta''[\dot{V}_2\dot{V}_1 + \bar{\nu}^2 r^{-2}\cos\eta\,V_2V_1]$ $- is_r s_\theta''[\dot{V}_2\dot{U}_1 + \bar{\nu}^2 r^{-2}\cos\eta\,V_2U_1]$ |
| Rayleigh → Love | $-s_\phi'' s_\theta'[\dot{W}_2\dot{V}_1 + \bar{\nu}^2 r^{-2}\cos\eta\,W_2V_1]$ $+ is_r s_\phi''[\dot{W}_2\dot{U}_1 + \bar{\nu}^2 r^{-2}\cos\eta\,W_2U_1]$ |
| Love → Rayleigh | $-is_r s_\phi'[\dot{U}_2\dot{W}_1 + \bar{\nu}^2 r^{-2}\cos\eta\,U_2W_1]$ $- s_\theta'' s_\phi'[\dot{V}_2\dot{W}_1 + \bar{\nu}^2 r^{-2}\cos\eta\,V_2W_1]$ |
| Love → Love | $s_\phi' s_\phi''[\dot{W}_2\dot{W}_1 + \bar{\nu}^2 r^{-2}\cos\eta\,W_2W_1]$ |
| | For $\Omega_\gamma^{(5)}$ |
| Rayleigh → Rayleigh | $s_r^2[\dot{U}_2\dot{U}_1 + \nu_2 r^{-1}\cos\eta\,U_2\dot{V}_1]$ $+ i\nu_1 r^{-1}s_r s_\theta'[\dot{U}_2U_1 + \nu_2 r^{-1}\cos\eta\,U_2V_1]$ $- is_r s_\theta''[\dot{V}_2\dot{U}_1 + \nu_2 r^{-1}\cos\eta\,V_2\dot{V}_1]$ $+ \nu_1 r^{-1}s_\theta'' s_\theta'[\dot{V}_2U_1 + \nu_2 r^{-1}\cos\eta\,V_2V_1]$ |

Table 16.2. *(continued)*

| | |
|---|---|
| Rayleigh → Love | $+is''_\phi s_r \dot{W}_2\dot{U}_1 + i\nu_2 r^{-1}s''_\phi s_r \cos\eta W_2\dot{V}_1 - \nu_1 r^{-1}s''_\phi s'_\theta \dot{W}_2 U_1$ $- \bar{\nu}^2 r^{-2}s''_\phi s'_\theta \cos\eta W_2 V_1$ |
| Love → Rayleigh | $+i\nu_2 r^{-1}s_r s''_\phi \sin\eta V_2\dot{W}_1 - i\bar{\nu}^2 r^{-2}s_r s'_\theta \sin\eta U_2 W_1$ $- \bar{\nu}^2 r^{-2}s'_\theta s''_\theta \sin\eta V_2 W_1 - \nu_2 r^{-1}s_r^2 \sin\eta U_2\dot{W}_1$ |
| Love → Love | $-i\nu_2 r^{-1}s''_\phi s_r \sin\eta W_2\dot{W}_1 + \bar{\nu}^2 r^{-2}s'_\theta s''_\phi \sin\eta W_2 W_1$ |
| | For $\Omega_\gamma^{(6)}$ |
| Rayleigh → Rayleigh | $s_r\dot{U}_2\dot{A}_1 - i\nu_2 r^{-1}s''_\theta U_2\dot{A}_1 + \nu_1 r^{-1}s_r \cos\eta\dot{V}_2 A_1$ $- i\bar{\nu}^2 r^{-2}s''_\theta \cos\eta V_2 A_1$ |
| Rayleigh → Love | $\nu_1 r^{-1}s_r \sin\eta\dot{W}_2 A_1 - i\bar{\nu}^2 r^{-2}s''_\theta \sin\eta W_2 A_1$ |
| Love → Rayleigh | $-is'_\phi s_r\dot{U}_2\dot{W}_1 - \nu_2 r^{-1}s''_\theta s'_\phi U_2\dot{W}_1$ $- \cos\eta s'_\phi[i\nu_1 r^{-1}s_r\dot{V}_2 W_1 + \bar{\nu}^2 r^{-2}s''_\theta V_2 W_1]$ |
| Love → Love | $- \sin\eta s'_\phi[i\nu_1 r^{-1}s_r\dot{W}_2 W_1 + \bar{\nu}^2 r^{-2}s''_\theta W_2 W_1]$ |

The ratio between these two also defines the direction of maximum splitting for SKS-waves, since $\tan\psi = G_s/G_c$. Thus, if we have a dense set of splitting observations, this can be used to constrain the ratio $a_1/a_2$. The depth dependence of anisotropy can then be recovered by inverting the local phase velocity. Partial derivatives for the coefficients $a_i$ are given by Montagner and Nataf [211].

A better way to combine SKS- and surface waves would be to use a parametrization in terms of $\epsilon$, $\gamma$ and $\delta$ and apply finite-frequency theory, since the frequencies of SKS-waves and surface waves differ considerably. A finite-frequency theory for surface wave phase velocity can be derived analogous to the treatment given in Chapter 11, where we used Born theory to model the phase angle deviation:

$$\delta\phi = \text{Im}\left(\frac{\delta u}{u}\right) \qquad \text{(11.2 again)}$$

with $\delta u$ given by (11.5). Interaction coefficients $\Omega$ for general anisotropy are given by Romanowicz and Snieder [294]. As argued earlier, such a general approach to the inverse problem is not very productive because of the lack of resolution. Panning and Nolet [250], give the interaction coefficient $\Omega$ in terms of $\epsilon$, $\delta$ and $\gamma$:

$$\Omega = \Omega_\epsilon\epsilon + \Omega_\delta\delta + \Omega_\gamma\gamma\,, \qquad (16.15)$$

where the $\Omega_x$ are listed in Table 16.2. A further reduction $\epsilon$ using the empirical correlations given by (16.13) may be desirable. The phase angle is perturbed as:

$$\delta\phi = \int (K^{\phi}_{\epsilon}\epsilon + K^{\phi}_{\delta}\delta + K^{\phi}_{\gamma}\gamma)\mathrm{d}^3\boldsymbol{x}\,, \qquad (16.16)$$

where $K^{\phi}_X$ is again:

$$K^{\phi}_X = \mathrm{Im}\left(\sum_{\text{mode 1}}\sum_{\text{mode 2}} N\Omega_X\mathcal{P}\right),$$

with $N$ and $\mathcal{P}$ given by (11.8) and (11.9), respectively. The terms for an isotropic perturbation can be added to this, or one can invert for anisotropic perturbations only after a best fit for an isotropic model has been obtained.

No experience has yet been obtained with the finite-frequency inversion of anisotropic parameters as described here. The inverse problem is nonlinear because the orientation $\hat{\boldsymbol{s}}$ of the symmetry axis adds two angles to the unknown parameters. Conceivably one can fix the azimuth using information from SKS splitting if that is available from a dense network. Alternatively, one can test the hypothesis that the LPO is in the direction predicted by contemporary plate motion and fix $\hat{\boldsymbol{s}}$ in that direction, or perturb it away using (16.11) and (16.12).

# 17

# Future directions

In this chapter we will take a brief look at several promising new developments. None of the new methods described here are as yet widely employed, and questions of viability remain for some of them. Some exciting new instruments are still in a stage of testing or exist only on the drawing board. The emphasis will be on showing the connection with the material presented in this book, rather than on a full development.

## 17.1 Beyond Born

Though the Born approximation improves on the ray-theoretical approximation, it also binds our hands considerably, since the amplitude of the perturbed wave $\delta u$ must be much smaller than that of the zero-order wavefield in order to justify the neglect of higher-order scattering (the repeated scattering of scattered waves). Especially at higher frequency the limitations of Born theory discussed in Section 7.5 may become prohibitive. Iterative methods that start with low frequencies and slowly increase the frequency content as the model becomes more complicated may offer some relief, if ray theory can be used to model the first-order perturbation, as is done in methods like PWI. Meier et al. [205] formulated a waveform inversion algorithm that combines 3D Born inversion with this strategy. It becomes potentially very powerful when combined with a 3D adjoint algorithm (see next section).

Encouraging progress has also been made to model the scattering of surface waves in a more complete manner. The earliest efforts to include multiple scattering go back more than twenty years. Kennett [158] used invariant embedding to model multiple forward and backward scattering in two-dimensional media. Maupin [202] and Odom [247] further improved this technique, and Bostock [30] extended it to three dimensions. Friederich [107] developed a computational technique that employs the relative dominance of forward scattering over backscattering to calculate a multiply scattered surface wave mode, including its Fréchet kernel. However, actual applications of such techniques are very few and they still have to find their way in routine applications of seismic tomography.

A completely different take on multiple scattering is offered by the method of radiative transfer, in which, rather than the full oscillatory signal, only the envelope of the signal is modelled. The theory uses the principle that energy must be conserved in a ray tube (see Section 4.4). In radiative transfer theory, the energy of the signal is modelled by summing seismic wave packets that travel a free path length in a smooth medium, following ray theory, until it is scattered – not unlike a photon in a cloud of dust particles. The free path length, the average length travelled between two scattering events, is related in a complicated way to the shape, strength and spatial distribution of the heterogeneities. For a clear introduction to radiative transfer theory as applicable in seismology see Margerin [192].

Margerin and Nolet [193, 194] and Shearer and Earle [309] used radiative transfer theory and Monte Carlo techniques to model the envelopes of seismograms, but the method has not yet been applied for true three-dimensional transmission tomography.

## 17.2 Adjoint methods

In Section 10.6 we presented a systematic method for the inversion of waveforms, using a 1D background model and linearized perturbations to the wavenumber $k_n$ of the surface wave modes summed to construct synthetic seismograms. Such simple waveform inversions are restricted to be linear, because a second iteration would have to start from a 3D background model. So far, no global tomographic Earth model is based on a nonlinear iterative inversion of data with frequencies beyond those of the normal mode regime. However, efforts in exploration seismics and more regional inversion studies have pointed the way to new techniques that may, sooner or later, be applicable in large-scale tomography.

The rapid growth in the power of computers now makes it possible to compute exact low- or even intermediate-frequency synthetic seismograms in fully 3D global models, using finite-difference or spectral element methods (e.g. Komatitsch et al. [165], Capdeville et al. [40]). Short periods around 1 Hz can even be used in the computation of regional seismograms. Mismatches between such predicted signals and observed waveforms contain information about the Earth's structure and need not be restricted to well-understood observables like delays in body wave arrivals or normal-mode splitting functions. To interpret such mismatches we need an inverse theory. The problem is inherently nonlinear, which poses the biggest challenge for seismic tomography where data volumes, as well as the number of parameters to be resolved, are very large.

The first attempts to formulate a nonlinear inverse approach for waveforms were by Tarantola's group in Paris in the 1980s [348, 350], who derived the Fréchet kernels using an adjoint method. Adjoint methods were developed outside of seismology as part of control theory (see, e.g., Wunsch [407]). Even though the actual numerical implementation may be far from trivial, the principle of adjoint methods is simple. Here we follow the brief review of such methods given by Nolet [238].

We assume we have an efficient numerical algorithm $\mathcal{A}$ that relates a model $\boldsymbol{m}$ and a displacement field $\boldsymbol{u}$ to a source $\boldsymbol{f}$:

$$\mathcal{A}[\boldsymbol{m}]\boldsymbol{u} = \boldsymbol{f} \,. \tag{17.1}$$

Though the solution $\boldsymbol{u}$ depends in a nonlinear way on the model $\boldsymbol{m}$, (17.1) is simply the wave equation in symbolic form, so the numerical operator $\mathcal{A}$ is linear in both $\boldsymbol{m}$ and $\boldsymbol{u}$. For example, comparison with the elastodynamic equation (2.8) shows that for elastic fields, $\mathcal{A}$ is represented by:

$$\mathcal{A}[\rho, c_{ijkl}]\, u_i = \rho \frac{\partial^2 u_i}{\partial t^2} - \sum_{jkl} \frac{\partial}{\partial x_j}\left( c_{ijkl} \frac{\partial u_l}{\partial x_k} \right) ,$$

or some discretized version of this equation. We shall assume from the outset that model and displacement fields are discretized so $\mathcal{A}$ is a matrix and we can work with ordinary dot products. Thus, the inverse problem can be formulated as:

$$\boldsymbol{u} = \boldsymbol{u}_{\text{obs}} \tag{17.2}$$

$$\text{while } \mathcal{A}[\boldsymbol{m}]\boldsymbol{u} = \boldsymbol{f} \,. \tag{17.3}$$

As usual, we adopt a least-squares approach to minimize the misfit between the observed $\boldsymbol{u}_{\text{obs}}$ and predicted $\boldsymbol{u}$. The condition (17.1) can be handled with the method of Lagrange multipliers (see Section 14.6). This means that we take the squared misfit of the predicted wavefield (17.2) and add a weighted misfit to (17.3), defining a penalty function $\mathcal{S}$:

$$\mathcal{S} = \frac{1}{2}|\boldsymbol{u} - \boldsymbol{u}_{\text{obs}}|^2 + \boldsymbol{v} \cdot (\mathcal{A}[\boldsymbol{m}]\boldsymbol{u} - \boldsymbol{f}) \;=\; \min \,. \tag{17.4}$$

The weights, or Lagrange multipliers, are arranged here in a new field $\boldsymbol{v}$, also called the adjoint field. Note that we obtain the original equation (17.1) by differentiating $\mathcal{S}$ with respect to the elements of $\boldsymbol{v}$ and setting these derivatives to zero. The adjoint algorithm $\mathcal{A}^\dagger$ is defined by:

$$\boldsymbol{v} \cdot \mathcal{A}[\boldsymbol{m}]\boldsymbol{u} = \mathcal{A}^\dagger[\boldsymbol{m}]\boldsymbol{v} \cdot \boldsymbol{u} \,. \tag{17.5}$$

If $\mathcal{A}$ is cast in the form of a matrix, the adjoint is simply the complex conjugate transpose of $\mathcal{A}$.

If we aim for a minimum misfit, and impose that $\mathcal{S}$ is stationary with respect to changes in $\boldsymbol{u}$ we find the backprojection for the adjoint or dual field $\boldsymbol{v}$:

$$\mathcal{A}^{\dagger}[\boldsymbol{m}]\boldsymbol{v} = \boldsymbol{u}_{\text{obs}} - \boldsymbol{u} \,. \tag{17.6}$$

The variation $\delta\mathcal{S}$ due to a model perturbation is found from the linearity of $\mathcal{A}$:

$$\delta\mathcal{S} = \boldsymbol{v} \cdot \mathcal{A}[\delta\boldsymbol{m}]\boldsymbol{u} \,. \tag{17.7}$$

Before discretization, the dot product in (17.7) represents a volume integral. As we saw in the development of Born theory in Chapter 7, the derivatives of the elastic constants can be removed by partial integration. Tarantola [350] used this to write $\mathcal{A}[\delta\boldsymbol{m}]$ as a linear function of the model perturbation $\boldsymbol{m}$. Designating the new operator by $\mathcal{B}[\boldsymbol{u}]$ with adjoint $\mathcal{B}^{\dagger}[\boldsymbol{u}]$:

$$\mathcal{A}[\delta\boldsymbol{m}]\boldsymbol{u} = \mathcal{B}[\boldsymbol{u}]\delta\boldsymbol{m} \,, \tag{17.8}$$

from which:

$$\delta\mathcal{S} = \mathcal{B}^{\dagger}[\boldsymbol{u}]\boldsymbol{v} \cdot \delta\boldsymbol{m} \,, \tag{17.9}$$

which yields a gradient $\mathcal{B}^{\dagger}[\boldsymbol{u}]\boldsymbol{v}$ after the dual field $\boldsymbol{v}$ has been found through the backprojection (17.6). This gradient can be used to adapt $\boldsymbol{m}$. We re-calculate $\boldsymbol{u}$ for the new model and repeat the search until a satisfactory fit has been found. If the operators are in the frequency domain, the complex conjugation changes $\mathrm{e}^{\mathrm{i}\omega t}$ into $\mathrm{e}^{-\mathrm{i}\omega t}$, which is equivalent to reversing the time direction $t \rightarrow -t$. Equivalently, if $\mathcal{A}$ contains a finite difference operator for the time derivative, transposing it conforms to interchanging the order of time indices, again reversing the time direction. Thus, the adjoint really represents a projection *back in time*.

In its original form the adjoint method was not overly succesful. This may be because the method was ahead of its time and had to wait for more CPU power to become available, but also because it was intended for reflection seismology, which is inherently more nonlinear than transmission tomography: even a small shift in the location of a reflector may significantly perturb phases. Pratt [266] and Pratt and Shipp [267] used a similar method for a 2D operator in the frequency domain and applied it successfully in crosshole transmission tomography. This method has obtained impressive results in a 2D blind test by Brenders and Pratt [32] on synthetic acoustic data.

The operator $\mathcal{A}$ need not necessarily be a numerical algorithm. Jin et al. [149] based their inversion method on a ray-theoretical expression like (7.23) and labelled their method 'ray-Born' inversion. Ribodetti et al. [275] extended ray-Born inversion to include attenuation, but this has so far only been applied to seismograms from laboratory experiments.

Nor is the adjoint method limited to waveforms. Gee and Jordan [110] recognize that the output from a short time window segment bandfiltered with a Gaussian filter around a central frequency $\omega_s$ can be characterized by an amplitude $A_s$, a phase delay $t_p$ and a group delay $t_g$:

$$u_s(\omega) \approx A_s \exp\left(-\frac{(\omega-\omega_s)^2}{2\sigma_s^2}\right) \exp[\mathrm{i}\omega_s t_p + \mathrm{i}(\omega-\omega_s)t_g]\,. \qquad (17.10)$$

The 'generalized data functionals' $A_s$, $t_p$ and $t_g$ are determined by minimizing the misfit to a synthetic seismogram that is windowed and filtered in the same way, or by minimizing the difference in cross-correllograms as in (7.12). If the waveform is a body wave, or a surface wave mode, first-order perturbation theory yields the more common linearized constraints for a tomographic model. Tromp et al. [368] and Liu and Tromp [187] extend this method to arbitrary selections from the seismogram and apply the adjoint method to construct Fréchet kernels at finite frequency.

Since the adjoint method avoids the storage of Fréchet kernels, it is more memory efficient than the inversion of the Born integral. The computational efficiency increases with the number of receivers for each source (Chen et al. [52]).

The backprojection of waveform misfits brings transmission tomography very close to reflection seismics, in which the signal itself is the 'misfit', i.e. the difference with the zero prediction for a model with no reflector at all. Indeed, stacking 'backprojected' seismograms has been applied by Lay et al. [175] and van der Hilst et al. [371] to map layering near the core–mantle boundary. The connection is a fascinating one but outside the scope of this book. To work for reflector imaging, a large number of sensors is a prerequisite. This is a concern that also plays a role in transmission tomography, and will be the topic of the next section.

## 17.3 Global coverage of seismic sensors

The widened sensitivity of finite-frequency kernels alone cannot make up for the fact that large volumes of the Earth's mantle remain very sparsely sampled by seismic waves.

Only dense arrays can provide the data density needed to resolve features of geological interest in the lithosphere or upper mantle. Temporary deployments of seismic stations are a proven method to enhance coverage in regions of interest, but the US Array component of EarthScope (Figure 17.1) brings this to a new level of technical sophistication (Levander and Nolet, [182]).

Such intensive deployments are not possible in the oceans, though, where deployments of ocean-bottom seismographs are not only more expensive, but where data retrieval cannot be done in real time. Simons et al. [317] report on an exciting

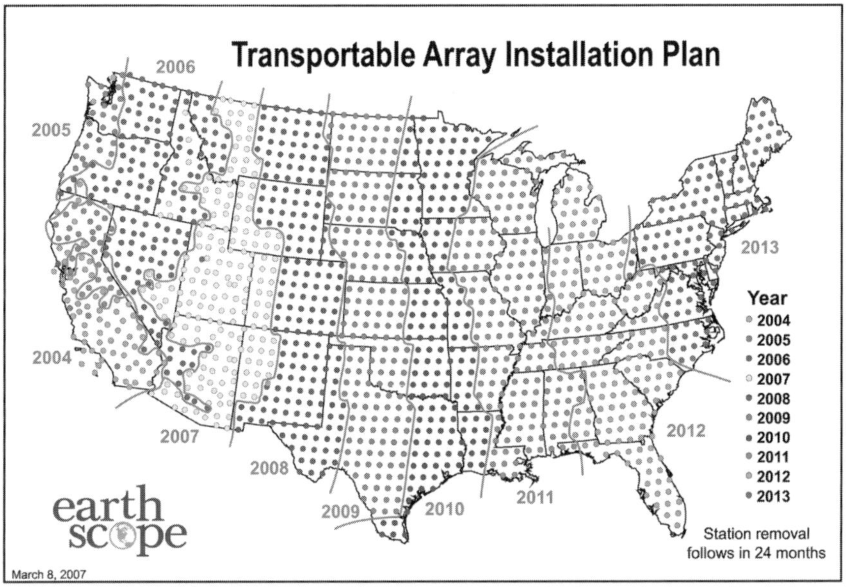


Fig. 17.1. The projected locations of the NSF-sponsored USArray deployment. Each of the sensors remains in place for two years. First sensors were installed in 2005, and the array moves eastwards to be completed in 2015. Figure courtesy Jim Fowler (IRIS).

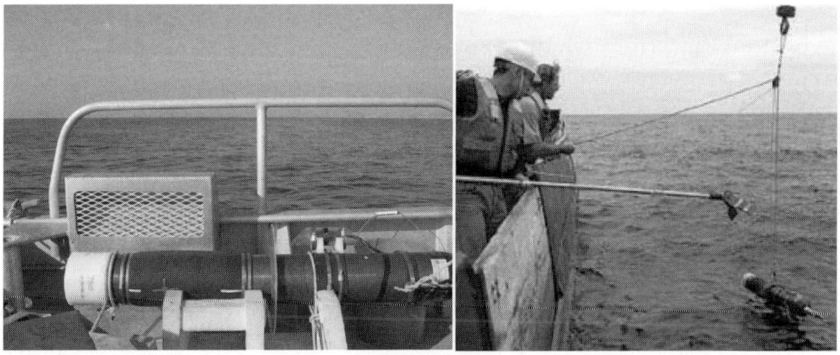

Fig. 17.2. A SOLO underwater float equipped with a hydrophone (left) being launched for testing (right). Photo courtesy Frederik Simons.

experiment with floats – autonomous sensors drifting below the sea surface – which indicates that it may be possible to record P-wave signals from strong earthquakes, despite the high noise level (Figure 17.2). During its maiden voyage, a prototype successfully recorded a P-wave at a distance of 46° from an earthquake of magnitude 6 while drifting 700 m below the sea surface. The floats control their depth level by inflating or deflating a bladder, and are designed to surface every few days to transmit information via satellite. Their trajectory is imposed by the prevailing ocean currents at depth, which are not yet well known, but a minor amount of steering is possible during the descent phase if the float has wings like a glider, and this can be used to steer the float away from the coast. A small flotilla of such instruments could provide an ever-changing array of sensors in oceanic regions.

## 17.4 Helioseismology and astroseismology

Because space missions require as much as 15 years of preparation, the increase in data flow and quality in the near future is somewhat easier to foresee than that in terrestrial seismology.

By combining forces with another very active field, that of the detection of exoplanets, astronomers are certain that many new data will be forthcoming: on December 27, 2006, a number of European countries and Brazil launched the CoRoT satellite, and the first data are indicating that it may be possible to observe and identify eigenfrequencies of a large number of variable stars for $\ell$ and $m$ as high as 3. A new space mission, Kepler, is scheduled to start in 2009 and a number of ground-based observatories are likely to increase the data volume dramatically in the next decade: Las Cumbres, a global network of 58 telescopes; SONG (for Stellar Oscillations Network Group) with 8 sites of four 1-metre telescopes each; and SIAMOIS, an Antarctica-based telescope scheduled to start operating around 2011.

The next mission from which helioseismology is likely to benefit greatly is the Solar Dynamics Observatory (SDO), planned to start observing in 2009, which will carry the Helioseismic and Magnetic Imager (HMI).† The spacecraft carries a 16-megapixel CCD and the data will include continuous Dopplergrams over the full solar disk as well as high resolution images of active targets. Currently, there is no certain follow-up for the SDO, though there are advanced plans for a Solar Orbiter mission after 2015 that would include the polar regions of the Sun in its coverage.

† See `http://hmi.stanford.edu`

# References


[1] Akaike, H. A new look at the statistical model identification. *IEEE Trans. on Autom. Control*, **AC-19**:716–723, 1974. {*272*}

[2] Aki, K., Christoffersson, A., and Husebye, E.S. Determination of the three-dimensional seismic structure of the lithosphere. *J. Geophys. Res.*, **82**:277–296, 1977. {*1*}

[3] Aki, K., Husebye, E.S., Christoffersson, A., and Powell, C. Three-dimensional seismic velocity anomalies in the crust and upper mantle under the USGS California seismic array. *EOS Trans. Am. Geophys. Un.*, **56**:1145, 1974. {*1*}

[4] Aki, K. and Lee, H.K. Determination of three-dimensional anomalies under a seismic array using first P arrival times from local earthquakes: 1. A homogeneous initial model. *J. Geophys. Res.*, **81**:4381–4399, 1976. {*1*}

[5] Aki, K. and Richards, P.G. *Quantitative Seismology, 2nd edn.* Univ. Science Books, Sausalito, CA, second ed., 2002. {*11, 16, 64, 92*}

[6] Allen, R.M. et al. The thin hot plume beneath Iceland. *Geophys. J. Int.*, **137**:51–63, 1999. {*146, 150*}

[7] Allen, R.M. et al. Imaging the mantle beneath Iceland using integrated seismological techniques. *J. Geophys. Res.*, **107**:2325 doi:10.1019/2001JB000595, 2002. {*5*}

[8] Ando, M. and Ishikawa, Y. Observations of shear-wave velocity polarization anisotropy beneath Honshu, Japan: two masses with different polarization in the upper mantle. *J. Phys. Earth*, **30**:191–199, 1982. {*289*}

[9] Babuska, V. and Cara, M. *Seismic Anisotropy in the Earth.* Kluwer Acad. Publ., Dordrecht, 1991. {*291*}

[10] Backus, G. Inference from inadequate and inaccurate data II. *Proc. NAS*, **65**:282–287, 1970. {*281*}

[11] Backus, G. and Gilbert, J.F. The rotational splitting of the free oscillations of the Earth. *Proc. Nat. Acad. Sci.*, **47**:362–371, 1961. {*158*}

[12] Backus, G. and Gilbert, J.F. Numerical applications of a formalism for geophysical inverse problems. *Geophys. J. Roy. astr. Soc.*, **13**:247–276, 1967. {*2, 158, 281*}

[13] Backus, G. and Gilbert, J.F. The resolving power of gross Earth data. *Geophys. J. Roy. astr. Soc.*, **16**:169–205, 1968. {*2, 281*}

[14] Backus, G. and Gilbert, J.F. Uniqueness on the inversion of inaccurate gross Earth data. *Phil. Trans. Roy. Soc. Lond.*, **A266**:123, 1970. {*2, 281*}

[15] Bagaini, C. Performance of time delay estimators. *Geophys.*, **70**:V109–V120, 2005. {*106*}

[16] Bahcall, J.N., Basu, S., and Pinsonneault, M.H. How uncertain are solar neutrino predictions? *Phys. Lett. B*, **433**:1–8, 1998. {*7*}

[17] Barber, C.B., Dobkin, D.P., and Huhdanpaa, H.T. The Quickhull algorithm for convex hulls. *ACM Trans. on Mathematical Software*, **22**:469–483, 1996. {*224*}

[18] Becker, T.W., Chevrot, S., Schulte-Pelkum, V., and Blackman, D.K. Statistical properties of seismic anisotropy predicted by upper mantle geodynamic models. *J. Geophys. Res.*, **111**:B08309, 2006. {*300*}

[19] Berger, J., Davis, P., and Ekström, G. Ambient Earth noise: A survey of the Global Seismograph Network. *J. Geophys. Res.*, **109**:doi:10.1029/2004JB003408, 2004. {*109*}

[20] Beucler, E., Stutzmann, E., and Montagner, J.-P. Surface wave higher-mode phase velocity measurements using a roller-coaster-type algorithm. *Geophys. J. Int.*, **155**:289–307, 2003. {*191*}

[21] Bhattacharyya, J., Masters, G., and Shearer, P.M. Global lateral variations of shear wave attenuation in the upper mantle. *J. Geophys. Res.*, **101**:22273–22289, 1996. {*145*}

[22] Bijwaard, H. and Spakman, W. Tomographic evidence for a narrow whole mantle plume below Iceland. *Earth Planet. Sci. Lett.*, **166**:121–126, 1999. {*5*}

[23] Bijwaard, H. and Spakman, W. Non-linear global P-wave tomography by iterated linearized inversion. *Geophys. J. Int.*, **141**:71–82, 2000. {*117*}

[24] Birch, A.C. and Kosovichev, A.G. Travel time sensitivity kernels. *Sol. Phys.*, **192**:193–201, 2000. {*9*}

[25] Blackman, D.K., Orcutt, J.A., Forsyth, D.W., and Kendall, J.M. Seismic anisotropy in the mantle beneath an oceanic spreading centre. *Nature*, **366**:675–677, 1993. {*289*}

[26] Böhm, G., Galuppo, P., and Vesnaver, A. 3D adaptive tomography using Delaunay triangles and Voronoi polygons. *Geophys. Prosp.*, **48**:723–744, 2000. {*226*}

[27] Bolton, H. and Masters, G. Travel times of P and S from global digital seismic networks: Implications for the relative variation of P and S velocity in the mantle. *J. Geophys. Res.*, **106**:13527–13540, 2001. {*99, 254*}

[28] Boschi, L. Measures of resolution in global body wave tomography. *Geophys. Res. Lett.*, **30**:doi:10.1029/2003GL018222, 2003. {*279*}

[29] Boschi, L., Becker, T.W., Soldati, G., and Dziewonski, A.M. On the relevance of Born theory in global seismic tomography. *Geophys. Res. Lett.*, **33**:L06302, 2006. {*231*}

[30] Bostock, M.G. Reflection and transmission of surface waves in laterally varying media. *Geophys. J. Int.*, **109**:411–436, 1992. {*306*}

[31] Bostock, M.G., Hyndman, R.D., Rondenay, S.J., and Peacock, S. An inverted continental Moho and the serpentinization of the forearc mantle. *Nature*, **417**:536–538, 2002. {*4*}

[32] Brenders, A.J. and Pratt, R.G. Full waveform tomography for lithospheric imaging: results from a blind test in a realistic crustal model. *Geophys. J. Int.*, **168**:133–151, 2007. {*309*}

[33] Browaeys, J.T. and Chevrot, S. Decomposition of the elastic tensor and geophysical applications. *Geophys. J. Int.*, **159**:667–678, 2004. {*293*}

[34] Brune, J.N. Travel times, body waves and normal modes of the Earth. *Bull. Seismol. Soc. Am.*, **54**:2099–2128, 1964. {*185*}

[35] Brune, J.N., Nafe, J.E., and Alsop, L.E. The polar phase shift of surface waves on a sphere. *Bull. Seismol. Soc. Am.*, **51**:247–257, 1961. {*213*}

[36] Butler, R. P-wave travel time and amplitude in western North America. *Nature*, **306**:677–678, 1983. {*145*}

[37] Calvet, M. and Chevrot, S. Travel time sensitivity kernels for PKP phases in the mantle. *Phys. Earth Planet. Inter.*, **153**:21–31, 2005. {*139*}

[38] Campillo, M. and Paul, A. Long range correlations in diffuse seismic coda. *Science*, **299**:547–549, 2003. {*91, 111*}

[39] Capdeville, Y. An efficient Born normal mode method to compute sensitivity kernels and synthetic seismograms in the Earth. *Geophys. J. Int.*, **163**:639–646, 2005. {*208*}

[40] Capdeville, Y., Larmat, C., Vilotte, J.-P., and Montagner, J.-P. A new coupled spectral element and modal solution method for global seismology: A first application to the scattering induced by a plume-like anomaly. *Geophys. Res. Lett.*, **29**:10.1029/2001GL013747, 2002. {*307*}

[41] Cara, M. Regional variations of higher Rayleigh-mode phase velocities: a spatial filtering method. *Geophys. J. Roy. astr. Soc.*, **54**:439–460, 1978. {*191*}

[42] Carter, G.C. Coherence and time delay estimation. *Proc. IEEE*, **75**:236–255, 1987. {*102*}

[43] Cazenave, A., Souriau, A., and Dominh, K. Global coupling of Earth surface topography with hotspots, geoid and mantle heterogeneities. *Nature*, **340**:54–57, 1989. {*3*}

[44] Censor, Y. Row-action methods for huge and sparse systems and their applications. *SIAM Rev.*, **23**:444–466, 1981. {*273*}

[45] Červený, V. *Seismic Ray Theory*. Cambridge University Press, Cambridge, UK, 2001. {*11*}

[46] Červený, V. and Hron, F. The ray series method and dynamic ray tracing system for three-dimensional inhomogeneous media. *Bull. Seismol. Soc. Am.*, **70**:47–77, 1980. {*50, 52*}

[47] Červený, V., Molotkov, I.A., and Pšenčík, I. *Ray Method in Seismology*. Univerzita Karlova, Prague, 1977. {*25*}

[48] Challis, L. and Sheard, F. The Green of Green functions. *Phys. Today*, **Dec.**:41–46, 2003. {*33*}

[49] Chapman, C.H. A new method for computing synthetic seismograms. *Geophys. J. Roy. astr. Soc.*, **54**:481–518, 1978. {*104*}

[50] Chapman, C.H. *Fundamentals of seismic wave propagation*. Cambridge University Press, Cambridge, UK, 2004. {*11*}

[51] Chapman, C.H., Chu, J.-Y., and Lyness, D.G. The WKBJ seismogram algorithm. In D. Doornbos, ed., *Seismological Algorithms*, pages 47–74. Acad. Press, London, 1988. {*104*}

[52] Chen, P., Jordan, T.H., and L.Zhao. Full three-dimensional tomography: a comparison between the scattering-integral and adjoint-wavefield methods. *Geophys. J. Int.*, **170**:175–181, 2007. {*310*}

[53] Chevrot, S. Finite-frequency vectorial tomography: a new method for high-resolution imaging of upper mantle anisotropy. *Geophys. J. Int.*, **165**:641–657, 2006. {*293, 297–299*}

[54] Chevrot, S. and Zhao, L. Multiscale finite frequency Rayleigh wave tomography of the Kaapvaal craton. *Geophys. J. Int.*, **169**:201–215, 2007. {*226*}

[55] Chiao, L.-Y. and Kuo, B.-Y. Multiscale seismic tomography. *Geophys. J. Int.*, **145**:517–527, 2001. {*226*}

[56] Chou, D.-Y. Acoustic imaging of solar active regions. *Sol. Phys.*, **192**:241–259, 2000. {*238*}

[57] Choy, G.L. and Richards, P.G. Pulse distortion and Hilbert transformation in multiply reflected and refracted body waves. *Bull. Seismol. Soc. Am.*, **65**:55–70, 1975. {*108*}

[58] Christensen-Dalsgaard, J. *Lecture notes on Stellar oscillations.* `http://whome.phys.au.dk/%7Ejcd/oscilnotes/print-chap-full.pdf`, Aarhus University, Denmark, 2003. {*xii, 7, 163*}

[59] Claerbout, J.F. Synthesis of a layered medium from its acoustic transmission response. *Geophys.*, **33**:264–269, 1968. {*113*}

[60] Claerbout, J.F. *Basic Earth Imaging*. Draft available from `http://sepwww.stanford.edu/ftp/prof/`, 2006. {*273*}

[61] Clayton, R.W. and Comer, R.P. A tomographic analysis of mantle heterogeneities from body wave travel time data. *EOS, Trans. Am. Geophys. Union*, **64**:776 (abstr.), 1983. {*4*}

[62] Clévédé, E., Mégnin, Ch., Romanowicz, B., and Lognonné, Ph. Seismic waveform modeling and surface wave tomography in a three-dimensional Earth: asymptotic and non-asymptotic approaches. *Phys. Earth Planet. Inter.*, **119**:37–56, 2000. {*208*}

[63] Coates, R.T. and Chapman, C.H. Ray perturbation theory and the Born approximation. *Geophys. J. Int.*, **100**:379–392, 1990. {*8, 140*}

[64] Constable, S.C., Parker, R.L., and Constable, C.G. Occam's inversion: a practical algorithm for generating smooth models from electromagnetic sounding data. *Geophys.*, **52**:289–300, 1987. {*271*}

[65] Corbard, T., Gizon, L., and Roth, M. Editor's note. *Astron. Nachr.*, **328**:203, 2007. {*xii*}

[66] Cox, A.N. and Guzik, J.A. Theoretical prediction of an observed solar g-mode. *Astrophys. J.*, **613**:L169–L171, 2004. {*164*}

[67] Crampin, S. Distinctive particle motion of surface waves as a diagnostic of anisotropy of layering. *Geophys. J. Roy. astr. Soc.*, **40**:177–186, 1975. {*290*}

[68] Crampin, S. and Peacock, S. A review of shear-wave splitting in the compliant crack-critical anisotropic Earth. *Wave Motion*, **41**:59–77, 2005. {*291*}

[69] Creager, K.C. and Jordan, T.H. Slab penetration in the lower mantle. *J. Geophys. Res.*, **89**:3031–3049, 1984. {*4*}

[70] Dahlen, F.A. The normal modes of a rotating, elliptical Earth - II. Near-resonance multiplet coupling. *Geophys. J. Roy. astr. Soc.*, **18**:397–436, 1969. {*158*}

[71] Dahlen, F.A. Elastic dislocation theory for a self-gravitating elastic configuration with an initial static stress field. *Geophys. J. Roy. astr. Soc.*, **28**:357–383, 1972. {*158*}

[72] Dahlen, F.A. Elastic dislocation theory for a self-gravitating elastic configuration with an initial static stress field. II Energy release. *Geophys. J. Roy. astr. Soc.*, **31**:469–484, 1973. {*158*}

[73] Dahlen, F.A. The spectra of unresolved split normal mode multiplets. *Geophys. J. Roy. astr. Soc.*, **58**:1–33, 1979. {*158*}

[74] Dahlen, F.A. Finite-frequency sensitivity kernels for boundary topography perturbations. *Geophys. J. Int.*, **162**:525–540, 2005. {*154, 239*}

[75] Dahlen, F.A. and Baig, A.M. Fréchet kernels for body-wave amplitudes. *Geophys. J. Int.*, **150**:440–466, 2002. {*141, 151*}

[76] Dahlen, F.A., Hung, S.-H., and Nolet, G. Fréchet kernels for finite-frequency traveltimes – I. Theory. *Geophys. J. Int.*, **141**:157–174, 2000. {*10, 54, 116*}

[77] Dahlen, F.A. and Nolet, G. Comment on the paper On sensitivity kernels for wave equation transmission tomography by de Hoop and van der Hilst. *Geophys. J. Int.*, **163**:949–951, 2005. {*138*}

[78] Dahlen, F.A. and Tromp, J. *Theoretical Global Seismology*. Princeton Univ. Press, Princeton NJ, 1998. {*11, 16, 53, 158, 161, 172, 181, 202*}
[79] de Hoop, M.V. and van der Hilst, R.D. On sensitivity kernels for wave equation transmission tomography. *Geophys. J. Int.*, **160**:621–633, 2005. {*138*}
[80] Deal, M.M. and Nolet, G. Nullspace shuttles. *Geophys. J. Int.*, **124**:372–380, 1996. {*288*}
[81] Deal, M.M. and Nolet, G. Slab temperature and thickness from seismic tomography 2. Izu-Bonin, Japan and Kuril subduction zones. *J. Geophys. Res.*, **104**:28803–28812, 1999. {*288*}
[82] Deal, M.M., Nolet, G., and van der Hilst, R.D. Slab temperature and thickness from seismic tomography, 1. Method and application to Tonga. *J. Geophys. Res.*, **104**:28789–28802, 1999. {*288*}
[83] Debayle, E. and Kennett, B.L.N. Anisotropy in the Australian upper mantle from Love and Rayleigh waveform inversion. *Earth Planet. Sci. Lett.*, **184**:339–351, 2000. {*301*}
[84] Debayle, E. and Sambridge, M. Inversion of massive surface wave data sets: Model construction and resolution assessment. *J. Geophys. Res.*, **109**:B02316, 2004. {*227*}
[85] Deubner, F.-L. Observations of low wavenumber nonradial eigenmodes of the Sun. *Astr. Astrophys.*, **44**:371–375, 1975. {*7*}
[86] Deuss, A. and Woodhouse, J.H. Theoretical free-oscillation spectra: the importance of wide band coupling. *Geophys. J. Int.*, **146**:833–842, 2001. {*169*}
[87] Deuss, A. and Woodhouse, J.H. Iteration method to determine the eigenvalues and eigenvectors of a target multiplet including full mode coupling. *Geophys. J. Int.*, **159**:326–332, 2004. {*169*}
[88] Duvall, T.L., D'Silva, S., Jefferies, S.M., Harvey, J.W., and Schou, J. Downflows under sunspots detected by helioseismic tomography. *Nature*, **379**:235–237, 1996. {*114*}
[89] Duvall, T.L., Jefferies, S.M., Harvey, J.W., and Pomerantz, M.A. Time-distance helioseismology. *Nature*, **362**:430–432, 1993. {*7*}
[90] Dziewonski, A.M. Mapping the lower mantle: Determination of lateral heterogeneity in P velocity up to degree and order 6. *J. Geophys. Res.*, **89**:5929–5952, 1984. {*3, 221*}
[91] Dziewonski, A.M. and Anderson, D.L. Preliminary reference Earth model. *Phys. Earth Planet. Inter.*, **25**:297–356, 1981. {*234*}
[92] Dziewonski, A.M. and Gilbert, J.F. The effect of small, aspherical perturbations on travel times and a re-examination of the corrections for ellipticity. *Geophys. J. Roy. astr. Soc.*, **44**:7–17, 1976. {*235*}
[93] Dziewonski, A.M., Hager, B.H., and O'Connell, R.J. Large-scale heterogeneities in the lower mantle. *J. Geophys. Res.*, **82**:239–255, 1977. {*2*}
[94] Dziewonski, A.M. and Woodhouse, J.H. An experiment in the systematic study of global seismicity: centroid-moment tensor solutions for 201 moderate and large earthquakes of 1981. *J. Geophys. Res.*, **88**:3247–3271, 1983. {*247*}
[95] Efron, B. Estimating the error rate of a prediction rule: Improvement on cross-validation. *J. Am. Stat. Ass.*, **78**:316–331, 1983. {*288*}
[96] Ekström, G. and Dziewonski, A.M. The unique anisotropy of the Pacific upper mantle. *Nature*, **394**:168–172, 1998. {*290*}
[97] Ellsworth, W.L. and Koyagani, R.Y. Three-dimension crust and upper mantle structure of the Kilauea volcano, Hawaii. *J. Geophys. Res.*, **82**:5379–5394, 1977. {*2*}

[98] Engdahl, E.R., van der Hilst, R.D., and Buland, R. Global teleseismic earthquake relocation with improved travel times and procedures for depth determination. *Bull. Seismol. Soc. Am.*, **88**:722–743, 1998. {*100, 239*}

[99] Farra, V. and Madariaga, R. Seismic waveform modeling in heterogeneous media by ray perturbation theory. *J. Geophys. Res.*, **92**:2697–2712, 1987. {*140, 146*}

[100] Favier, N. and Chevrot, S. Sensitivity kernels for shear wave splitting in transverse isotropic media. *Geophys. J. Int.*, **153**:213–228, 2003. {*143, 297*}

[101] Favier, N., Chevrot, S., and Komatitsch, D. Near-field influence on shear wave splitting and traveltime sensitivity kernels. *Geophys. J. Int.*, **156**:467–482, 2004. {*298*}

[102] Ferreira, A.M.G. and Woodhouse, J.H. Source, path and receiver effects on seismic surface waves. *Geophys. J. Int.*, **168**:109–132, 2007. {*53*}

[103] Forsyth, D.W. The early structural evolution and anisotropy of the oceanic upper mantle. *Geophys. J. Roy. astr. Soc.*, **43**:103–162, 1975. {*290*}

[104] Foulger, G.R. et al. The seismic anomaly beneath Iceland extends down to the mantle transition zone and no deeper. *Geophys. J. Int.*, **142**:F1, 2000. {*5*}

[105] Franklin, J.N. Well posed stochastic extention of ill-posed linear problems. *J. Math. Anal. Appl.*, **31**:682–716, 1970. {*267*}

[106] Fréchet, J. Sismogènese et doublets sismiques. *Thèse d'Etat, Grenoble*, 1985. {*108*}

[107] Friederich, W. Propagation of seismic shear and surface waves in a laterally heterogeneous mantle by multiple forward scattering. *Geophys. J. Int.*, **136**:180–204, 1999. {*306*}

[108] García, R.A. et al. Tracking solar gravity modes: the dynamics of the solar core. *Science*, **316**:1591, 2007. {*164*}

[109] Gautier, S., Nolet, G., and Virieux, J. Finite-frequency tomography in a crustal environment: application to the western part of the Gulf of Corinth. *Geophys. Prosp.*, in press 2008. {*138*}

[110] Gee, L.S. and Jordan, T.H. Generalized seismological data functionals. *Geophys. J. Int.*, **111**:363–390, 1992. {*310*}

[111] Giardini, D., Li, X., and Woodhouse, J.H. Three-dimensional structure of the Earth from splitting in free-oscillation spectra. *Nature*, **325**:405–411, 1987. {*172*}

[112] Gilbert, J.F. Excitation of normal modes of the Earth by earthquake sources. *Geophys. J. Roy. astr. Soc.*, **22**:223–226, 1971. {*158*}

[113] Gilbert, J.F. Differential kernels for group velocity. *Geophys. J. Roy. astr. Soc.*, **49**:649–660, 1976. {*199*}

[114] Gilbert, J.F. The representation of seismic displacements in terms of traveling waves. *Geophys. J. Roy. astr. Soc.*, **44**:275–280, 1976. {*189*}

[115] Gilbert, J.F. and Backus, G. Propagator matrices in elastic wave and vibration problems. *Geophysics*, **31**:326–332, 1966. {*140*}

[116] Gilbert, J.F. and Dziewonski, A.M. An application of normal mode theory to the retrieval of structural parameters and source mechanisms from seismic spectra. *Phil. Trans. R. Soc. Lond.*, **A278**:187–269, 1975. {*158*}

[117] Gizon, L. and Birch, A.C. Time-distance helioseismology: the forward problem for random distributed sources. *Astrophys. J.*, **571**:966–986, 2002. {*114, 124*}

[118] Gizon, L. and Birch, A.C. Local helioseismology. *Living Rev. Solar Phys.*, `www.livingreviews.org/lrsp-2005-6`, 2005. {*xii, 7, 34*}

[119] Golub, G.H., Heath, M., and Wahba, G. Generalized cross-validation as a method for choosing a good ridge parameter. *Technometrics*, **21**:215–223, 1979. {*288*}

[120] Gomberg, J.S., Priestley, K.F., Masters, G., and Brune, J.N. The structure of the crust and upper mantle of northern Mexico. *Geophys. J. Roy. astr. Soc.*, **94**:1–20, 1988. {*215*}

[121] Got, J.-L., Fréchet, J., and Klein, F.W. Deep fault plane geometry inferred from multiple relative relocation beneath the south flank of Kilauea. *J. Geophys. Res.*, **99**:15375–15386, 1994. {*108*}
[122] Grand, S.P. A possible bias in travel time measurements reported to ISC. *Geophys. Res. Lett.*, **17**:17–20, 1990. {*99, 137*}
[123] Grand, S.P. Mantle shear structure beneath the Americas and surrounding oceans. *J. Geophys. Res.*, **99**:11591–11621, 1994. {*5*}
[124] Grand, S.P. Mantle shear-wave tomography and the fate of subducting slabs. *Phil. Trans. R. Soc. Lond.*, **A360**:2475–2491, 2002. {*vi*}
[125] Grand, S.P., van der Hilst, R.D., and Widiyantoro, S. Global seismic tomography: a snapshot of convection in the Earth. *GSA Today*, **7**:1–7, 1997. {*5*}
[126] Gudmundsson, O., Davies, J.H., and Clayton, R.W. Stochastic analysis of global travel time data: mantle heterogeneity and random errors in the ISC data. *Geophys. J. Int.*, **102**:25–44, 1990. {*99, 272*}
[127] Haddon, R.A.W. and Husebye, E.S. Joint interpretation of P-wave time and amplitude anomalies in terms of lithospheric heterogeneities. *Geophys. J. Roy. astr. Soc.*, **55**:19–43, 1978. {*146*}
[128] Hager, B.H., Clayton, R.W., Richards, M.A., Comer, R.P., and Dziewonski, A.M. Lower mantle heterogeneity, dynamic topography and the geoid. *Nature*, **313**:541–545, 1985. {*3*}
[129] Harris, F. On the use of windows for harmonic analysis with the discrete Fourier transform. *Proc. Instr. Electr. Eng.*, **66**:51–83, 1978. {*215*}
[130] Hearn, T.M. Anisotropic Pn tomography in the western United States. *JGR*, **101**:8403–8414, 1996. {*290*}
[131] Helbig, K. and Schoenberg, M. Anomalous polarizations of elastic waves in transversely isotropic media. *J. Acoust. Soc. Am.*, **81**:1235–1245, 1987. {*291*}
[132] Hess, H.H. Seismic anisotropy of the uppermost mantle under the oceans. *Nature*, **203**:629–631, 1964. {*289*}
[133] Hindman, B.W., Gizon, L., Duvall, T.L., Haber, D.A., and Toomre, J. Comparison of solar subsurface flows assessed by ring and time-distance analysis. *Astrophys. J.*, **613**:1253–1262, 2004. {*190*}
[134] Ho-Liu, P., Kanamori, H., and Clayton, R.W. Applications of attenuation tomography to Imperial Valley and Coso-Indian Wells region, southern California. *J. Geophys. Res.*, **93**:10501–10520, 1988. {*145*}
[135] Houser, C., Masters, G., Shearer, P.M., and Laske, G. Shear and compressional velocity models of the mantle from cluster analysis of long-period waveforms. *Geophys. J. Int.*, **174**:195–212, 2008. {*104, 245, 287*}
[136] Huang, C., Jin, W., and Liao, X. A new nutation model of a non-rigid Earth with ocean and atmosphere. *Geophys. J. Int.*, **146**:126–133, 2001. {*234*}
[137] Hudson, J. Scattered waves in the coda of P. *J. Geophys.*, **43**:359–374, 1977. {*8*}
[138] Hung, S.-H., Dahlen, F.A., and Nolet, G. Fréchet kernels for finite-frequency travel times – II. Examples. *Geophys. J. Int.*, **141**:175–203, 2000. {*119, 122, 144*}
[139] Hung, S.-H., Dahlen, F.A., and Nolet, G. Wavefront healing: a banana-doughnut perspective. *Geophys. J. Int.*, **146**:289–312, 2001. {*117, 143*}
[140] Ianniello, J.P. Time delay estimation via cross-correlation in the presence of large estimation errors. *IEEE Trans. Acoust. Speech Signal Proc.*, **30**:998–1003, 1982. {*102*}
[141] Inoue, H., Fukao, Y., Tanabe, K., and Ogata, Y. Whole mantle P-wave travel time tomography. *Phys. Earth Planet. Inter.*, **59**:294–328, 1990. {*5*}

[142] Ishii, M. and Tromp, J. Normal mode and free air gravity constraints on lateral variations in velocity and density of Earth's mantle. *Science*, **285**:1231–1236, 1999. {*5, 158, 281*}
[143] Ishii, M. and Tromp, J. Constraining large-scale mantle heterogeneity using mantle and inner-core sensitive normal modes. *Phys. Earth Planet. Inter.*, **146**:113–124, 2004. {*5, 158*}
[144] Jackson, D.D. Interpretation of inaccurate, insufficient and inconsistent data. *Geophys. J. Roy. astr. Soc.*, **28**:97–109, 1972. {*2*}
[145] Jackson, D.D. The use of a priori data to resolve non-uniqueness in linear inversion. *Geophys. J. Roy. astr. Soc.*, **57**:137–157, 1979. {*4, 267*}
[146] Jeffreys, H. On travel times in seismology. *Bur. Centr. Seism. Trav. S.*, **14**:3–36 (reprinted in The Collected Papers of Sir Harold Jeffreys, Vol. 2, Gordon and Breach 1973), 1936. {*257*}
[147] Jeffreys, H. *The Earth (4th ed.)*. Cambridge University Press, Cambridge, 1954. {*245*}
[148] Jeffreys, H. The damping of S waves. *Nature*, **208**:675, 1965. {*90*}
[149] Jin, S., Madariaga, R., Virieux, J., and Lambaré, G. Two-dimensional asymptotic iterative inversion. *Geophys. J. Int.*, **108**:575–588, 1992. {*9, 132, 309*}
[150] Johnsonbaugh, R. *Discrete Mathematics*. Macmillan, New York, 1984. {*44*}
[151] Jordan, T.H. A procedure for estimating lateral variations from low-frequency eigenspectra data. *Geophys. J. Roy. astr. Soc.*, **52**:441–455, 1978. {*158, 206*}
[152] Jordan, T.H., Puster, P., Glatzmaier, G.A., and Tackley, P.J. Comparisons Between Seismic Earth Structures and Mantle Flow Models Based on Radial Correlation Functions. *Science*, **261**:1427–1431, 1993. {*232*}
[153] Jordan, T.H. and Sipkin, S.A. Estimation of the attenuation operator for multiple ScS waves. *Geophys. Res. Lett.*, **4**:167–170, 1977. {*145*}
[154] Jung, H. and Karato, S. Water-induced fabric transitions in Olivine. *Science*, **293**:1460–1463, 2001. {*291*}
[155] Kak, A.C. and Slaney, M. *Principles of Computerized Tomography*. SIAM, 2001. {*135*}
[156] Kárason, H. and van der Hilst, R.D. Tomographic imaging of the lowermost mantle with differential times of refracted and diffracted core phases PKP,$P_{diff}$. *J. Geophys. Res.*, **106**:6569–6587, 2001. {*138*}
[157] Keilis-Borok, V.I. *Seismic surface waves in a laterally inhomogeneous Earth*. Kluwer, Dordrecht, 1989. {*180*}
[158] Kennett, B.L.N. Guided wave propagation in laterally varying media, I. Theoretical development. *Geophys. J. Roy. astr. Soc.*, **79**:235–255, 1984. {*306*}
[159] Kennett, B.L.N. *The Seismic Wavefield Vol. I: Introduction and theoretical development*. Cambridge University Press, Cambridge, UK, 2001. {*11*}
[160] Kennett, B.L.N. and Engdahl, E.R. Travel times for global earthquake location and phase association. *Geophys. J. Int.*, **105**:429–465, 1991. {*245*}
[161] Kennett, B.L.N., Engdahl, E.R., and Buland, R. Constraints on seismic velocities in the Earth from traveltimes. *Geophys. J. Int.*, **122**:108–124, 1995. {*245*}
[162] Kennett, B.L.N. and Gudmundsson, O. Ellipticity corrections for seismic phases. *Geophys. J. Int.*, **127**:40–48, 1996. {*236*}
[163] Kennett, B.L.N. and Nolet, G. The interaction of the S-wave field with upper mantle heterogeneity. *Geophys. J. Int.*, **101**:751–762, 1990. {*197, 208*}
[164] Kim, S. and Cook, R. 3D traveltime computation using second-order ENO scheme. *Geophysics*, **64**:1867–1876, 1999. {*48*}

[165] Komatitsch, D., Ritsema, J., and Tromp, J. The spectral-element methods, Beowulf computing, and global seismology. *Science*, **298**:1737–1742, 2002. {*307*}
[166] Kosovichev, A.G., Duvall, T.L., and Scherrer, P.H. Time-distance inversion methods and results. *Sol. Phys.*, **192**:159–176, 2000. {*7*}
[167] Kosovichev, A.G. and Zharkova, V.V. X-ray flare sparks quake inside the Sun. *Nature*, **393**:317, 1998. {*7*}
[168] Kuhn, J.R., Bush, R.I., Scheick, X., and Scherrer, P.H. The Sun's shape and brightness. *Nature*, **392**:155–157, 1998. {*234*}
[169] Kuo, B.-Y., Garnero, E.J., and Lay, T. Tomographic inversion of S-SKS times for shear velocity heterogeneity in $D''$ : Degree 12 and hybrid models. *J. Geophys. Res.*, **105**:28139–28157, 2000. {*221*}
[170] Larose, E., Khan, A., Nakamura, Y., and Campillo, M. Lunar subsurface investigated from correlation of seismic noise. *Geophys. Res. Lett.*, **32**:L16201, 2005. {*91*}
[171] Laske, G. and Masters, G. Constraints on global phase velocity maps from long period polarization data. *J. Geophys. Res.*, **101**:16059–16075, 1996. {*216*}
[172] Laske, G. and Masters, G. Surface-wave polarization data and global anisotropic structure. *Geophys. J. Int.*, **132**:508–520, 1998. {*290, 301*}
[173] Lawrence, J.F. and Wysession, M.E. Seismic evidence for subducted-transported water in the lower mantle. In S. Jacobsen and S. van der Lee, eds., *Earth's deep water cycle*, vol. 168 of *Geophys. Monogr.*, pages 251–262. AGU, Washington D.C., 2006. {*145*}
[174] Lay, T. and Helmberger, D.V. Body wave amplitude patterns and upper mantle attenuation variations across North America. *Geophys. J. Roy. astr. Soc.*, **66**:691–726, 1981. {*145*}
[175] Lay, T., Hernlund, J., Garnero, E.J., and Thorne, M.S. A Post-Perovskite Lens and $D''$ Heat Flux Beneath the Central Pacific. *Science*, **314**:1272–1276, 2006. {*310*}
[176] Lebedev, S., Nolet, G., and Meier, T. Automated multimode inversion of surface and S waveforms. *Geophys. J. Int.*, **162**:951–964, 2005. {*197*}
[177] Lebedev, S. and van der Hilst, R.D. Global upper-mantle tomography with the automated multimode inversion of surface and S wave forms. *Geophys. J. Int.*, page subm., 2007. {*197*}
[178] Leighton, R.B., Noyes, R.W., and Simon, G.W. Velocity fields in the solar atmosphere, I. Preliminary report. *Astrophys. J.*, **135**:474–499, 1962. {*7*}
[179] Leonard, M. Comparison of manual and automatic onset time picking. *Bull. Seismol. Soc. Am.*, **90**:1384–1390, 2000. {*97*}
[180] Leonard, M. and Kennett, B.L.N. Multicomponent autoregressive techniques for the analysis of seismograms. *Phys. Earth Planet. Inter.*, **113**:247–263, 1999. {*97*}
[181] Lerner-Lam, A. and Jordan, T.H. Earth structure from fundamental and higher-mode waveform analysis. *Geophys. J. Roy. astr. Soc.*, **75**:759–797, 1983. {*191*}
[182] Levander, A. and Nolet, G. Perspectives on array seismology and USArray. In A. Levander and G. Nolet, eds., *Seismic Earth*, vol. 157 of *Geophysical Monogr.*, pages 1–6. AGU, Washington DC, 2005. {*310*}
[183] Leveque, J.-J., Rivera, L., and Wittlinger, G. On the use of the checker-board test to assess the resolution of tomographic inversions. *Geophys. J. Int.*, **115**:313–318, 1993. {*285*}
[184] Levshin, A.L., Ritzwoller, M.H., and Shapiro, N.M. The use of crustal higher modes to constrain crustal structure across Central Asia. *Geophys. J. Int.*, **160**:961–972, 2005. {*191*}

[185] Li, X. and Romanowicz, B. Comparison of global waveform inversions with and without considering cross branch coupling. *Geophys. J. Int.*, **121**:695–709, 1995. {*9, 207, 208*}

[186] Li, X. and Tanimoto, T. Waveforms of long period body waves in a slightly aspherical Earth. *Geophys. J. Int.*, **112**:92–102, 1993. {*9, 207, 208*}

[187] Liu, Q. and Tromp, J. Finite-frequency kernels based on adjoint methods. *Bull. Seismol. Soc. Am.*, **96**:2383–2397, 2006. {*310*}

[188] Loris, I., Nolet, G., Daubechies, I., and Dahlen, F.A. Tomographic inversion using $\ell_1$-norm regularization of wavelet coefficients. *Geophys. J. Int.*, **170**:359–379, 2007. {*226, 271*}

[189] Love, A.E.H. *Some problems of geodynamics*. Cambridge University Press, Cambridge (repr. by Dover, 1967), 1911. {*158*}

[190] Luo, Y. and Schuster, G.T. Wave-equation travel time tomography. *Geophysics*, **56**:645–653, 1991. {*9, 123*}

[191] Luscombe, J.H. and Luban, M. Simplified recursive algorithm for Wigner 3j and 6j symbols. *Phys. Rev. E*, **57**:7274–7277, 1998. {*172*}

[192] Margerin, L. Introduction to radiative transfer of seismic waves. In A. Levander and G. Nolet, eds., *Seismic Earth*, vol. 157 of *AGU Geophys. Monogr. Ser.*, pages 229–252. 2005 (for figures uncontaminated by AGU's typesetters see: `http://www-lgit.obs.ujf-grenoble.fr/users/lmarger/`). {*307*}

[193] Margerin, L. and Nolet, G. Multiple scattering of high-frequency seismic waves in the deep Earth: modeling and numerical examples. *J. Geophys. Res.*, **108**:10.029/2002JB001974, 2003. {*307*}

[194] Margerin, L. and Nolet, G. Multiple scattering of high-frequency seismic waves in the deep Earth: PKP precursor analysis and inversion for mantle granularity. *J. Geophys. Res.*, **108**:doi:10.1029/2003JB002455, 2003. {*91, 307*}

[195] Marquering, H., Dahlen, F.A., and Nolet, G. Three-dimensional sensitivity kernels for finite-frequency travel times: the banana-doughnut paradox. *Geophys. J. Int.*, **137**:805–815, 1999. {*9, 123, 138*}

[196] Marquering, H., Nolet, G., and Dahlen, F.A. Three-dimensional waveform sensitivity kernels. *Geophys. J. Int.*, **132**:521–534, 1998. {*9, 128, 138*}

[197] Marquering, H. and Snieder, R. Surface-wave mode coupling for efficient forward modelling and inversion of body-wave phases. *Geophys. J. Int.*, **120**:186–208, 1995. {*207*}

[198] Masters, G., Johnson, S., Laske, G., and Bolton, H. A shear velocity model of the mantle. *Phil. Trans. R. Soc. Lond.*, **A354**:1385–1410, 1996. {*121, 221*}

[199] Masters, G., Jordan, T.H., Silver, P.G., and Gilbert, J.F. Aspherical earth structure from spheroidal mode oscillations. *Nature*, **298**:609–613, 1982. {*3, 202*}

[200] Masters, G., Laske, G., Bolton, H., and Dziewonski, A.M. The relative behaviour of shear velocity, bulk sound speed and compressional velocity in the mantle: implications for chemical and thermal structure. In S. Karato, A. Forte, R. C. Liebermann, G. Masters, and L. Stixrude, eds., *Earth's Deep Interior*, pages 63–88. AGU, Washington DC, 2000. {*246*}

[201] Masters, G., Laske, G., and Gilbert, J.F. Autoregressive estimation of the splitting matrix of free oscillation multiplets. *Geophys. J. Int.*, **141**:25–42, 2000. {*158, 169, 174, 175*}

[202] Maupin, V. Surface waves across 2-D structures: a method based on coupled local modes. *Geophys. J. Int.*, **93**:173–185, 1988. {*306*}

[203] McCarthy, D.D. and Babcock, A.K. The length of day since 1656. *Phys. Earth Planet. Inter.*, **44**:281–292, 1986. {*253*}

[204] McNamara, D.E. and Buland, R. Ambient noise levels in the continental United States. *Bull. Seismol. Soc. Am.*, **94**:1517–1527, 2004. {*109*}

[205] Meier, T., Lebedev, S., Nolet, G., and Dahlen, F.A. Diffraction tomography using multimode surface waves. *J. Geophys. Res.*, **102**:8255–8267, 1997. {*306*}

[206] Menke, W. Lateral inhomogeneities in P-velocity under the Tarbella array of the Lesser Himalayas of Pakistan. *Bull. Seismol. Soc. Am.*, **67**:725–734, 1979. {*2*}

[207] Menke, W. Case studies of seismic tomography and earthquake location in a regional context. In A. Levander and G. Nolet, eds., *Seismic Earth: Array Analysis of Broadband Seismograms*, pages 7–36. AGU, Washington DC, 2005. {*225*}

[208] Mitchell, B.J., Cheng, C.C., and Stauder, W. A three-dimensional velocity model of the lithosphere beneath the New Madrid seismic zone. *Bull. Seismol. Soc. Am.*, **62**:1061–1074, 1977. {*2*}

[209] Mochizuki, E. Free oscillations and surface waves in a spherical Earth. *Geophys. Res. Lett.*, **13**:1478–1481, 1986. {*204*}

[210] Montagner, J.-P. and Jobert, N. Vectorial tomography – II. Application to the Indian Ocean. *Geophys. J. Roy. astr. Soc.*, **94**:309–344, 1988. {*243*}

[211] Montagner, J.-P. and Nataf, H.-C. A simple method for inverting the azimuthal anisotropy of surface waves. *J. Geophys. Res.*, **91**:511–520, 1986. {*304*}

[212] Montagner, J.-P. and Tanimoto, T. Global upper mantle tomography of seismic velocities and anisotropy. *J. Geophys. Res.*, **96**:20337–20351, 1991. {*290, 301*}

[213] Montelli, R., Nolet, G., and Dahlen, F.A. Comment on 'Banana-doughnut kernels and mantle tomography' by van der Hilst and de Hoop. *Geophys. J. Int.*, **167**:1204–1210, 2006. {*242*}

[214] Montelli, R., Nolet, G., Dahlen, F.A., and Masters, G. A catalogue of deep mantle plumes: new results from finite-frequency tomography. *Geochem. Geophys. Geosys. (G3)*, **7**:Q11007, 2006. {*6, 231*}

[215] Montelli, R., Nolet, G., Dahlen, F.A., Masters, G., Engdahl, E.R., and Hung, S.-H. Finite frequency tomography reveals a variety of plumes in the mantle. *Science*, **303**:338–343, 2004. {*5, 226, 247, 259*}

[216] Mooney, W.D., Laske, G., and Masters, G. CRUST5.0: A global crustal model at $5° \times 5°$. *J. Geophys. Res.*, **103**:727–747, 1998. {*241*}

[217] Moore, B.J. Seismic ray theory for lithospheric structures with slight lateral variations. *Geophys. J. Roy. astr. Soc.*, **63**:671–689, 1980. {*146*}

[218] Moore, M.M., Garnero, E.J., Lay, T., and Williams, Q. Shear wave splitting and waveform complexity for lowermost mantle structures with low-velocity lamellae and transverse isotropy. *J. Geophys. Res.*, **109**:B02319, 2004. {*291*}

[219] Morelli, A. and Dziewonski, A.M. Topography of the core-mantle boundary and lateral homogeneity of the liquid core. *Nature*, **325**:678–683, 1987. {*98, 99*}

[220] Morita, Y. and Hamaguchi, H. Automatic detection of onset time of seismic waves and its confidence interval using the autoregressive model fitting. *Zisin*, **37**:281–293, 1984. {*97*}

[221] Moser, T.J. Shortest path calculations of seismic rays. *Geophys.*, **56**:59–67, 1991. {*25, 43*}

[222] Moser, T.J., Nolet, G., and Snieder, R. Ray bending revisited. *Bull. Seismol. Soc. Am.*, **82**:259–288, 1992. {*44*}

[223] Munk, W., Worcester, P., and Wunsch, C. *Ocean Acoustic Tomography*. Cambridge University Press, Cambridge, UK, 1995. {*3, 34, 36*}

[224] Nakanishi, I. and Yamaguchi, K. A numerical experiment on nonlinear image reconstruction from first-arrival times for two-dimensional island arc structure. *J. Phys. Earth*, **34**:195–201, 1986. {*43*}

[225] Nataf, H.-C., Nakanishi, I., and Anderson, D.L. Measurement of mantle wave velocities and inversion for lateral heterogeneity and anisotropy, III. Inversion. *J. Geophys. Res.*, **91**:7261–7307, 1986. {*290*}

[226] Neele, F. and de Regt, H. Imaging upper-mantle discontinuity topography using underside-reflection data. *Geophys. J. Int.*, **137**:91–106, 1999. {*152, 239*}

[227] Neele, F., VanDecar, J.C., and Snieder, R. A formalism for including amplitude data in tomographic inversions. *Geophys. J. Int.*, **115**:482–496, 1993. {*146*}

[228] Neele, F., VanDecar, J.C., and Snieder, R. The use of P-wave amplitude data in joint inversions with travel times for upper-mantle velocity structure. *J. Geophys. Res.*, **98**:12033–12054, 1993. {*145, 146*}

[229] Neumann-Denzau, G. and Behrens, J. Inversion of seismic data using tomographic reconstruction techniques for investigation of laterally inhomogeneous media. *Geophys. J. Roy. astr. Soc.*, **79**:305–316, 1984. {*4*}

[230] Nichols, D.E. Maximum energy arrival traveltimes calculated in the seismic frequency band. *Geophysics*, **61**:253–263, 1996. {*48*}

[231] Nissen-Meyer, T., Dahlen, F.A., and Fournier, A. Spherical-earth Fréchet sensitivity kernels. *Geophys. J. Int.*, **168**:1051–1066, 2007. {*139*}

[232] Nolet, G. *Higher modes and the determination of upper mantle structure*. Ph.D. Thesis, Utrecht University, 1976. {*198*}

[233] Nolet, G. Inversion and resolution of linear tomographic systems (abstract). *EOS Trans. Am. Geophys. Un.*, **64**:775–776, 1983. {*4, 274*}

[234] Nolet, G. Solving or resolving inadequate and noisy tomographic systems. *J. Comp. Phys.*, **61**:463–482, 1985. {*4, 274, 284*}

[235] Nolet, G. Seismic wave propagation and seismic tomography. In G. Nolet, ed., *Seismic Tomography*, pages 1–23. Reidel, Dordrecht, 1987. {*8, 269*}

[236] Nolet, G. Partitioned waveform inversion and 2D structure under the NARS array. *J. Geophys. Res.*, **95**:8513–8526, 1990. {*194*}

[237] Nolet, G. Imaging the deep earth: technical possibilities and theoretical limitations. In A. Roca, ed., *Proc. XXIIth Assembly ESC*, pages 107–115. Barcelona, 1991. {*8*}

[238] Nolet, G. A general view of the seismic inverse problem. In E. Boschi, G. Ekström, and A. Morelli, eds., *Seismic modeling of Earth structure*, pages 1–27. Editrice Compositori, Bologna, 1996. {*308*}

[239] Nolet, G. and Dahlen, F.A. Wave front healing and the evolution of seismic delay times. *J. Geophys. Res.*, **105**:19043–19054, 2000. {*118, 119, 146*}

[240] Nolet, G., Dahlen, F.A., and Montelli, R. Traveltimes and amplitudes of seismic waves: a re-assessment. In A. Levander and G. N. eds., *Array analysis of broadband seismograms*, vol. 157 of *Geophys. Monograph Ser.*, pages 37–48. 2005. {*122, 129*}

[241] Nolet, G. and Kennett, B.L.N. Normal mode representation of multiple-ray reflections in a spherical Earth. *Geophys. J. Roy. astr. Soc.*, **53**:219–226, 1978. {*112, 185*}

[242] Nolet, G. and Montelli, R. Optimal parameterization of tomographic models. *Geophys. J. Int.*, **161**:1–8, 2005. {*223, 226, 227*}

[243] Nolet, G. and Panza, G.F. Array analysis of seismic surface waves: limits and possibilities. *Pure Appl. Geophys.*, **114**:776–790, 1976. {*191*}

[244] Nolet, G., van Trier, J., and Huisman, R. A formalism for nonlinear inversion of seismic surface waves. *Geophys. Res. Lett.*, **13**:26–29, 1986. {*192*}

[245] Nowack, R.L. and Lutter, W.J. Linearized rays, amplitude and inversion. *Pure Appl. Geophys.*, **128**:401–421, 1988. {*146*}

[246] Nowack, R.L. and Lyslo, J.A. Fréchet derivatives for curved interfaces in the ray approximation. *Geophys. J. Int.*, **97**:497–509, 1989. {*146*}
[247] Odom, R.I. A coupled mode examination of irregular waveguides including the continuum spectrum. *Geophys. J. Roy. astr. Soc.*, **86**:425–453, 1986. {*306*}
[248] Osher, S. and Sethian, J. Fronts propagating with curvature dependent speed: algorithms based on Hamilton-Jacobi formulations. *J. Comp. Phys.*, **79**:12–49, 1988. {*48*}
[249] Paige, C.C. and Saunders, M.A. LSQR: An algorithm for sparse, linear equations and sparse least squares. *A.C.M. Trans. Math. Softw.*, **8**:43–71, 1982. {*274*}
[250] Panning, M. and Nolet, G. Surface wave tomography for azimuthal anisotropy in a strongly reduced parameter space. *Geophys. J. Int.*, doi:10.1111/j.1365–246X.2008.03833.x, in press. {*294, 302, 304*}
[251] Panning, M. and Romanowicz, B. A three-dimensional radially anisotropic model of shear velocity in the whole mantle. *Geophys. J. Int.*, **167**:361–379, 2006. {*290, 293, 294*}
[252] Park, J. Synthetic seismograms from coupled free oscillations: Effects of lateral structure and rotation. *J. Geophys. Res.*, **91**:6441–6464, 1986. {*158, 168*}
[253] Park, J. Roughness constraints in surface wave tomography. *Geophys. Res. Lett.*, **16**:1329–1332, 1989. {*208*}
[254] Park, J. and Yu, Y. Seismic determination of elastic anisotropy and mantle flow. *Science*, **261**:1159–1162, 1993. {*290*}
[255] Parker, R.L. Understanding inverse theory. *Ann. Rev. Earth Planet Sci.*, **5**:35–64, 1977. {*281*}
[256] Parker, R.L. *Geophysical Inverse Theory*. Princeton Univ. Press, Princeton, 1994. {*281*}
[257] Paulssen, H. and Stutzmann, E. On PP-P differential travel time measurements. *Geophys. Res. Lett.*, **23**:1833–1836, 1996. {*108*}
[258] Pedersen, H.A. Impacts of non-plane waves on two-station measurements of phase velocities. *Geophys. J. Int.*, **165**:279–287, 2006. {*208*}
[259] Pekeris, C.L., Alterman, A., and Jarosch, H. Rotational multiplets in the spectrum of the Earth. *Phys. Rev.*, **122**:1692–1700, 1961. {*158*}
[260] Pekeris, C.L. and Jarosch, H. The free oscillations of the Earth. In *Contributions in Geophysics in honor of Beno Gutenberg*, pages 171–192. Pergamon Press, New York, NY, 1958. {*158*}
[261] Pijpers, F.P. and Thompson, M.J. The SOLA method for helioseismic inversion. *Astron. Astrophys.*, **281**:231–240, 1994. {*283*}
[262] Pintore, S., Quintiliani, M., and Franceschi, D. Teseo: a vectoriser of historical seismograms. *Comp. Geosci.*, **31**:1277–1285, 2005. {*93*}
[263] Podvin, P. and Lecomte, I. Finite-difference computation of travetimes in very contrasted velocity models: A massively parallel approach and its associated tools. *Geophys. J. Int.*, **105**:271–284, 1991. {*47*}
[264] Polak, E. and Ribiere, G. Note sur la convergence de methodes de directions conjugees. *Fr. Inf. Rech. Oper.*, **16-R1**:35–43, 1969. {*193*}
[265] Poupinet, G., Ellsworth, W.L., and Fréchet, J. Monitoring velocity variations in the crust using earthquake doublets: an application to the Calaveras fault, California. *J. Geophys. Res.*, **89**:5719–5713, 1984. {*108*}
[266] Pratt, R.G. Seismic waveform inversion in the frequency domain, Part I: Theory and verification in a physical scale model. *Geophysics*, **64**:888–901, 1999. {*309*}
[267] Pratt, R.G. and Shipp, R.M. Seismic waveform inversion in the frequency domain, Part 2: Fault delineation in sediments using crosshole data. *Geophysics*, **64**:902–914, 1999. {*309*}

[268] Pratt, T.L., Templeton, M.E., Frost, R., and Shafer, A.P. Variations in short-period geophone responses in temporary seismic arrays. *Seism. Res. Lett.*, **77**:377–388, 2006. {*253*}

[269] Press, W.H., Teukolsky, S.A., Vettering, W.T., and Flannery, B.P. *Numerical Recipes*. Cambridge University Press, Cambridge, UK, second ed., 1992. {*41, 44, 273*}

[270] Qin, F., Luo, Y., Olsen, K.B., Chai, W., and Schuster, G.T. Finite-difference solution of the eikonal equation along expanded wavefronts. *Geophysics*, **57**:478–487, 1992. {*47*}

[271] Rawlinson, N. and Sambridge, M. Multiple reflection and transmission phases in complex layered media using a multistage fast marching method. *Geophysics*, **69**:1338–1350, 2004. {*47*}

[272] Rayleigh, Lord. *The Theory of Sound (1877)*. Repr. by Dover, New York, NY, 1945. {*166*}

[273] Reid, F.J.L., Woodhouse, J.H., and van Heijst, H.-J. Upper mantle attenuation and velocity structure from measurements of differential S phases. *Geophys. J. Int.*, **145**:615–630, 2001. {*145*}

[274] Revenaugh, J. A scattered-wave image of subduction beneath the Transverse Ranges. *Science*, **268**:1888–1892, 1995. {*209*}

[275] Ribodetti, A., Gaffet, S., Operto, S., Virieux, J., and Saracco, G. Asymptotic waveform inversion for unbiased velocity and attenuation measurements: numerical tests and application for Vesuvius lava sample analysis. *Geophys. J. Int.*, **158**:353–371, 2004. {*309*}

[276] Ribodetti, A., Operto, S., Virieux, J., Lambaré, G., Valero, H.P., and Gibert, D. Asymptotic viscoacoustic diffraction tomography of ultrasonic laboratory data: a tool for rock property analysis. *Geophys. J. Int.*, **140**:324–340, 2000. {*9*}

[277] Richards, P.G. and Frasier, C.W. Scattering of elastic waves from depth-dependent inhomogeneities. *Geophysics*, **41**:441–458, 1976. {*92*}

[278] Rickett, J.E. and Claerbout, J.F. Calculation of the Sun's acoustic impulse response by multi-dimensional spectral factorization. *Sol. Phys.*, **192**:203–210, 2000. {*7*}

[279] Ritsema, J., Rivera, L., Komatitsch, D., Tromp, J., and van Heijst, H.-J. Effects of crust and mantle heterogeneity on PP/P and SS/S amplitude ratios. *Geophys. Res. Lett.*, **29**:10.1029/2001GL013831, 2002. {*146*}

[280] Ritsema, J., van Heijst, H.-J., and Woodhouse, J.H. Complex shear wave velocity structure imaged beneath Africa and Iceland. *Science*, **286**:1925–1928, 1999. {*282*}

[281] Ritsema, J., van Heijst, H.-J., and Woodhouse, J.H. Global transition zone tomography. *J. Geophys. Res.*, **109**:B02302, 2004. {*282*}

[282] Ritzwoller, M.H., Masters, G., and Gilbert, J.F. Observations of anomalous splitting and their interpretation in terms of aspherical structure. *J. Geophys. Res.*, **91**:10203–10228, 1986. {*172*}

[283] Rodi, W.L., Glover, P., Li, T.M.C., and Alexander, S.S. A fast, accurate method for computing group-velocity partial derivatives for Rayleigh modes. *Bull. Seismol. Soc. Am.*, **65**:1105–1114, 1975. {*189*}

[284] Röhm, A.H.E., Bijwaard, H., Spakman, W., and Trampert, J. Effects of arrival time errors on traveltime tomography. *Geophys. J. Int.*, **142**:270–276, 2000. {*254*}

[285] Röhm, A.H.E., Trampert, J., Paulssen, H., and Snieder, R. Bias in arrival times deduced from ISC Bulletin. *Geophys. J. Int.*, **137**:163–174, 1999. {*99, 254*}

[286] Romanowicz, B. Seismic structure of the upper mantle beneath the United States by three dimensional inversion of body wave arrival times. *Geophys. J. Roy. astr. Soc.*, **57**:479–506, 1979. {*2*}

[287] Romanowicz, B. A study of large scale variations of P velocity in the upper mantle beneath western Europe. *Geophys. J. Roy. astr. Soc.*, **63**:217–232, 1980. {*2*}

[288] Romanowicz, B. Seismic tomography of the Earth's mantle. *Ann. Rev. Earth Planet Sci.*, **19**:77–99, 1991. {*4*}

[289] Romanowicz, B. Anelastic tomography: a new perspective on upper-mantle thermal structure. *Earth Planet. Sci. Lett.*, **128**:113–121, 1994. {*5*}

[290] Romanowicz, B. Attenuation tomography of the Earth's mantle: a review of current status. *Pure Appl. Geophys.*, **153**:257–272, 1998. {*84*}

[291] Romanowicz, B. Global mantle tomography: progress status in the past 10 years. *Ann. Rev. Earth Planet Sci.*, **31**:303–328, 2003. {*4, 6*}

[292] Romanowicz, B. and Mitchell, B.J. Q of the Earth from crust to core. In *Treatise of Geophysics*. Elsevier, New York, in press, 2007. {*84*}

[293] Romanowicz, B. and Roult, G. First-order asymptotics for the eigenfrequencies of the earth and application to the retrieval of large-scale lateral variations and structure. *Geophys. J. Roy. astr. Soc.*, **87**:209–240, 1986. {*213*}

[294] Romanowicz, B. and Snieder, R. A new formalism for the effect of lateral heterogeneity on normal modes and surface waves - II: General anisotropic perturbations. *Geophys. J. Roy. astr. Soc.*, **93**:91–100, 1988. {*304*}

[295] Rowe, C.A., Aster, R.C., Borchers, B., and Young, C.J. An automatic, adaptive algorithm for refining phase picks in large seismic data sets. *Bull. Seismol. Soc. Am.*, **92**:1660–1674, 2002. {*104*}

[296] Rubin, A.M., Gillard, D., and Got, J.-L. Streaks of microearthquakes along creeping faults. *Nature*, **400**:635–641, 1999. {*108*}

[297] Ruff, L. Multi-trace deconvolution with unknown trace scale factors: omnilinear inversion of P and S waves for source time functions. *Geophys. Res. Lett.*, **16**:1043–1046, 1989. {*104*}

[298] Saito, M. DISPER80: A subroutine package for the calculation of seismic normal-mode solutions. In D. Doornbos, ed., *Seismological Algorithms*, pages 294–319. Acad. Press, New York, 1988. {*163*}

[299] Sambridge, M., Braun, J., and McQueen, H. Geophysical parametrization and interpolation of irregular data using natural neighbours. *Geophys. J. Int.*, **122**:837–857, 1995. {*224*}

[300] Sambridge, M. and Gudmundsson, O. Tomographic systems of equations with irregular grids. *Geophys. J. Int.*, **103**:773–781, 1998. {*224, 225*}

[301] Sanders, C., Ho-Liu, P., and Rinn, D. Anomalous shear wave attenuation in the shallow crust beneath the Coso volcanic region, California. *J. Geophys. Res.*, **93**:3321–3338, 1988. {*145*}

[302] Sato, H. and Fehler, M.C. *Seismic wave propagation and scattering in the heterogeneous Earth*. Springer, New York, 1997. {*59*}

[303] Savage, M. Seismic anisotropy and mantle deformation: what have we learned from shear wave splitting? *Rev. Geophys.*, **37**:65–106, 1999. {*291, 296*}

[304] Scales, J.A., Gersztenkorn, A., and Treitel, S. Fast $\ell_p$ solutions of large sparse linear systems: Application to seismic travel time tomography. *J. Comp. Phys.*, **75**:314–333, 1988. {*261*}

[305] Sengupta, M.K. and Toksöz, M.N. Three-dimensional model of seismic velocity variation in the Earth's mantle. *Geophys. Res. Lett.*, **3**:84–86, 1976. {*2*}

[306] Shapiro, N.M., Campillo, M., Stehly, L., and Ritzwoller, M.H. High resolution surface wave tomography from ambient seismic noise. *Science*, **307**:1615–1618, 2005. {*91, 111*}

[307] Shearer, P.M. *Introduction to Seismology*. Cambridge University Press, Cambridge, UK, 1999. {*xi*}

[308] Shearer, P.M. Improving global seismic event locations using source-receiver reciprocity. *Bull. Seismol. Soc. Am.*, **91**:594–603, 2001. {*245*}

[309] Shearer, P.M. and Earle, P.S. The global short-period wavefield modelled with a Monte Carlo seismic phonon method. *Geophys. J. Int.*, **158**:1103–1117, 2004. {*91, 280, 307*}

[310] Shearer, P.M. and Masters, G. Global mapping of topography on the 660-km discontinuity. *Nature*, **355**:791–796, 1992. {*5*}

[311] Shin, C., Min, D.-H., Marfurt, K.J., Lin, H.Y., Yang, D., Cha, Y., Ko, S., Yoon, K., Ha, T., and Hong, S. Traveltime and amplitude calculations using the damped wave solution. *Geophysics*, **67**:1637–1646, 2002. {*48*}

[312] Sieminski, A., Liu, Q., Trampert, J., and Tromp, J. Finite-frequency sensitivity of surface waves to anisotropy based adjoint methods. *Geophys. J. Int.*, **168**:1153–1174, 2007. {*297*}

[313] Sigloch, K. and Nolet, G. Measuring finite-frequency body wave amplitudes and travel times. *Geophys. J. Int.*, **167**:271–287, 2006. {*103, 105, 147, 247, 248*}

[314] Silver, P.G. and Chan, W.W. Implications for continental structure and evolution from seismic anisotropy. *Nature*, **335**:34–39, 1988. {*290*}

[315] Silver, P.G. and Jordan, T.H. Fundamental spheroidal mode observations of aspherical heterogeneity. *Geophys. J. Roy. astr. Soc.*, **64**:605–634, 1981. {*202*}

[316] Simons, F.J., Dahlen, F.A., and Wieczorek, M.A. Spatiospectral concentration on a sphere. *SIAM Rev.*, **48**:504–536, 2006. {*232*}

[317] Simons, F.J., Nolet, G., Babcock, J.M., Davis, R.E., and Orcutt, J.A. A future for drifting seismic networks. *EOS Trans. AGU*, **31**:305–307, 2006. {*3, 97, 310*}

[318] Simons, F.J., van der Hilst, R.D., Montagner, J.-P., and Zielhuis, A. Multimode Rayleigh wave inversion for heterogeneity and azimuthal anisotropy of the Australian upper mantle. *Geophys. J. Int.*, **151**:738–754, 2002. {*301*}

[319] Simons, F.J., van der Hilst, R.D., and Zuber, M.T. Spatiospectral localization of isostatic coherence anisotropy in Australia and its relation to seismic anisotropy: Implications for lithospheric deformation. *J. Geophys. Res.*, **108**:2250, 2003. {*232*}

[320] Sipkin, S.A. and Jordan, T.H. Frequency dependence of $Q_{ScS}$. *Bull. Seismol. Soc. Am.*, **69**:1055–1079, 1979. {*150*}

[321] Skarsoulis, E.K. and Cornuelle, B.D. Travel-time sensitivity kernels in ocean acoustic tomography. *J. Acoust. Soc. Am.*, **116**:227–238, 2004. {*3, 135*}

[322] Sleeman, R. and van Eck, T. Robust automatic P-phase picking: an on-line implementation in the analysis of broadband seismogram recordings. *Phys. Earth Planet. Inter.*, **113**:265–275, 1999. {*97*}

[323] Smith, G.P. and Ekström, G. Improving event locations using a three-dimensional Earth model. *Bull. Seismol. Soc. Am.*, **86**:788–796, 1996. {*245*}

[324] Smith, M.L. and Dahlen, F.A. The azimuthal dependence of Love and Rayleigh wave propagation in a slightly anisotropic medium. *J. Geophys. Res.*, **78**:3321–3333, 1973. {*301*}

[325] Snieder, R. 3-D linearized scattering of surface waves and a formalism for surface wave holography. *Geophys. J. Roy. astr. Soc.*, **84**:581–605, 1986. {*8*}

[326] Snieder, R. The influence of topography on the propagation and scattering of surface waves. *Phys. Earth Planet. Inter.*, **44**:226–241, 1986. {*210, 243*}

[327] Snieder, R. Large-scale waveform inversions of surface waves for lateral heterogeneity, 1. Theory and numerical examples. *J. Geophys. Res.*, **93**:12055–12066, 1988. {*8, 209*}

[328] Snieder, R. Large-scale waveform inversions of surface waves for lateral heterogeneity, 2. Application to surface waves in Europe and the Mediterranean. *J. Geophys. Res.*, **93**:12067–12080, 1988. {*209*}

[329] Snieder, R. On the connection between ray theory and scattering theory for surface waves. In N. Vlaar, G. Nolet, M. Wortel, and S. Cloetingh, eds., *Mathematical Geophysics*, pages 76–81. Reidel, Dordrecht, 1988. {*8*}

[330] Snieder, R. Extracting the Green's function from the correlation of coda waves: A derivation based on stationary phase. *Phys. Rev. E*, **69**:046610, 2004. {*111*}

[331] Snieder, R. and Aldridge, D.F. Perturbation theory for travel times. *J. Acoust. Soc. Am.*, **98**:1565–1569, 1995. {*25, 117*}

[332] Snieder, R. and Chapman, C.H. The reciprocity properties of geometrical spreading. *Geophys. J. Int.*, **132**:89–95, 1998. {*141*}

[333] Snieder, R. and Lomax, A. Wavefield smoothing and the effect of rough velocity perturbations on arrival times and amplitudes. *Geophys. J. Int.*, **125**:796–812, 1996. {*9*}

[334] Snieder, R. and Nolet, G. Linearized scattering of surface waves on a spherical Earth. *J. Geophys.*, **61**:55–63, 1987. {*8, 181, 199, 210*}

[335] Soldati, G. and Boschi, L. Whole Earth tomographic models: a resolution analysis. *Eos Trans. AGU*, Fall Meeting Suppl. S13D–1092, 2004. {*231*}

[336] Solomon, S.C. and Toksöz, M.N. Lateral variation of attenuation of P and S waves beneath the United States. *Bull. Seismol. Soc. Am.*, **60**:819–838, 1970. {*145*}

[337] Sommerfeld, A. *Partial differential equations in physics*. Acad. Press, New York, 1949. {*178*}

[338] Spakman, W. and Nolet, G. Imaging algorithms, accuracy and resolution in delay-time tomography. In N. V. et al., ed., *Mathematical Geophysics*, pages 155–187. Reidel, Hingham, Mass, 1988. {*4, 266, 269*}

[339] Spakman, W., van der Lee, S., and van der Hilst, R.D. Travel-time tomography of the European-Mediterranean mantle down to 1400 km. *Phys. Earth Planet. Inter.*, **79**:3–74, 1993. {*5*}

[340] Spakman, W., Wortel, M.J.R., and Vlaar, N.J. The Hellenic subduction zone: a tomographic image and its geodynamic implications. *Geophys. Res. Lett.*, **15**:60–63, 1988. {*4*}

[341] Spencer, C. and Gubbins, D. Travel-time inversion for simultaneous earthquake location and velocity structure determination in laterally varying media. *Geophys. J. Roy. astr. Soc.*, **63**:95–116, 1980. {*246*}

[342] Stark, Ph.B. and Nikolayev, D.I. Towards tubular tomography. *J. Geophys. Res.*, **98**:8095–8106, 1993. {*137*}

[343] Stein, S. and Wysession, M. *An introduction to seismology, earthquakes and Earth structure*. Blackwell, Malden, USA, 2003. {*xi*}

[344] Stork, C. and Clayton, R.W. Linear aspects of tomographic velocity analysis. *Geophys.*, **56**:483–495, 1991. {*239*}

[345] Su, W.-J. and Dziewonski, A.M. Simultaneous inversion for 3-D variations in shear and bulk velocity in the mantle. *Phys. Earth Planet. Inter.*, **100**:135–156, 1997. {*5*}

[346] Tanimoto, T. Formalism for traveltime inversion with finite frequency effects. *Geophys. J. Int.*, **121**:103–110, 1995. {*9*}

[347] Tanimoto, T. Geometrical approach to surface wave finite frequency effects. *Geophys. Res. Lett.*, **30**:doi:10.1029/2003GL017475, 2003. {*204, 205*}

[348] Tarantola, A. Inversion of seismic reflection data in the acoustic approximation. *Geophys.*, **49**:1259–1266, 1984. {*308*}

[349] Tarantola, A. *Inverse problem theory*. Elsevier, Amsterdam, 1987. {*267*}

[350] Tarantola, A. Theoretical background for the inversion of seismic waveforms, including elasticity and attenuation. *Pure Appl. Geophys.*, **128**:365–399, 1988. {*4, 308, 309*}

[351] Tarantola, A. *Inverse problem theory and methods for model parameter estimation.* SIAM, (available on line from `www.ipgp.jussieu.fr/~tarantola/Files/Professional/Books/index.html`), 2005. {*256*}

[352] Tarantola, A. and Nercessian, A. Three-dimensional inversion without blocks. *Geophys. J. Roy. astr. Soc.*, **76**:299–306, 1984. {*4*}

[353] Tarantola, A. and Valette, B. Generalized nonlinear inverse problems solved using the least squares criterion. *Rev. Geophysics*, **20**:219–232, 1982. {*4*}

[354] Taylor, S.R. and Toksöz, M.N. Three-dimensional crust and upper mantle structure of the northeastern United States. *J. Geophys. Res.*, **84**:7627–7644, 1979. {*2*}

[355] Thomas, P.D. Geodetic arc lengths on the reference ellipsoid to second-order terms in the flattening. *J. Geophys. Res.*, **70**:3331–3340, 1965. {*243*}

[356] Thomsen, L. Weak elastic anisotropy. *Geophysics*, **51**:1954–1966, 1986. {*294*}

[357] Thomson, C. Ray theoretical amplitude inversion for laterally varying structure below NORSAR. *Geophys. J. Roy. astr. Soc.*, **74**:525–558, 1983. {*146*}

[358] Thomson, D.J. Spectrum estimation and harmonic analysis. *IEEE Proc.*, **70**:1055–1096, 1982. {*216*}

[359] Thurber, C.H. Earthquake location and 3D crustal structure in the Coyote Lake area, California. *J. Geophys. Res.*, **88**:8226–8236, 1983. {*222*}

[360] Thurber, C.H. Seismic detection of the summit magma complex of Kilauea volcano, Hawaii. *Science*, **223**:165–167, 1984. {*2*}

[361] Tian, Y., Hung, S.-H., Nolet, G., Montelli, R., and Dahlen, F.A. Dynamic ray tracing and traveltime corrections for global seismic tomography. *J. Comp. Phys.*, **226**:672–687, 2007. {*142, 144, 236*}

[362] Tian, Y., Montelli, R., Nolet, G., and Dahlen, F.A. Computing traveltime and amplitude sensitivity kernels in finite-frequency tomography. *J. Comp. Phys.*, **226**:2271–2288, 2007. {*143, 144*}

[363] Tibuleac, I.M., Nolet, G., Michaelson, C., and Koulakov, I. P wave amplitudes in a 3-D Earth. *Geophys. J. Int.*, **155**:1–10, 2003. {*148, 248*}

[364] Tikhonov, A.N. Solution of incorrectly formulated problems and the regularization method. *Dokl. Akad. Nauk. SSSR*, **151**:501–504, 1963. {*266*}

[365] Trampert, J. and Snieder, R. Model estimations biased by truncated expansions: possible artifacts in seismic tomography. *Science*, **271**:1257–1260, 1996. {*219*}

[366] Trampert, J. and van Heijst, H.-J. Global azimuthal anisotropy in the transition zone. *Science*, **296**:1297–1299, 2002. {*282*}

[367] Trampert, J. and Woodhouse, J.H. Global anisotropic phase velocity maps for fundamental mode surface waves between 40 and 150 s. *Geophys. J. Int.*, **154**:154–165, 2003. {*301*}

[368] Tromp, J., Tape, C., and Liu, Q. Seismic tomography, adjoint methods, time reversal and banana-doughnut kernels. *Geophys. J. Int.*, **160**:195–216, 2005. {*156, 310*}

[369] Turin, G.L. An introduction to matched filters. *IEEE Trans. Inform. Theory*, **6**:311–329, 1960. {*100*}

[370] Ulrich, R.K. The five-minute oscillations on the solar surface. *Astrophys. J.*, **162**:993–1002, 1970. {*7*}

[371] van der Hilst, R.D., de Hoop, M.V., Wang, P., Shim, S.-H., Ma, P., and Tenorio, L. Seismostratigraphy and thermal structure of Earth's core-mantle boundary region. *Science*, **315**:1813–1817, 2007. {*310*}

[372] van der Hilst, R.D., Engdahl, E.R., Spakman, W., and Nolet, G. Tomographic imaging of subducted lithosphere below northwest Pacific island arcs. *Nature*, **353**:37–43, 1991. {*5*}

[373] van der Hilst, R.D., Widiyantoro, S., and Engdahl, E.R. Evidence for deep mantle circulation from global tomography. *Nature*, **386**:578–584, 1997. {*5*}

[374] van der Lee, S. and Frederiksen, A. Surface wave tomography applied to the North American upper mantle. In A. Levander and G. Nolet, eds., *Seismic Earth*, vol. 157 of *Geophys. Monogr.*, pages 67–80. AGU, Washington DC, 2005. {*vi*}

[375] van der Lee, S. and Nolet, G. Upper mantle S velocity structure of North America. *J. Geophys. Res.*, **102**:22815–22838, 1998. {*184, 197*}

[376] van der Lee, S., Paulssen, H., and Nolet, G. Variability of P660S phases as a consequence of topography of the 660 km discontinuity. *Phys. Earth Planet. Inter.*, **86**:147–164, 1994. {*146*}

[377] van der Sluis, A. and van der Vorst, H.A. Numerical solution of large, sparse linear algebraic systems arising from tomographic problems. In G. Nolet, ed., *Seismic Tomography*, pages 49–84. Reidel, Dordrecht, 1987. {*273*}

[378] van Heijst, H.-J. and Woodhouse, J.H. Measuring surface-wave overtone phase velocities using a mode branch stripping technique. *Geophys. J. Int.*, **131**:209–230, 1997. {*191*}

[379] van Trier, J. and Symes, W. Upwind finite-difference computation of travel times. *Geophysics*, **56**:812–821, 1991. {*47*}

[380] VanDecar, J.C. and Crosson, R. Determination of teleseismic arrival times using multi-channel cross-correlation and least squares. *Bull. Seismol. Soc. Am.*, **80**:150–159, 1990. {*106*}

[381] VanDecar, J.C. and Snieder, R. Obtaining smooth solutions to large, linear, inverse problems. *Geophysics*, **59**:818–829, 1994. {*269*}

[382] Vasco, D.W., Johnson, L.R., and Marques, O. Resolution, uncertainty, and whole Earth tomography. *J. Geophys. Res.*, **108**:doi:10.1029/2001JB000412, 2003. {*278*}

[383] Vidale, J. Finite-difference calculation of travel times. *Bull. Seismol. Soc. Am.*, **78**:2062–2076, 1988. {*47*}

[384] Vinnik, L.P., Farra, V., and Romanowicz, B. Azimuthal anisotropy in the Earth from observations of SKS at GEOSCOPE and NARS broadband stations. *Bull. Seismol. Soc. Am.*, **79**:1542–1558, 1989. {*290*}

[385] Virieux, J. Seismic ray tracing. In E. Boschi, G. Ekström, and A. Morelli, eds., *Seismic modelling of Earth structure*, pages 223–304. Editrice Compositori, Bologna, 1996. {*138*}

[386] Virieux, J., Garnier, N., Blanc, E., and Dessa, J.-X. Paraxial ray tracing for atmospheric wave propagation. *Geophys. Res. Lett.*, **31**:L20106, 2004. {*34*}

[387] Vlaar, N.J. On the excitation of the Earth's seismic normal modes. *Pure Appl. Geophys.*, **114**:863–875, 1976. {*170*}

[388] Waldhauser, F. and Ellsworth, W.L. A double-difference earthquake location algorithm: method and application to the Northern Hayward Fault, California. *Bull. Seismol. Soc. Am.*, **90**:1353–1368, 2000. {*108*}

[389] Wang, Z. and Dahlen, F.A. Spherical-spline parameterization of three-dimensional Earth. *Geophys. Res. Lett.*, **22**:3099–3102, 1995. {*223*}

[390] Wapenaar, K. and Fokkema, J. Green's function representations for seismic interferometry. *Geophys.*, **71**:S133–S146, 2006. {*111*}

[391] Warren, L.M. and Shearer, P.M. Mapping lateral variations in upper mantle attenuation by stacking P and PP spectra. *J. Geophys. Res.*, **107**:doi:10.1029/2001JB001195, 2002. {*145, 248*}

[392] Wessel, P. and Smith, W. H. F. New version of the Generic Mapping Tools released. *Eos Trans. AGU*, **76**:329, 1995. {*224*}

[393] Wielandt, E. First-order asymptotic theory of the polar phase shift of Rayleigh waves. *Pure Appl. Geophys.*, **118**:1214–1227, 1980. {*213*}

[394] Wielandt, E. On the validity of the ray approximation for interpreting delay times. In G. Nolet, ed., *Seismic Tomography*, pages 85–98. Reidel, Dordrecht, 1987. {*8, 119*}

[395] Wiggins, R. A. General linear inverse problem - Implication of surface waves and free oscillations for Earth structure. *Rev. Geophys. Space Phys.*, **10**:251–285, 1972. {*2*}

[396] Wolfe, C.J., Bjarnason, I.T., VanDecar, J.C., and Solomon, S.C. Seismic structure of the Iceland plume. *Nature*, **385**:245–247, 1997. {*5*}

[397] Woodhouse, J.H. Surface waves in a laterally varying structure. *Geophys. J. Roy. astr. Soc.*, **37**:461–490, 1974. {*180*}

[398] Woodhouse, J.H. The calculation of eigenfrequencies and eigenfunctions of the free oscillations of the Earth and the Sun. In D. Doornbos, ed., *Seismological Algorithms*, pages 321–370. Acad. Press, New York, 1988. {*163*}

[399] Woodhouse, J.H. and Dahlen, F.A. The effect of a general spherical perturbation on the free oscillation of the Earth. *Geophys. J. Roy. astr. Soc.*, **53**:335–354, 1978. {*158, 170*}

[400] Woodhouse, J.H. and Dziewonski, A.M. Mapping the upper mantle: three dimensional modelling of the Earth structure by inversion of seismic waveforms. *J. Geophys. Res.*, **89**:5953–5986, 1984. {*3, 158, 207*}

[401] Woodhouse, J.H. and Girnius, T.P. Surface waves and free oscillations in a regionalized Earth model. *Geophys. J. Roy. astr. Soc.*, **68**:653–673, 1982. {*8, 158, 203*}

[402] Woodhouse, J.H. and Wong, Y. Amplitude, phase and path anomalies of mantle waves. *Geophys. J. Roy. astr. Soc.*, **87**:753–773, 1986. {*158*}

[403] Woodward, M.J. Wave-equation tomography. *Geophys.*, **57**:15–26, 1992. {*9, 135*}

[404] Woodward, R.L. and Masters, G. Global upper mantle structure from long-period differential travel times. *J. Geophys. Res.*, **96**:6351–6378, 1991. {*103, 120*}

[405] Wu, R.S. Seismic wave scattering. In D. James, ed., *The Encyclopedia of Solid Earth Geophysics*, pages 1166–1187. Van Nostrand, New York, 1989. {*58*}

[406] Wu, R.S. and Aki, K. Scattering characteristics of elastic waves by an elastic heterogeneity. *Geophys.*, **50**:582–595, 1985. {*60*}

[407] Wunsch, C. Tracer inverse problems. In D. Anderson and J. Willebrand, eds., *Oceanic circulation models: combining data and dynamics*, pages 1–78. Kluwer Acad. Publ., Dordrecht, 1989. {*308*}

[408] Yang, T. and Shen, Y. Frequency-dependent crustal correction for finite-frequency seismic tomography. *Bull. Seismol. Soc. Am.*, **96**:2441–2448, 2006. {*242*}

[409] Yomogida, K. Fresnel zone inversion for lateral heterogeneities in the Earth. *Pure Appl. Geophys.*, **138**:391–406, 1992. {*9*}

[410] Yomogida, K. and Aki, K. Amplitude and phase data inversions for phase velocity anomalies in the Pacific Ocean basin. *Geophys. J. Int.*, **88**:161–204, 1987. {*209*}

[411] Yoshizawa, K. and Kennett, B.L.N. Determination of influence zone for surface wave paths. *Geophys. J. Int.*, **149**:440–453, 2002. {*206*}

[412] Yoshizawa, K. and Kennett, B.L.N. Sensitivity kernels for finite-frequency surface waves. *Geophys. J. Int.*, **162**:910–926, 2005. {*218*}

[413] Zhang, H. and Thurber, C.H. Double-difference tomography: the method and its application to the Hayward fault, California. *Bull. Seismol. Soc. Am.*, **93**:1875–1889, 2003. {*108*}

[414] Zhao, A., Zhang, Z., and Teng, J. Minimum travel time tree algorithm for seismic ray tracing: improvement in efficiency. *J. Geophys. Eng.*, **1**:245–251, 2004. {*5*}

[415] Zhao, D., Xu, Y., Wiens, D.A., Dorman, L., Hildebrand, J., and Webb, S. Depth extent of the Lau back-arc spreading center and its relation to subduction processes. *Science*, **278**:254–257, 1997. {*117*}

[416] Zhao, L. and Dahlen, F.A. Mode-sum to ray-sum transformation in a spherical and aspherical Earth. *Geophys. J. Int.*, **126**:389–412, 1996. {*9, 60, 77*}

[417] Zhao, L. and Jordan, T.H. Sensitivity of frequency dependent traveltimes to laterally heterogeneous, anisotropic structure. *Geophys. J. Int.*, **133**:683–704, 1998. {*9*}

[418] Zhao, L., Jordan, T.H., and Chapman, C.H. Three-dimensional Fréchet kernels for seismic delay times. *Geophys. J. Int.*, **141**:558–576, 2000. {*9, 138*}

[419] Zhou, Y., Dahlen, F.A., and Nolet, G. Three-dimensional sensitivity kernels for surface wave observables. *Geophys. J. Int.*, **158**:142–168, 2004. {*181, 210, 216–218*}

[420] Zhou, Y., Dahlen, F.A., Nolet, G., and Laske, G. Finite-frequency effects in global surface wave tomography. *Geophys. J. Int.*, **163**:1087–1111, 2005. {*214, 218, 243*}

[421] Zhou, Y., Nolet, G., and Dahlen, F.A. Surface sediment effect on teleseismic P wave amplitude. *J. Geophys. Res.*, **108**:doi:10.1029/2002JB002331, 2003. {*248, 249*}

[422] Zollo, A., Auria, L.D, Matteis, R. De, Herrero, A., Virieux, J., and Gasparini, P. Bayesian estimation of 2-D P-velocity models from active seismic arrival time data: imaging of the shallow structure of Mt Vesuvius (Southern Italy). *Geophys. J. Int.*, **151**:566–582, 2002. {*272*}

# Author index

Numbers refer to the corresponding numbers in the bibliography.

# General index